Exploratory Data Analysis Using Fisher Information

B. Roy Frieden and Robert A. Gatenby (Eds)

Exploratory Data Analysis Using Fisher Information

 Springer

B. Roy Frieden, BS, MS, PhD
College of Optical Sciences
University of Arizona
Tucson
USA

Robert A. Gatenby, BSE, MD
Radiology Department
Arizona Health Sciences Center
Tucson
USA

Cover image: DNA nebula or DNA molecule? See Chap. 5, Sec. 5.1.7. Image courtesy of NASA/JPL-Caltech, and with thanks also to Professor Mark Morris, UCLA.

British Library Cataloguing in Publication Data
A catalogue record for this book is available from the British Library

ISBN 978-1-84996-615-3   e-ISBN 978-1-84628-777-0

© Springer-Verlag London Limited 2007
Softcover reprint of the hardcover 1st edition 2007

9 8 7 6 5 4 3 2 1

Springer Science+Business Media
springer.com

Contents

Contributor Details

H. Cabezas

Heriberto Cabezas is the Chief of the Sustainable Environments Branch, a multidisciplinary research group of approximately 20 scientists and engineers at the National Risk Management Research Laboratory, Office of Research and Development (ORD), US Environmental Protection Agency. Dr. Cabezas serves as Chair of the Environmental Division at the American Institute of Chemical Engineers. He serves as his laboratory's designated representative to the Council for Chemical Research. Dr. Cabezas was the recipient of the Sustainable Technology Division 1997 Individual Achievement Award, the 1998 EPA Science Achievement Award in Engineering, and the 2003 ORD Diversity Leadership Award. He received the PhD in chemical engineering from the University of Florida in 1985 in the area of thermodynamics and statistical mechanics. His areas of research interest include complex sustainable systems, chemical process design for pollution prevention, design of environmentally benign substitute-solvents, and thermodynamics. His publications include over 40 peer-reviewed technical publications, many conference proceedings, guest editorship of the *Journal of Chromatography* and the journal *Resources, Conservation and Recycling*, various trade publications, invited lectures at national and international conferences, presentations at national and international conferences, and invited research seminars at academic, private, and government institutions. Dr. Cabezas is a member of the American Institute of Chemical Engineers and the American Association for the Advancement of Science (AAAs).

Affiliation and contact details:
Dr. Heriberto Cabezas
US Environmental Protection Agency
26 West Martin Luther King Drive
Cincinnati, OH 45268
E-mail: cabezas.heriberto@epa.gov

J.L. D'Anna

Joseph D'Anna is Managing Director and Head of Equity-Fund Derivatives Structuring for the Americas at Rabobank International, New York. Formerly, he was a Senior Financial Engineer in Société Générale's Equity Derivatives group, also serving as Chief Investment Officer of Catalyst Re, SG's reinsurance subsidiary. He joined SG via Constellation Financial Management (CFM), a structured finance boutique acquired by SG in 2003. At CFM he was Managing Director of Strategic Capital Raising and Alternative Risk Transfer, where he directed the design and placement of $1.3 billion in notes backed by mutual fund and annuity derivatives. At Bank of America, he analyzed financial market data, validated derivative models, and helped set bankwide standards. He is also an NASD registered representative and CFA Charterholder. Joe holds Bachelor of Science degrees in math and physics from the University of Redlands and a PhD in physics from UC Santa Barbara.

Affiliation and contact details:
Dr. Joseph L. D'Anna
Managing Director, Head of Equity-Fund Derivatives Structuring
North America Rabobank International, Inc.
245 Park Avenue, 37th floor, NY 10167
E-mail: Joseph.D'Anna@Rabobank.com

B.D. Fath

Brian Fath is an Assistant Professor in the Department of Biological Sciences at Towson University in Towson, Maryland. He teaches courses in ecosystem ecology, environmental biology, human ecology, and networks, and has taught short courses in ecological management and modeling in Portugal, Denmark, and Croatia. His research is in the area of ecological network analysis, sustainability, and integrated assessment. He has published more than 45 papers in referred journals, conferences, and book chapters. He has a summer appointment as a visiting research scientist in the Dynamic Systems Program at the International Institute of Applied Systems Analysis in Laxenburg, Austria, where he has organized a series of workshops and supervised student summer research. He is currently an associate editor for the international journal *Ecological Modelling and Systems Ecology*, and coordinating editor for a forthcoming *Encyclopedia of Ecology*.

Affiliation and contact details:
Professor Brian D. Fath
Biology Department
Towson University
Towson, MD 21252
E-mail: bfath@towson.edu

B.R. Frieden

Roy Frieden is an Emeritus Professor in the College of Optics, University of Arizona, Tucson. He consults for government agencies in statistical optics (NASA, James Webb 2nd generation space telescope) and in scientific uses of Fisher information. The latter include the monitoring of ecologies for the Environmental Protection Agency, as well as the analysis of functional and cancerous growth for the University of Arizona Medical Center. Other such uses of the information are listed in Section 1.7 of this book, and include the parallel behavior of biological, cosmological, and social systems as information transducers, and the invention of optimal capital investment programs in economics. Past research successes in optics have been the invention of laser beam shaping techniques (currently used in laser fusion energy research), the proposed use of e.m. stealth technology for the US navy, three-dimensional imaging theory, maximum entropy image restoration, and imagery whose resolution much-exceeds the diffraction limit. The author has published well over 100 peer-reviewed papers in scientific journals as well as 20 invited chapters and 5 books. He is a fellow of the AAAS, Optical Society of America, and SPIE.

Affiliation and contact details:
Professor B. Roy Frieden
College of Optics, University of Arizona
Tucson, AZ 85721
E-mail: roy.frieden@optics.arizona.edu

R.A. Gatenby
Robert Gatenby is a Professor of radiology and applied mathematics at the University of Arizona, and has served as Chair of the Radiology Department since July 1, 2005. His research interests are focused on integrating mathematical models and empirical techniques for purposes of understanding tumor invasion and carcinogenesis. Nonlinear cellular and microenvironmental dynamics dominate these processes. Dr. Gatenby received the BSE in engineering from Princeton University, and an MD from the University of Pennsylvania. He completed his residency in radiology at the Hospital of the University of Pennsylvania. His interest in tumor biology began during his early career while working at the Fox Chase Cancer Center in Philadelphia. At the University of Arizona, Dr. Gatenby works with an eclectic group of researchers at the interface of mathematics and medicine. This includes young physicians and life scientists who are interested in the use of mathematical modeling techniques, as well as applied mathematicians, theoretical physicists, aerospace engineers, and optical scientists who wish to apply their skills to biology.

Affiliation and contact details:
Dr. Robert A. Gatenby
Chair, Department of Radiology
University Medical Center
University of Arizona
Tucson, AZ 85726
E-mail: rgatenby@uph.org

R.J. Hawkins
Raymond J. Hawkins is Executive Vice President, Interest Rate Risk Management, at Countrywide Bank, NA The bank is a member of the Countrywide Financial

Corporation family of companies. Dr. Hawkins is also an Adjunct Professor in the College of Optical Sciences of the University of Arizona. At Countrywide he provides oversight of the bank's strategies that address interest rate risk as it impacts investments, loan portfolios, time deposits, escrow deposits, and other bank activities. He entered the financial services profession in 1993 in the fixed-income derivatives group at Salomon Brothers. Since that time, Hawkins has held a variety of positions with organizations including Bank of America, Bear Stearns, and Barclay's Global Investors. These were as a strategist, specializing in risk management and securities valuation. Prior to joining Countrywide Bank he served as Managing Partner, Director of Derivatives Trading and Research for Mulsanne Capital Management. At the University of Arizona, Dr. Hawkins leverages his earlier work in guided-wave electromagnetics and integrated photonics in a research program that centers on applications of information theory and statistical physics to the emerging field of econophysics. Prior to 1993 Hawkins served as a team leader and researcher in computational photonics at Lawrence Livermore National Laboratory. Dr. Hawkins was formally educated at Columbia University in the city of New York, beginning with Columbia College and ending with a doctorate for work done in condensed matter physics. He holds a Chartered Financial Analyst designation, and has published extensively in physics, econometrics, and econophysics.

Affiliation and contact details:
Dr. Raymond J. Hawkins
Executive Vice President
Interest Rate Risk Management
Countrywide Bank
225 West Hillcrest Drive, MS TO-20
Thousand Oaks, CA 91360
E-mail: Raymond_Hawkins@Countrywide.Com

A.L. Mayer
Audrey Mayer is a senior researcher at the University of Tampere, Finland, where she is working on two issues: applying ecologically based indices to sustainable forest management, and measuring impacts on biodiversity of global trade in forest products. Prior to this appointment, she was an ecologist at the US Environmental Protection Agency (Office of Research and Development, National Risk Management Research Laboratory), working in multidisciplinary groups in the Sustainable Environments Branch on theoretical and applied sustainability tools, particularly using concepts from dynamic regimes theory. She received the PhD in ecology and evolutionary biology from the University of Tennessee, Knoxville, and completed postdoctoral positions at the University of Cincinnati and the USEPA. Her research spans several areas, including multidisciplinary sustainability indices influenced by information theory, managing impacts of impervious surface on aquatic communities at the watershed scale, and bird conservation, particularly land use/land cover effects at multiple spatial scales on terrestrial bird communities.

Affiliation and contact details:
Dr. Audrey L. Mayer
Research Centre Synergos
School of Economics and Business Administration
University of Tampere
33100 Tampere, Finland
E-mail: audrey.mayer@uta.fi

Christopher Pawlowski
received his Ph.D. in mechanical engineering from the University of California, Berkeley. His dissertation dealt with hierarchical control applied to environmental systems for space applications. He worked at NASA Ames for two years on market-based methods for power management of a space station, and then accepted a postdoctoral position to work on environmental modeling and sustainability at the National Risk Management Research Laboratory of the US EPA. He is currently with R.D. Zande and Associates, an environmental engineering firm.

Affiliation and contact details:
Dr. Christopher Pawlowski
R.D. Zande and Associates
11500 Northlake Dr., Soite 150
Cincinnati, OH 45249
33100 Tampere, Finland
E-mail: cw_pawlowski@yahoo.com

A. Plastino
Angelo Plastino is an Emeritus Professor at the National University of La Plata, Argentina. His main research areas are quantum mechanics, thermodynamics, statistical mechanics, information physics, and information theory. In these areas he has published about 390 refereed articles in science journals since 1961. Awards and Honors: Director of the La Plata Physics Institute, 1998–present; President of the National University of La Plata, 1986–1992; Honorary Professor of the University of Buenos Aires; Doctor Honoris Causa of the University of Pretoria; Member of the Mexican Academy of Sciences; Member of the Brazilian Academy of Sciences; Advisory Editor of the Elsevier Journal *Physica A*.

Affiliation and contact details:
Professor Angelo Plastino
La Plata Physics Institute – UNLP
C. C. 727 – 1900 La Plata
Argentina
E-mail: plastino@fisica.unlp.edu.ar

A.R. Plastino
Angel R. Plastino is a Full Professor at the University of Pretoria, South Africa. His main research areas are quantum mechanics, statistical physics, the physics

of information, and physics education. In these areas he has published about 100 refereed articles in science journals since 1992.

Affiliation and contact details:

Professor Angelo Ricardo Plastino
Physics Department
University of Pretoria
Pretoria 0002, South Africa
E-mail: arplastino@maple.up.ac.za, vdfsarp9@uib.es

B.H. Soffer

Bernard Soffer has 50 years of experience in physical research, beginning at MIT with the infrared spectroscopy of single crystals. At Korad Corporation, he invented the continuously tunable dye laser. He performed research characterizing laser parameters and the spectroscopy of several successful new solid-state lasers. He did fundamental studies of the Verneuil crystal growth processes. He discovered the first dye Q-switching materials for the giant pulse ruby and Nd lasers, and researched the physics of these materials including their dynamics and photochemistry. As head of the Signal Processing Department, Hughes Research Laboratories, he was responsible for many programs in image and signal processing (restoration, compression, pattern recognition, fusion), communication, sparse antenna arrays, radar and neural networks. He invented a programmable optical neural network using nonlinear materials for the holographic storage of weights and interconnections, resulting in a device with as many neurons as in the brain of a bee. Most recently, he has been investigating fundamental issues in physics from the point of view of information theory. He has published more than 75 peer-reviewed papers, in addition to scores of talk summaries, including invited talks, in conference proceedings, and several contributed book chapters. He holds 37 patents. He was for eight years an associate editor of *Applied Optics*, and continues to serve on the editorial board of *Neural Networks*.

Affiliation and contact details:

Bernard H. Soffer
University of California, Los Angeles
Department of Electrical Engineering
Correspondence: 665 Bienveneda Avenue
Pacific Palisades, CA 90272
E-mail: bsoffer@ucla.edu

R.C. Venkatesan

Ravi Venkatesan received the PhD in computational sciences from the University of Illinois at Urbana-Champaign. He has worked at the Centre for Development of Advanced Computing and the Centre for Advanced Technology, both in the Department of Atomic Energy of India. He has had assignments at the Australian National University, INRIA (Sophia-Antipolis, France), and the Russian Academy of Sciences. He is presently with Systems Research Corporation in India. Ravi has authored 39 journal publications and numerous conference proceedings. His areas of activity are in quantum computing, pattern recognition (quantum and SVD clustering), information theoretic cryptosystems, game theory, computer vision, and image processing. These include scientific data mining, the design of invariant and symplectic numerical schemes, and the design of symbolic-numeric computing environments.

Affiliation and contact details:

Dr. Ravi C. Venkatesan
Computational Sciences Division
Systems Research Corporation
Aundh, Pune 411007, India
E-mail: ravi@systemsresearchcorp.com

M.I. Yolles

Maurice Yolles is a professor of management systems at Liverpool John Moores University, based in the Business School. His doctorate, completed more than a generation ago, was in mathematical social theory, in particular the formal dynamics of peace and conflict. His research book on management systems was published in 1999 and his new book *Organisations as Complex Systems* is due out shortly. He has published more than 140 papers in refereed journals, conferences, and book chapters, mostly in managerial cybernetics and its development in social collectives, international joint alliance theory, and human resource management. He is editor of the international journal Organisational Transformation and Social Change. He is also the vice president of the International Society of Systems Science. His main teaching area is in change and knowledge management, and he heads the Centre for Creating Coherent Change and Knowledge. Within this context he has also been involved in, and run, a number of international research and development projects for the EU under various programs within countries experiencing transformational change. This includes involvement in TEMPUS projects in Central and Eastern European countries. He has also lectured and run organizational change programs in China.

Affiliation and contact details:

Professor Maurice Yolles
School of Business Information
Liverpool John Moores University
Liverpool, L3 5UZ, UK
E-mail: M.Yolles@livjm.ac.uk

1
Introduction to Fisher Information: Its Origin, Uses, and Predictions

B. ROY FRIEDEN

This is a book on new approaches to the exploratory analysis of an unknown system. The approaches are based upon uses of Fisher information and, in particular, on a unified approach to science called EPI (extreme physical information). Both theory and applications are given. The scope of systems that can be so analyzed is quite broad, in fact encompassing *all statistically based phenomena*. The new approaches grow out of analyzing how these systems convey Fisher information to *an observer*. The result is a novel program of exploratory data analysis (EDA), whose inputs are real or *gedanken* data, and whose outputs are the natural laws governing the systems. Examples are the Schrodinger wave equation and the quarter-power laws of biology.

The aim of this chapter is to introduce Fisher information and to develop those of its properties that are later used. The chapters that follow cover an eclectic range of applications of this EDA program. These applications are independent and, so, may be read in any order.

The universe consists of systems whose parts move, collide, attract, repel, etc., each other. Traditional methods of analyzing these systems are *phenomenological* in nature. Thus, the motions of mechanical particles are analyzed by directly addressing phenomena of force and motion; the interactions among charges, currents, and magnets are analyzed by assessing the effects of Coulomb and magnetic forces, etc. These approaches generally regard phenomena as basically distinct. Although the approaches usually work, as we learned in elementary physics courses, it gives the impression that the phenomena are fundamentally unrelated. This is of course not the case. The aim of physics is to unify, and various well-known approaches to unification[1] have been found (see, e.g., Section 1.4.7).

In particular, there is a way to understand these phenomena that emphasizes *their commonalities* rather than differences. It is almost a tautology to remark that we know that systems behave in such-and-such a way because we *observe* them to do so. These observations can of course be either direct or indirect, i.e., by real-time viewing or by inference from other observations. In turn, whatever their

---

[1] This is in the sense of a unified conceptual approach to *all physics*; not in the sense of specifically unifying gravitational and quantum effects (a famously unresolved problem).

nature, observations that faithfully describe a system must contain *information about* the system. The traditional way of using these observations is as data inputs to the phenomenological methods of analysis mentioned above.

However, there is another way to do the analysis. This arises out of considering instead the nature of *the information* that is conveyed by the data. This is irrespective of the *particular* effect that is under observation and, so, provides a unified point of view. Analyzing the system on the basis of information flow, rather than phenomenologically as usual, amounts to a *paradigm shift*.

To review, *all physical effects*, regardless of their nature, require observation, and observation requires *a flow of information*. The flow is from the source effect to the observer. Thus, the act of measurement, rather than the specific data, is fundamental to the approach. Herein lies the paradigm shift. But, by what mechanism does the act of measurement provide information about the source effect? How does the concept of information allow an observer to acquire quantitative knowledge about an unknown system?

As has been known since at least Heisenberg, any measurement perturbs the state of the system under measurement. In fact, this perturbation turns out to *reveal* to the observer the law governing formation of the data. This is expressed below by the EPI principle, which permits calculation of the law. In this way, each exploratory application of the EPI principle to data amounts to *a process of discovery of the law governing that data*. The laws uncovered in this way by EPI range from those of quantum mechanics and cosmology to those of economics and the life sciences. The many applications taken up in this book serve to bring the reader up to date on both the process of discovery and how it is accomplished in practice. Overall, an operational point of view is taken. Its aim is to provide the analyst with a new tool of EDA for discovering the laws governing *unknown systems*.

In this way, the traditional phenomenological approach, which is piecemeal in scope and ad hoc in nature, is replaced with an information-based approach of a general nature. Moreover, the information-based approach often gives an answer that applies to a broader class of systems than does the corresponding phenomenon-based approach (see Sections 1.4.7 and 8.3). However, EPI does not get you something for nothing. Each application of the approach requires some *prior knowledge* about the system. This is in the form of a generalized invariance property, as described below. Searching for and trying out candidate invariance properties that fit the application is the "exploratory" phase of this EDA approach.

Getting down to specifics, all systems, including living systems and inanimate scientific effects, are defined by statistical laws. Examples are the system likelihood law—exemplified by a wave function of quantum mechanics—and the system input–output law or data law—exemplified by a power law of biology. Knowledge of these laws is central to understanding the systems. Given an unknown system, Fisher information may be used as an effective tool for estimating these laws. In subsequent chapters, it is shown how physical, biological, ecological, economic, image-forming, and even sociological laws have been so estimated. By its nature, Fisher information applies to all statistically repeatable scenarios where data are taken.

Thus, the concept of *information* is finally, nowadays, achieving its long-standing goal of providing a unifying concept for understanding our world. As will be seen, this is both the world outside and the world within. The particular information due to R.A. Fisher is evolving into a particularly powerful concept for this purpose. This is on two distinct levels: the philosophical and the practical. The information provides both a philosophy of science and a framework for calculating *all its laws*. What is required is prior knowledge of an invariance principle that describes the data-taking scenario (see below). Hence, the information has acted to unify science on both the *conceptual* and *perceptual* levels.

A detailed summary of past uses of the Fisher approach is given in Table 1.1, where its uses in subsequent chapters are sketched as well.

Any bona fide physical theory must make predictions. To demonstrate the utility of Fisher information in analyzing unknown systems, some of its predictions are summarized at the chapter's end.

1.1. Mathematical Tools

This introduction to Fisher methods requires some basic mathematical tools, which are next reviewed. An operational view is taken. As likewise in the rest of the book, the question of "how to" is kept paramount.

1.1.1. Partial Derivatives

The concept of the partial derivative is essential to this work. It is important to distinguish it from the ordinary or total derivative, particularly in the context of the Euler–Lagrange Eq. (1.6). The distinction is easy to show. Consider a given function

$$\mathcal{L} = \dot{x}^2 - \sin(ax) + 2t^3, \quad x = x(t), \tag{1.1}$$

where the dot denotes a derivative d/dt. This function has *partial* derivatives

$$\frac{\partial \mathcal{L}}{\partial \dot{x}} = 2\dot{x}, \quad \frac{\partial \mathcal{L}}{\partial x} = -a\cos(ax) \quad \text{and} \quad \frac{\partial \mathcal{L}}{\partial t} = 6t^2. \tag{1.2}$$

This is to be compared with the *total* derivative $d\mathcal{L}/dt$ which, by the chain rule, obeys

$$\frac{d\mathcal{L}}{dt} = \frac{\partial \mathcal{L}}{\partial t} + \frac{\partial \mathcal{L}}{\partial \dot{x}}\frac{d\dot{x}}{dt} + \frac{\partial \mathcal{L}}{\partial x}\frac{dx}{dt} = 6t^2 + 2\dot{x}\ddot{x} - a\cos(ax)\dot{x} \tag{1.3}$$

by the dot notation.

Also, from the first Eq. (1.2),

$$\frac{d}{dt}\left(\frac{\partial \mathcal{L}}{\partial \dot{x}}\right) = 2\ddot{x}. \tag{1.4}$$

Operationally, the partial derivative is simpler to take than the total derivative. For example, in constructing the partial derivative $\partial \mathcal{L}/\partial \dot{x}$, one simply scans Eq. (1.1) in $\mathcal{L}$ for instances of $\dot{x}$, and then differentiates these alone. Thus the term $\dot{x}^2$ is differentiated, while the other terms $2t^3$ and $\sin(ax)$ are ignored. Or, in constructing $\partial\mathcal{L}/\partial x$, only the term $-\sin(ax)$ is differentiated, the terms $2t^3$ and $\dot{x}^2$ now ignored; etc. for $\partial\mathcal{L}/\partial t$.

1.1.2. Euler–Lagrange Equations

Note that function $\mathcal{L}$ in Eq. (1.1) is a function of functions. This is typically denoted as $\mathcal{L}[x(t), \dot{x}(t), t]$ or $\mathcal{L}[x, \dot{x}, t]$, and called a "Lagrangian." Suppose that x is the coordinate defining a system, such as the position x of a given snowflake in a storm. Function $x(t)$ is often called the "trajectory" of the system.

1.1.2.1. Extremum Problem

Suppose that a trajectory $x(t)$ is sought that gives an extreme value, i.e., either maximum or minimum or point of inflection, for the integral of the Lagrangian $\mathcal{L}$ over some finite time interval (a, b). That is, an optimized trajectory $x(t)$ is sought that satisfies

$$\int_a^b dt\,\mathcal{L}[x, \dot{x}, t] = \text{extrem.} \tag{1.5}$$

(For simplicity, we do not give a different notation, for example $x_s(t)$, to the solution.) The principle of EPI, introduced below, is of this form.

1.1.2.2. Solution

The general solution $x(t)$ to (1.5) for *any* well-behaved Lagrangian $\mathcal{L}$ is known to obey the differential equation [1]

$$\frac{d}{dt}\left(\frac{\partial\mathcal{L}}{\partial\dot{x}}\right) = \frac{\partial\mathcal{L}}{\partial x}. \tag{1.6}$$

This is derived in many elementary texts, such as [2] or [3]. Equation (1.6) is called the Euler–Lagrange equation after its originators. In particular, for our Lagrangian (1.1), the use of Eqs. (1.2) and (1.4) in (1.6) gives a solution

$$2\ddot{x} + a\cos(ax) = 0. \tag{1.7}$$

An E–L solution such as (1.7) may be solved for its unknown trajectory, here $x(t)$, using any convenient table of solutions to differential equations [4]. The result is almost always *a family* of solutions. Oftentimes this is of advantage to a user who has prior information or bias about the type of solution $x(t)$ to be sought. An example would be a tendency for $x(t)$ to approach zero as t increases, rather than to be oscillatory. This would give preference to an exponential solution rather than a sinusoidal solution. Also, because of the second-order nature of Eq. (1.7), two boundary conditions are needed to uniquely fix a solution from the family. We do not pursue further the particular solution to (1.7). Our main point here is that the problem (1.5) of optimization usually admits of a solution, in the form of a known differential equation (1.6) representing a family of solutions.

In the same way, a multidimensional, multicomponent problem of optimization

$$\int dt \, \mathcal{L}[\mathbf{x}, \dot{\mathbf{x}}, \mathbf{t}] = \text{extrem.}, \tag{1.8}$$

with coordinates $\mathbf{t} = (t_1, \ldots, t_M)$, $dt = dt_1 \cdots dt_M$, and components $\mathbf{x} = x_1, \ldots, x_N$, has a solution [1–3]

$$\sum_{m=1}^{M} \frac{d}{dt_m} \left(\frac{\partial \mathcal{L}}{\partial x_{nm}} \right) = \frac{\partial \mathcal{L}}{\partial x_n}, \quad n = 1, \ldots, N, \tag{1.9}$$

where $x_{nm} \equiv \partial x_n / \partial t_m$. Note that undefined integration limits, as in (1.8), generally go from $-\infty$ to $+\infty$.

1.1.2.3. Nature of Extremum

We will have occasion to require that the extremum in principle (1.5) be a minimum in particular. A well-known necessary condition for the minimum is that the second partial derivative obey

$$\frac{\partial^2 \mathcal{L}}{\partial \dot{x}^2} > 0. \tag{1.10}$$

This is called "Legendre's condition" [1] for a minimum. A maximum is, instead, indicated if the second derivative is negative.

In the same way, for multiple functions x_n, $n = 1, \ldots, N$ of a single coordinate t, a necessary condition for a minimum is that the matrix of elements $[\partial^2 \mathcal{L}/\partial \dot{x}_j \partial \dot{x}_k]$, $j, k = 1, \ldots, N$ independently, be positive definite. That is,

$$\det \left[\frac{\partial^2 \mathcal{L}}{\partial \dot{x}_j \partial \dot{x}_k} \right] > 0 \tag{1.11}$$

for all subminors of the matrix. The condition for a minimum in the more general scenario of multiple coordinates t_m will not be of interest to us.

1.1.2.4. Building in Constraints

There are scenarios where the user knows that any legitimate solution $x(t)$ to the general optimization problem (1.5) must also obey certain side conditions, in the form of constraint equations

$$\int_a^b dt f_n(x, \dot{x}, t) = F_n, \quad n = 1, \ldots, N. \tag{1.12}$$

Here the kernel functions f_n and data F_n are known. The solution to this problem is obtained by simply forming a new Lagrangian

$$\tilde{\mathcal{L}} = \mathcal{L} + \sum_{n=1}^{N} \lambda_n f_n(x, \dot{x}, t), \quad \lambda_n = \text{const.} \tag{1.13}$$

The λ_n are called Lagrange multipliers and are initially unknown. The procedure is to use $\tilde{\mathcal{L}}$ in place of $\mathcal{L}$ in the Euler–Lagrange Eq. (1.6). The result is a new family of solutions $x(t)$. Any such solution contains the unknown constants λ_n. One solution is chosen on the basis of prior knowledge, and then back substituted into the N constraint Eqs. (1.12). The result is N equations in the N unknown constants λ_n, which is usually sufficient for their solution.

Likewise, in the more general scenario of multiple coordinates and components, there can be generally K constraints upon each component x_n,

$$\int dt\, f_{nk}(\mathbf{x}, \dot{\mathbf{x}}, t) = F_{nk}, \quad n = 1, \ldots, N; \ k = 1, \ldots, K \tag{1.14}$$

Here the kernel functions $f_{nk}(\mathbf{x},\dot{\mathbf{x}},t)$ and constraint values F_{nk} are given. Then the effective Lagrangian used in Eq. (1.9) is

$$\tilde{\mathcal{L}} = \mathcal{L} + \sum_{nk}^{N} \lambda_{nk} f_{nk}(\mathbf{x}, \dot{\mathbf{x}}, t), \quad \lambda_{nk} = \text{const.} \tag{1.15}$$

Back substitution of the new solution $\mathbf{x}$ into the multiple-coordinate constraint Eqs. (1.14) now results in NK equations in the NK unknowns λ_{nk}, which usually have a well-defined solution.

1.1.2.5. Example: MFI Lagrangian

The MFI or *minimum Fisher information* approach (see below) to estimating a probability law utilizes a Lagrangian

$$\mathcal{L} = 4\dot{x}^2 \quad \text{where } f_{nk} \equiv f_1 = x^2 t^b, \ b = \text{const.}$$

Let the reader show that the net Lagrangian is then

$$\tilde{\mathcal{L}} = 4\dot{x}^2 + \lambda_1 x^2 t^b,$$

and the Euler–Lagrange solution (1.6) is the differential equation

$$8\ddot{x} = 2\lambda_1 x t^b.$$

1.1.3. Dirac Delta Function

An oft-used concept is the Dirac delta function $\delta(t)$. This has the defining properties

$$\int dt\, \delta(t) = 1, \quad \text{with } \delta(t) = 0 \ \text{ for } t \neq 0. \tag{1.16}$$

In words, despite having zero width, $\delta(t)$ has finite area under it. This can only mean that it has an infinite spike at the origin, $\delta(0) \to \infty$. An example is an infinitely tall and narrow rectangle function. That is, in the limit as $\epsilon \to 0$,

$$\delta(t) = \frac{1}{\epsilon} \quad \text{for} \ -\frac{\epsilon}{2} \le t \le +\frac{\epsilon}{2} \quad \text{and} \ \ 0 \text{ otherwise.} \tag{1.17a}$$

Another, much used, representation is the Fourier one

$$\delta(t) = (2\pi)^{-1} \int dt'\, \exp(itt'), \quad i = \sqrt{-1}. \tag{1.17b}$$

The representation (1.17a) is handy in deriving the "sifting" property

$$\int dt' f(t')\delta(t - t') = \lim_{\epsilon \to 0} \frac{1}{\epsilon} \int_{t-\epsilon/2}^{t+\epsilon/2} dt' f(t') = \lim_{\epsilon \to 0} \frac{1}{\epsilon} f(t)\epsilon = f(t). \quad (1.18)$$

This is probably the most-often used property of the Dirac delta function. Another useful identity is

$$\delta(ax) = \frac{\delta(x)}{|a|}, \quad a = \text{const.} \quad (1.19)$$

1.1.4. Unitary Transformation

This concept plays a key role in uses of EPI. Suppose that a (now) general trajectory $x(t)$ has a "length" L_x^2 of

$$L_x^2 \equiv \int dt |x(t)|^2 \equiv \int dt x^*(t)x(t), \quad (1.20)$$

where the asterisk denotes a complex conjugate. Suppose that $x(t)$ relates to a trajectory $y(t')$ in a new space t' defined as

$$y(t') \equiv \int dt x(t)U(t, t'), \quad (1.21)$$

in terms of a transformation kernel $U(t, t')$. What condition must $U(t, t')$ obey in order that length be preserved under the transformation, i.e.,

$$L_y^2 = L_x^2. \quad (1.22)$$

By definitions (1.20) and (1.21),

$$L_y^2 \equiv \int dt' y^*(t')y(t') = \int dt' \int dt x^*(t)U^*(t, t') \int dt'' x(t'')U(t'', t') \quad (1.23)$$

$$= \int dt x^*(t) \int dt'' x(t'') \int dt' U^*(t, t')U(t'', t') \quad (1.24)$$

after rearranging orders of integration. Now if

$$\int dt' U^*(t, t')U(t'', t') = \delta(t'' - t), \quad (1.25)$$

a Dirac delta function, then (1.24) becomes

$$L_y^2 = \int dt x^*(t) \int dt'' x(t'')\delta(t'' - t) = \int dt x^*(t)x(t) = L_x^2, \quad (1.26)$$

where the second equality is by the sifting property (1.18) of the Dirac delta function, and the third equality is by definition (1.20). Condition (1.25) is the defining condition of a unitary transformation. Essentially its cross-correlation is an infinitely tall and narrow spike.

The unitary condition (1.25) is satisfied, for example, by a *Fourier transformation F*,

$$U(t, t') = (2\pi)^{-1/2} \exp(itt') \equiv F(t, t'). \qquad (1.27)$$

As a check, directly

$$\int dt' U^*(t, t') U(t'', t') = \int dt' (2\pi)^{-1/2} \exp(-itt')(2\pi)^{-1/2} \exp(it''t')$$

$$= (2\pi)^{-1} \int dt' \exp[it'(t'' - t)] = \delta(t'' - t). \qquad (1.28)$$

The latter is by the delta function representation (1.17b).

Any number of other unitary transformation kernels can be found similarly. Unitary transformations are at the heart of the EPI principle described later.

Given the preceding mathematical tools, we can now proceed to the definition of Fisher information.

1.2. A Tutorial on Fisher Information

The EPI principle [3] utilizes two Fisher information quantities, termed I and J. In the following subsections we develop the various forms of I. Following this, in Section 1.3 we turn to information J.

1.2.1. Definition and Basic Properties

Fisher information arises naturally out of the classical problem of estimation [2, 3, 5]. The manner by which it arises is described next. The derivation is simple, direct, and impressively general. In fact it is sufficiently general to suggest its exploratory use in analyzing any new scientific effect. The derivation simply rests upon the process of observation. By the scientific method, any effect must be observed for it to be known. The effect–observation process is as follows.

A system is specified by a parameter value a. This is unknown. An observer collects data $\mathbf{y} \equiv y_1, \ldots, y_N$ from the system, with the aim of *estimating a*. His assumption is that the data contain "information," in some as-yet undefined way, about *the value of a*. The data relate to the parameter value through a "likelihood law" $p(\mathbf{y}|a)$, which may or may not be known. This is the probability of the $\mathbf{y}$ if a (i.e., data $\mathbf{y}$ in the presence of a). The aim is to quantify the information in the data $\mathbf{y}$ *about a*, i.e., define it in terms of properties of the system.

1.2.1.1. Comparison with Shannon Information

It is instructive to digress for a moment and compare this problem with the one that defines *Shannon information*. Given data $\mathbf{y}$, the Shannon observer wants, by comparison, to infer from the data whether an event a *occurred*. This is a very *different* problem from that posed above. For this purpose he first forms the ad hoc ratio $P(a|\mathbf{y})/P(a)$, where $P(a|\mathbf{y})$ is the probability that a occurred conditional upon knowledge of data $\mathbf{y}$, and $P(a)$ is the probability of the event a without such knowledge. He then takes the logarithm of the ratio, and calls this the Shannon information. This definition is indeed well suited to the problem it addresses. The extent to which an unknown event a is more probable in the presence of data than in its absence, i.e. the value of the ratio $P(a|\mathbf{y})/P(a)$, does effectively "measure" the amount of evidence or information there is *in* $\mathbf{y}$ *about* the occurrence of a. A drawback is that this definition is, of course, an *ad hoc* construction whereas, by comparison, the Fisher definition (below) is *derived*. The only assumptions of the derivation are (i) the legitimacy of the mean-squared error as an error measure and (ii) the use of an unbiased estimator. Thus, on the basis of ad hoc assumptions, Fisher has the advantage.

A further point of comparison is that the *occurrence* of a and the *size* of a are two very different things. Hence, a system that carries high information (i.e., Shannon information) about the former does not necessarily carry high information (Fisher information) about the latter. Therefore the two informations should be *expected* to obey generally different expressions. (See also Section 1.2.2.12.) We return now to the main development.

1.2.1.2. System Input–Output Law

The system has a random component so that in the presence of a fixed value of a the data $\mathbf{y}$ randomly vary from one set of observations to another. For example, the system might obey an input-output law

$$\mathbf{y} \equiv \mathbf{y}(a, \mathbf{x}) = a + \mathbf{x}, \tag{1.29}$$

where $\mathbf{x} \equiv x_1, \ldots, x_N$ defines the fluctuations of the system. Then for any fixed a there are randomly many possible sets of data $\mathbf{y}$. This behavior defines the system likelihood law $p(\mathbf{y}|a)$, described previously. Intuitively, for the data $\mathbf{y}$ to contain information about a, they cannot be independent of a, i.e., cannot obey $p(\mathbf{y}|a) = p(\mathbf{y})$. We return to this point below.

1.2.1.3. Unbiased Estimators

The observer wants to estimate the value of a. Call an estimate $\hat{a}$. All he has to go on are the data $\mathbf{y}$. Hence, he forms a function $\hat{a}(\mathbf{y})$, called the "estimator function," which he hopes is a good approximation to a. Suppose he can form estimator functions that are correct on average, i.e., that obey $\langle \hat{a}(\mathbf{y}) \rangle = a$ or, equivalently,

$$\langle (\hat{a}(\mathbf{y}) - a) \rangle \equiv \int d\mathbf{y}\, (\hat{a}(\mathbf{y}) - a)\, p(\mathbf{y}|a) = 0. \tag{1.30}$$

The average is over all possible sets of data $\mathbf{y}$ in the presence of a. Estimators $\hat{a}(\mathbf{y})$ obeying (1.30) are called "unbiased," borrowing a term from experimental physics where "unbiased experiments" likewise give correct outputs on average. Amazingly, this unbiasedness condition (1.30) *alone* will lead directly to the definition of Fisher information. It also will *derive the mean-squared error* as the particular error measure that is physically meaningful to the problem. Because these results are so powerful and, also, form the basis for the entire EPI approach, we show the derivation in full. It is quite simple.

The first step is to differentiate (1.30) $\partial/\partial a$. This entails differentiating the integrand, giving

$$\int d\mathbf{y}\,(\hat{a}(\mathbf{y}) - a)\frac{\partial p}{\partial a} - \int d\mathbf{y}\,p = 0, \quad p \equiv p(\mathbf{y}|a). \tag{1.31a}$$

Using the identity $\partial p/\partial a = p\,\partial \ln p/\partial a$ and the normalization property $\int d\mathbf{y}\,p = 1$ in (1.31a) gives

$$\int d\mathbf{y}\,(\hat{a}(\mathbf{y}) - a)\frac{\partial \ln p}{\partial a}p = 1. \tag{1.31b}$$

Reversing the equality and factoring the integrand appropriately gives

$$1 = \int d\mathbf{y}[(\hat{a}(\mathbf{y}) - a)\sqrt{p}]\left[\frac{\partial \ln p}{\partial a}\sqrt{p}\right]. \tag{1.32}$$

The $\sqrt{p}$ item in each factor is key, as it turns out.

1.2.1.4. Use of Schwarz Inequality

Squaring (1.32) and using the Schwarz inequality [1] gives

$$1^2 = 1 \le \int d\mathbf{y}[(\hat{a}(\mathbf{y}) - a)\sqrt{p}]^2 \int d\mathbf{y}\left[\frac{\partial \ln p}{\partial a}\sqrt{p}\right]^2. \tag{1.33a}$$

Squaring out the integrands,

$$1 \le \int d\mathbf{y}\left[(\hat{a}(\mathbf{y}) - a)^2 p\right]\int d\mathbf{y}\left[\left(\frac{\partial \ln p}{\partial a}\right)^2 p\right]. \tag{1.33b}$$

Each of these integrals is now an expectation. The first is

$$\int d\mathbf{y}[(\hat{a}(\mathbf{y}) - a)^2 p] \equiv \langle(\hat{a}(\mathbf{y}) - a)^2\rangle \equiv e^2, \tag{1.34}$$

the mean-squared error in the estimate. The second is

$$\int d\mathbf{y}\left[\left(\frac{\partial \ln p}{\partial a}\right)^2 p\right] \equiv \left\langle\left(\frac{\partial \ln p}{\partial a}\right)^2\right\rangle \equiv I \equiv I(a), \quad p \equiv p(\mathbf{y}|a), \tag{1.35}$$

where I is defined to be the *Fisher information*. The form (1.35) is shown in Chapter 7 to be useful as a diagnostic tool for predicting ecosystem regime changes.

Note that the derivation step (1.31a) required the likelihood law to have a well-defined partial derivative $\partial p/\partial a$. Hence, this approach rules out, in particular, likelihood laws that depend upon only *discrete* values of a.

By the gradient $\partial/\partial a$ operation in (1.35), I effectively measures the gradient content of $p(\mathbf{y}|a)$. Thus, the slower p changes with the value of a, the lower is the net information value $I(a)$. This holds to the limit, so that if $p(\mathbf{y}|a) = p(\mathbf{y})$, which is independent of a, then $I(a) = 0$. As discussed previously in Section 1.2.1.2, this makes intuitive sense: If the data are independent of the unknown parameter, they certainly carry no information about it.

A less extreme scenario is where $p(\mathbf{y}|a)$ changes slowly with a. Then neighboring a values will tend to give similar data sets $\mathbf{y}$. Since the estimate $\hat{a} \equiv \hat{a}(\mathbf{y})$ depends entirely upon the data, it could not then accurately distinguish these a values. Therefore the error e^2 will tend to be large. Hence, I is low and the error is large; or vice versa, when I is high the error is small. Thus, I varies inversely with the error in the estimate of a. This indicates that I measures the amount of information that is present *about a*. The argument may be further quantified as follows.

1.2.1.5. Cramer–Rao Inequality

The use of Eqs. (1.34) and (1.35) in inequality (1.33b) gives a famous result,

$$e^2 I \geq 1 \qquad (1.36)$$

called the "Cramer–Rao inequality" or, simply, the "error inequality." It shows that the mean-squared error and the Fisher information obey a complementarity relation—if one is very small the other cannot also be very small. Or, taken another way, the *minimum possible* mean-squared error obeys

$$e^2_{min} = \frac{1}{I}. \qquad (1.37)$$

This quantifies the above indication that when the information level is high the error is low, and vice-versa. It holds as strictly an inverse relation in the minimum-error case.

The C–R inequality (1.36) has many applications to scientific laws in and of itself. It has been used to derive the Heisenberg uncertainty principle [3], an uncertainty principle of population biology (Chapter 8 and [3]), and a decision rule on predicting population cataclysms (Chapter 8).

1.2.1.6. Fisher Coordinates

The coordinates $\mathbf{y}, a$ of the system define its space or "arena" of activities. These are called "Fisher coordinates." These coordinates vary from one problem to another, and the first step in attacking a systems problem via EPI is to establish its coordinates. In relativistic scenarios, strictly speaking, these have to be four-vectors. Examples are space-time coordinates $(ict, x, y, z), i = \sqrt{-1}$, or equivalent covariant and contravariant coordinates as below. The use of three or less coordinates can lead to some sacrifice of accuracy, as in classical nonrelativistic physics.

Individual Fisher coordinates may be either real or imaginary. (Mixed cases have not yet been encountered.) There are useful applications where the Fisher coordinates are, in fact, a single imaginary time coordinate ict, with c the speed of light [3]. Does the C–R inequality hold in these cases? Repeating the derivation steps (1.30) to (1.36) shows Eq. (1.36) still holds, although now with the error e^2 and the information I real *and negative* numbers. The negative information, in particular, is useful in giving rise to the correct d'Alembertian forms of the wave equations derived by EPI in [3].

1.2.1.7. Efficient Estimation

It is important to consider when the minimum error (1.37) can actually be *achieved*. These are cases of likelihood laws $p(y|a)$ that allow the equality sign in (1.36) which, in turn, follows if the Schwarz inequality (1.33a) is likewise an equality. This occurs when the two "vectors" within the square brackets of (1.32) are parallel, i.e., when all $\mathbf{y}$ components of the vectors are proportional,

$$\frac{\partial \ln p}{\partial a}\sqrt{p} = k(a)\,(\hat{a}(\mathbf{y}) - a)\,\sqrt{p}. \tag{1.38}$$

The constant k of proportionality is in general any function of the constant a. Dividing through by $\sqrt{p}$ gives the final condition

$$\frac{\partial \ln p}{\partial a} = k(a)\,(\hat{a}(\mathbf{y}) - a), \quad p = (y/a). \tag{1.39}$$

The estimators $\hat{a}(\mathbf{y})$ that satisfy (1.39) are called "minimum error" or, simply, "efficient," since then $e^2 = e^2_{\min} = 1/I$. They also must obey the unbiasedness condition (1.30) assumed at the outset. From its invention, in about 1922, until recently, Fisher information has been mainly used to rate any proposed estimator function $\hat{a}(\mathbf{y})$ according to how close its mean-squared error e^2 is to the theoretical minimum value given by (1.37).

The efficiency condition (1.39) is used in Chapter 8 to predict the probability law on biological mass. Here, efficiency is assumed to be one aspect of Darwinian evolution, which tends to optimize population dynamics.

1.2.1.8. Examples of Tests for Efficiency

The reader can readily verify that a normal likelihood law

$$p(\mathbf{y}|a) = \left(\frac{1}{\sqrt{2\pi}}\right)^N \exp\left[-\frac{1}{2\sigma^2}\sum_{n=1}^{N}(y_n - a)^2\right] \tag{1.40}$$

can be put in the form (1.39), and hence allows efficient estimation of a. Hence, the ideal position a of a quantum harmonic oscillator, which has the likelihood law (1.40) in its ground state with $N = 1$, admits of efficient estimation. By comparison, a likelihood law

$$p(y|a) = A \sin^2 [n\pi(y - a)/L] \tag{1.41}$$

cannot be placed in the form (1.39). Equation (1.41) is the law for a freefield quantum particle in a box of length L with the origin in its middle. Strangely, the ideal position a of such a particle cannot, then, be efficiently estimated.

1.2.2. Alternative Forms of Fisher Information I

The various applications of EPI use alternative forms of *one general* Fisher information expression, Eqs. (1.49) or (1.58). These alternative forms are developed next. (See [3] for full details.) Their past applications to systems problems are mentioned.

1.2.2.1. Multiple Parameters and Data

Consider a more general measurement scenario $y = a + x$, where there are now N parameters $\mathbf{a} = a_1, \ldots, a_N$ to be estimated, using N data $\mathbf{y} = y_1, \ldots, y_N$ in the presence of N random quantities $\mathbf{x}$. In place of (1.35) defining a Fisher information scalar $I(a)$, there is now a Fisher information *matrix* $[F]$ with elements

$$F_{mn} \equiv \left\langle \frac{\partial \ln p}{\partial a_m} \frac{\partial \ln p}{\partial a_n} \right\rangle \equiv \int d\mathbf{y} \frac{\partial \ln p}{\partial a_m} \frac{\partial \ln p}{\partial a_n} p, \quad p \equiv p(\mathbf{y}|\mathbf{a}). \tag{1.42}$$

Note that the *trace* of the FI matrix is then

$$Tr[F] = \sum_{n=1}^{N} \int d\mathbf{y} \left(\frac{\partial \ln p}{\partial a_n} \right)^2 p. \tag{1.43}$$

Comparing this with Eq. (1.35) shows that (1.43) is merely the sum of N scalar informations $I(a_n)$.

The Cramer–Rao inequality governing the error in the nth parameter estimate is now [2]

$$e_n^2 = [F]_{nn}^{-1}, \tag{1.44}$$

the nth diagonal element in the inverse matrix to $[F]$.

But an information form (1.43) is not quite general enough for, in particular, *relativistic* applications. Here each scalar element a_n becomes effectively a *vector* $\mathbf{a}_n$ of length $M = 4$, with $\mathbf{a}_n \equiv a_{n0}, a_{n1}, \ldots, a_{n3}$, and similarly for elements x_n and y_n. Note that this scenario is so highly vectorized that it admits of no single scalar Fisher information quantity. Instead, a matrix $[F]$ of Fisher values is known. Stam [6] suggested, in effect, taking the sum down the diagonal of $[F]$ as the net Fisher measure. This is

$$Tr[F] = \sum_{n=1}^{N} \int d\mathbf{y} \, p \sum_{v=0}^{3} \left(\frac{\partial \ln p}{\partial a_{nv}} \right)^2 \equiv I. \tag{1.45}$$

Next, we take the viewpoint that *nature reveals itself out of potential states of maximum information* [7][2]. (Note: The *ansatz* of *maximum* information is generalized to *extreme* information preceding Eq. (1.63).) Information *I* is, in fact, *maximized* in the presence of independent data [3]. Here the likelihood law becomes a product [2]

$$p(\mathbf{y}|\mathbf{a}) = \prod_{n=1}^{N} p_n(\mathbf{y}_n|\mathbf{a}_n), \tag{1.46}$$

causing the individual informations in *n* to directly add. Using (1.46) in (1.45) gives

$$I = \sum_{n=1}^{N} \int d\mathbf{y}_n \, p_n \sum_{\nu=0}^{3} \left(\frac{\partial \ln p_n}{\partial a_{n\nu}} \right)^2 \tag{1.47}$$

$$= \sum_{n=1}^{N} \int d\mathbf{y}_n \frac{1}{p_n} \sum_{\nu=0}^{3} \left(\frac{\partial p_n}{\partial a_{n\nu}} \right)^2, \quad p_n \equiv p_n(\mathbf{y}_n|\mathbf{a}_n),$$

where the familiar formula for the derivative of a logarithm was used.

Note that, again, this is not the actual Fisher information for the *physical* data-taking scenario, where the data are *not* necessarily independent. Rather it is the information for an ideal scenario of potentially maximized information. This latter scenario, and the resulting information form (1.47), *is the one out of which the laws of science follow via EPI.*

Equation (1.47) gives the maximized information in an assumed scenario of rectangular coordinates **y**. The more general scenario of curved space is taken up below. Borrowing some language from *Shannon* information theory, the state (1.46) defines the "channel capacity" information (1.47) of the physical effect, the latter represented by $p(\mathbf{y}|\mathbf{a})$.

We now specialize to some cases of past application.

1.2.2.2. Shift-Invariant Cases

There are many applications where

$$p_n(\mathbf{y}_n|\mathbf{a}_n) \equiv p_{X_n}(\mathbf{y}_n - \mathbf{a}_n|\mathbf{a}_n) = p_{X_n}(\mathbf{x}_n), \quad \mathbf{x}_n \equiv \mathbf{y}_n - \mathbf{a}_n. \tag{1.48}$$

The first equality merely defines the probability density function (PDF) on fluctuations $\mathbf{y}_n - \mathbf{a}_n$ in the presence of fixed variables $\mathbf{a}_n$. The second states that the statistics of the fluctuations $\mathbf{y}_n - \mathbf{a}_n$ are the same for all absolute positions in $\mathbf{a}_n$ space. This holds for phenomena that have no absolute origins. For example, if a particle and its field source are in a box that is moved to a new room in the research laboratory, and the particle is observed for space-time fluctuations **x** after some arbitrary fixed time, the same law $|\psi(\mathbf{x})|^2$ of fluctuation holds in the original room. Likewise, an optical diffraction pattern tends to remain the same regardless of the lateral position of its point source, within a certain range of positions [8]; these

---

[2] This *ansatz* also fits in with the "knowledge game" in Section 1.6.3.

define what is called an "isoplanatic patch." Other phenomena, such as Poisson processes, *do not* obey shift invariance. The observer has to know a priori whether it is obeyed.

In a case (1.48), the general form (1.47) becomes, after changing integration variables to $\mathbf{x}_n$ and suppressing the now redundant subscript n for $\mathbf{x}_n$,

$$I = \sum_{n=1}^{N} \int d\mathbf{x} \frac{1}{p_{X_n}} \sum_{\nu=0}^{3} \left(\frac{\partial p_{X_n}}{\partial x_\nu} \right)^2, \quad p_{X_n} \equiv p_{X_n}(\mathbf{x}). \tag{1.49}$$

1.2.2.3. One-Dimensional Applications

In scenarios where a one-dimensional coordinate $\mathbf{x} \equiv x$ is observed, (1.49) becomes

$$I = \int dx \frac{1}{p} \left(\frac{\partial p}{\partial x} \right)^2, \quad p = p_X(x). \tag{1.50}$$

This is the usual textbook [5] shift-invariant form of the information, and probably the most used in applications. See, e.g., Chapter 2 (Frieden and Hawkins) on econophysics.

A variant on (1.50) is the expression of I as a *cross entropy* [9]

$$I = \lim_{\Delta x \to 0} \left(\frac{2}{\Delta x^2} \right) \int dx p(x) \log \frac{p(x)}{p(x + \Delta x)}. \tag{1.51a}$$

Taking the limit in (1.51a) results in a $0/0$ indeterminate form (note that $\log[p(x)/p(x)] = \log 1 = 0$). However, the indeterminacy is removed by straightforward use of l'Hôpital's rule—twice in succession because of the quadratic nature of the zero Δx^2 in the denominator. The result is the form (1.50). Hence the Fisher information is effectively the cross entropy between a PDF and its infinitesimally shifted version. The form (1.51a) is also quite useful physically. It may be used, e.g., to prove [3] the unidirectional behavior (1.75) of Fisher information.

1.2.2.4. No Shift-Invariance, Discrete Data

Here the data y_n are discrete. Consequently, in the definition (1.35), the indicated averaging becomes a weighted sum rather than the stated integral, as

$$I = \sum_{n=1}^{N} \left(\frac{\partial \ln p_n}{\partial a} \right)^2 p_n, \quad p_n \equiv p(y_n|a), \tag{1.51b}$$

$$\text{or} \quad I = \sum_{n=1}^{N} \frac{(\partial p_n/\partial a)^2}{p_n}$$

after use of the derivative of the logarithm again. This representation was used [3] to derive the general growth and transport laws of physics and biological growth processes (including DNA growth) and polymer growth processes. Note that this information is the exact Fisher information, and not the channel capacity as in Eqs. (1.47) and (1.49). Also, shift-invariance is not assumed.

1.2.2.5. Amplitude q-*Forms* of I

We return to the shift-invariant case (1.49). The division by $p_{X_n}(\mathbf{x})$ in the integrand is bothersome if $p_{X_n}(\mathbf{x})$ has zeros at certain points $\mathbf{x}$. It is not obvious whether these give finite or infinite contributions to the integral. To avoid this problem, it is convenient to reexpress I in terms of real *probability amplitudes*, defined as

$$p_{X_n}(\mathbf{x}) \equiv q_n^2(\mathbf{x}). \tag{1.52}$$

Fisher himself used probability amplitudes, although seemingly not for analysis but, rather, for display purposes. However, the significance of (1.52) cannot be understated. It predicts the probabilistic particle picture of quantum mechanics, where $\mathbf{x}$ is in particular the fluctuation in position of a particle (say, an electron). In this scenario (1.52) amounts to the Born hypothesis, which was a revolutionary *ansatz* in its day. Here, the result seems merely to be true by construction, i.e., trivially. (See Eq. (1.57) for the more general complex version.) But this is not, in fact, true. The same EPI theory that predicts the effect (1.52) also shows that the amplitudes q_n, ψ_n obey wave equations [3], which are the *essential* effects defining quantum mechanics.

Using (1.52) in (1.49) directly gives the q-*form* of I

$$I = 4 \sum_{n=1}^{N} \int d\mathbf{x} \sum_{\nu=0}^{3} \left(\frac{\partial q_n}{\partial x_\nu} \right)^2 \tag{1.53}$$

or, equivalently,

$$I = 4 \int d\mathbf{x} \sum_{n=1}^{N} \nabla q_n \cdot \nabla q_n, \quad \nabla \equiv \partial/\partial x_\nu, \ \nu = 0, 1, 2, 3. \tag{1.54}$$

The sum over ν in (1.53) is now implicit in the dot product. Most important, there is no longer a division in the integrand, lifting the ambiguity problem. The particular form (1.53) for I was used [3] to derive the Maxwell–Boltzmann velocity law.

Likewise, using (1.52) in (1.51b) gives a q-*form* of I for *discrete* random variables y_n,

$$I = 4 \sum_{n=1}^{N} \left(\frac{\partial q_n}{\partial a} \right)^2, \quad p_n \equiv q_n^2(y_n|a). \tag{1.54a}$$

Note that this is an exact Fisher information expression, whereas (1.53) and (1.54) are channel capacities as before.

1.2.2.6. Tensor *Form* of I

In relativistic problems, often a tensor [3] form

$$I = 4 \int d^4x \, q_{n,\lambda} \, q_n^{\cdot,\lambda}, \quad d^4x \equiv dx^0 \cdots dx^3 \tag{1.54b}$$

of (1.54) is handy. Here the x^ν, $\nu = 0$–3 are called "contravariant" coordinates, the x_λ, $\lambda = 0$–3 are called "covariant" coordinates, and these relate to ordinary time and space coordinates (t, x, y, z) as $x_0 \equiv ct$ with $(x_1, x_2, x_3) \equiv (x, y, z)$. Also, note the derivative notation $q_{n,\lambda} \equiv \partial q_n/\partial x^\lambda$ and $q_n{}^{,\lambda} \equiv \partial q_n/\partial x_\lambda$.

1.2.2.7. One-Dimensional q-*Form* of I

The one-dimensional $\mathbf{x} = x$, single component $N = 1$ case of (1.54) is much used,

$$I = 4 \int dx \left(\frac{\partial q}{\partial x}\right)^2, \quad p \equiv q^2. \tag{1.55}$$

Examples of past uses [3] of (1.55) are in deriving both the Boltzmann energy distribution and the *in-situ* cancer-growth law.

1.2.2.8. Complex Amplitude ψ-*Form* of I

In quantum mechanics complex, rather than real, probability amplitudes are used. As a result, quantum mechanics is most conveniently derived [3] out of an I that is expressed in terms of complex amplitudes ψ_n rather than real amplitudes q_n as in (1.53). To form such a replacement, define complex amplitudes ψ_n by packing the q_n as their successive real and imaginary parts,

$$\psi_n(\mathbf{x}) \equiv \frac{1}{\sqrt{N}}[q_{2n-1}(\mathbf{x}) + iq_{2n}(\mathbf{x})], \quad n = 1, \ldots, N/2. \tag{1.56}$$

Note that this gives, in sequence, $\psi_1 = N^{-1/2}(q_1 + iq_2)$, $\psi_2 = N^{-1/2}(q_3 + iq_4)$, ... From this, not only do we directly get

$$\sum_{n=1}^{N/2} \psi_n^* \psi_n = \frac{1}{N} \sum_{n=1}^{N} q_n^2 \equiv p(\mathbf{x}) \tag{1.57}$$

(imaginary terms dropping out), which is the well-known Born hypothesis of quantum mechanics, but also

$$I = 4N \int d\mathbf{x} \sum_{n=1}^{N/2} \nabla \psi_n^* \cdot \nabla \psi_n. \tag{1.58}$$

Note that Fisher coordinates $\mathbf{x}$ in (1.58) are still general. With in particular $\mathbf{x} = (ict, x, y, z)$, i.e., space-time, (1.58) becomes a key part of the Lagrangian that EPI provides in deriving the Klein–Gordon-, Dirac-, Rarita–Schwinger-, and Schroedinger wave equations of quantum mechanics [3]. Newton's laws of classical mechanics, including the Virial theorem, follow as well [3]. Information (1.58) is also used in Chapter 2 on econophysics, where optimum investment dynamics are developed. There a one-dimensional $\mathbf{x} \equiv K \equiv K(t)$ is the capital value of the system (say, company) at a time t, and an optimum program of investment $i(t) \equiv \dot{K}$ is sought. These define investment dynamics that can optimize capital gain.

1.2.2.9. Principal Value Integrals

Any PDF must be defined over a definite support region. This has an important ramification to uses of Fisher information.

In certain applications the amplitude function $q_n(\mathbf{x})$ has infinite gradients ∇q_n at isolated points $\mathbf{x}$. If these are included in the domain of integration in (1.54) the result will be $I = \infty$. Are these cases meaningful? Or, should these isolated points be skipped by taking a principal value integral [1]? Here is where further a priori information enters in. All such points should be included within the domain of integration if they do not lie on a boundary of the physically realizable points $\mathbf{x}$. We call such a boundary a "critical boundary." Alternatively, if they do lie on a critical boundary, they should be skipped.

For example, in a diffusion process, *all* positions $\mathbf{x} = (x, y, z)$ are, in principal, realizable. Therefore there is no critical boundary. Hence, all points should be *included in* the integration, giving infinite information (as we saw). In fact this makes sense—If such infinite gradients exist at a time t_0, after some passage of time the gradients *smooth out* by diffusion, thereby reducing the values of I to finite, and ever decreasing, values. This is consistent with the increased disorder predicted by the Fisher version (1.75) of the Second law.

An opposite example is where $\mathbf{x} = E$, the kinetic energy of a molecule in a gas. Points $E < 0$ can never be realized since they are unphysical (kinetic energy never being negative). Therefore, the point $E = 0$ now lies on a critical boundary and, hence, is *skipped* by taking a principal value integral. This was the case, e.g., in derivation of the Boltzmann energy distribution in [3].

1.2.2.10. I in Curved Space

Many EPI problems occur in a curved space, specified by a metric function $g_{ij}(\mathbf{x})$ of vector coordinates $\mathbf{x}$. Let the latter be sampled at discrete points $\mathbf{x}_n$, and denote the metric values $g_{ij}(\mathbf{x}_n) \equiv g_{ijn}$, $n = 1, \ldots, N$. In quantum gravity, a related metric $G_{ijkl}(\mathbf{x}_n) \equiv G_{ijkln}$ exists that allows the Fisher information about all space-time to be represented. For a spatially closed universe,

$$I = 4 \int d\mathbf{g} \sum_{ijkln} G_{ijkln} \frac{\partial \psi^*}{\partial g_{ijn}} \frac{\partial \psi}{\partial g_{kln}}, \quad d\mathbf{g} \equiv dg_{111} \cdots dg_{33N}. \quad (1.59)$$

Here, the

$$G_{ijkl}(\mathbf{x}) \equiv \frac{1}{2\sqrt{g}}(g_{ik}g_{jl} + g_{il}g_{jk} - g_{ij}g_{kl}), \quad g \equiv \det[g_{ij}]. \quad (1.60)$$

In the limit as the discrete points $\mathbf{x}_n$ become so closely spaced as to approach a continuum, the integral (1.59) becomes what is called a "functional integral," or a "path integral" [10] in the interpretation of Feynman. The use of information (1.59) in EPI gives rise to the Wheeler–DeWitt equation of quantum gravity [3].

1.2.2.11. *I* for *Gluon* Four-Position Measurement

The subject of quantum chromodynamics describes the strong force interaction within the nucleus. Its Lagrangian famously [11, 12] contains a term governing the complex *gluon* amplitude functions $Q_{nv}^\mu(x^\alpha)$, as

$$\mathcal{L}_G = 4N_G \int dx^\mu \, \partial^\mu Q_{nv}^\mu \partial_\mu Q_n^{\mu v}. \tag{1.61}$$

Here tensor notation [3] is used, where repeated indices signify a sum. According to EPI (below), the form (1.61) arises as the Fisher channel capacity for a gluon space-time measurement. This is as follows.

The Fisher information in a space-time measurement of a particle whose amplitude function ψ_{nv} has rank 2 obeys a rank 2 version of (1.54b),

$$I_G \equiv 4N_G \int d^4x \, \partial^\mu \psi_{nv} \partial_\mu \psi_n^v. \tag{1.62}$$

In fact, by the standard model of fundamental particles, a gluon *does* obey an amplitude function ψ_{nv} of rank 2. Thus, the information obtained in measuring its space-time is (1.62). Hence, if (1.61) arises out of Fisher theory applied to a gluon measurement, it must somehow be equivalent to (1.62). A complication is that the amplitude functions Q_{nv}^μ in (1.61) have instead rank 3, owing to the extra index μ. Also, in (1.62) the integration is over one four-space region, while in (1.61) it is over multiple one-space regions dx^μ. We next show that, nevertheless, *the Lagrangian term (1.61) is in fact a special case of information (1.62)*.

As we discuss below Eq. (1.75), in multicomponent problems the Fisher information *I* to be used in EPI is in a maximized state. Here, as one of the special aspects of chromodynamics, it must be generalized to an *extremized* state. Hence, we endeavor to find the solution ψ_{nv} that extremizes the information (1.62), irrespective of whether the extremum is a maximum or a minimum. As probability amplitudes, the ψ_{nv} must also obey normalization,

$$\int d^4x \, \psi_{nv} \psi_n^v = 1 \quad \text{for each } n, v \text{ (no summation)} \tag{1.63}$$

so we use this as a constraint on the extremization of (1.62). Using the approach of Eqs. (1.12) to (1.15), this is a problem

$$\int d^4x \, \partial^\mu \psi_{nv} \partial_\mu \psi_n^v + \lambda_{nv} \left(\int d^4x \, \psi_{nv} \psi_n^v - 1 \right) = \text{extrem.} \tag{1.64}$$

Note that the Lagrange multipliers λ_{nv} are (implicitly) summed over, as in (1.15). The solution obeys an Euler–Lagrange equation

$$\partial_\mu \left[\frac{\partial \mathcal{L}}{\partial(\partial_\mu \psi_n^v)} \right] = \frac{\partial \mathcal{L}}{\partial \psi_n^v}, \quad \partial_\mu \equiv \partial/\partial x^\mu, \quad n = 1\text{–}8, \quad v = 0\text{–}3. \tag{1.65}$$

The integrand of (1.64) is the Lagrangian

$$\mathcal{L} = \partial^\mu \psi_{nv} \partial_\mu \psi_n^v + \lambda_{nv} \psi_{nv} \psi_n^v. \tag{1.66}$$

Then the Euler-Lagrange solution (1.65) is

$$\Box\psi_{n\nu} - \lambda_{n\nu}\psi_{n\nu} = 0 \quad \text{(no summation)}, \tag{1.67}$$

$$n = 1\text{–}8, \quad \nu = 0\text{–}3,$$

$$\Box \equiv \partial_\mu \partial^\mu = \frac{1}{c^2}\frac{\partial^2}{\partial t^2} - \nabla^2.$$

These formally number $8 \times 4 = 32$ equations.

A particular solution $\psi_{n\nu}^P$, using separation of variables, may be sought. This has the product form

$$\psi_{n\nu}^P = Q_n^{0\nu}(x^0)Q_n^{1\nu}(x^1)Q_n^{2\nu}(x^2)Q_n^{3\nu}(x^3). \tag{1.68}$$

Hence, extremizing the information leads to independent data coordinates, as might have been expected. Trial substitution of (1.68) into (1.67) gives the usual result of separation of variables—one ordinary differential equation in each marginal amplitude function $Q_{n\nu}^\mu(x^\mu)$. These are easily solved (not shown).

We note at this point that for each each trio of indices n, λ, ν (no sums) the product

$$Q_{n\nu}^\lambda(x^\lambda)Q_n^{\lambda\nu}(x^\lambda) = p_{n\nu}^\lambda(x^\lambda), \tag{1.69}$$

a PDF (cf. Eq. (1.52)). These of course obey normalization. This is used next.

The information I_G for the solution (1.68) that extremizes the information is found by substituting (1.68) into the general information form (1.62). This gives, after the use of q-form identifications (1.69),

$$I_G = 4N_G \int dx^\mu \left[\int_{\beta\neq\mu} dx^\beta p_{n\nu}^\beta(x^\beta) \int_{\gamma\neq\beta,\mu} dx^\gamma p_{n\nu}^\gamma(x^\gamma) \right. \tag{1.70}$$

$$\left. \times \int_{\delta\neq\beta,\mu,\gamma} dx^\delta p_{n\nu}^\delta(x^\delta) \right] \partial^\mu Q_{n\nu}^\mu \partial_\mu Q_n^{\mu\nu}.$$

Notation: There is no sum on β, γ, δ, but there is an implied sum on μ. The lower-limit integration notation means that, for each μ, the coordinate indices β, γ, δ are to be the three *different* indices from among 0, 1, 2, 3 that do not equal μ. No permutation over indices β, γ, δ is to be performed. For example, with $\mu = 0$ the three integrals within the square brackets are with respect to coordinate indices 1, 2, and 3 in any one order.

By the normalization mentioned above, the three integrals in the square brackets in (1.70) collapse to *unity* for each choice of indices β, γ, δ. This leaves only the outer integral, which is the required Lagrangian term (1.61). QED.

In summary, the Lagrangian term (1.61) is the Fisher information

$$I_G = 4N_G \int dx^\mu \, \partial^\mu Q_{n\nu}^\mu \partial_\mu Q_n^{\mu\nu} \tag{1.71}$$

in a gluon space-time measurement, where the gluon is in an extreme (separable) information state (1.68). The general state is a weighted superposition of these.

1.2.2.12. Local vs Global Measures of Disorder

All of the above measures of the information I depend upon the local *slope values* of a given PDF or amplitude curve. Hence they depend upon the manner by which the points *locally* connect up. Such local measures have the obvious property that discontinuities in the ordinate values (p, q, Q, etc.) give rise to infinite slopes and, hence, to infinite values for the information. Thus, I is very sensitive to the local behavior of the system and, resultingly, is called a "local measure" of system disorder. By comparison, the entropy measure $-\int dx p(x) \ln p(x)$ manifestly does not depend upon the local slopes of $p(x)$ [or $q(x)$] and, hence, is not nearly as sensitive to local discontinuities or near-discontinuities. In fact the points x may be arbitrarily rearranged globally, without changing the value of the entropy. Hence, the entropy is a "global" measure of disorder, in comparison to the "local" measure provided by Fisher. The localized behavior of the Fisher measure is, in fact, what ultimately gives rise to EPI solutions that are differential equations, as compared to the algebraic equations that result from maximizing the entropy [3].

It is notable that these differential equations correctly define the fundamental laws of physics [3]. They also provide information-based definitions of basic physical concepts such as energy [3] or magnetic charge (Section 4.2.7). This property can be described succinctly as "it from I," in obvious parallel with the Wheeler slogan "it from bit" [13, 14]. However, the two are not the same, since "bit" describes entropy, and entropy is (as we saw in Section 1.2.1.1) definitely not Fisher information.

1.3. Introduction to Source Fisher Information J

Whereas information I is that in the data, information J is the level of Fisher information at its source, i.e., in the effect that is observed. It is also called the "bound information" (in the sense of being *bound to* the source). Any observation is the output of an information-flow process

$$J \underset{\text{messenger}}{\rightarrow} I. \tag{1.72}$$

Thus, the source is *not directly* observed. Instead, information of level J *about* the source is transported, by intermediary "messenger" particles, to the data space, at level $I \neq J$. Examples of messenger particles are photons in an optical microscope; or lactic acid ions in cancer growth (see below). These particles play a dual role in EPI, also providing its needed perturbation (see below).

1.3.1. EPI Zero Principle

It is presumed that all such flows are passive, so that I can never exceed J, or

$$I - \kappa J = 0, \quad 0 \leq \kappa \leq 1, \quad \kappa = \text{const}. \tag{1.73}$$

This is called the "EPI zero" principle. The passive nature of κ also ties in with the Second law, and with classical ideas of philosophy, as below.

In general, the observer, the messenger particle, and the observed effect, together define an *information channel*. Information I is that at the output or observer, and J is that at the input or source effect. The channel is the effect or system that is to be analyzed for its statistical likelihood law or functional input-output law.

1.3.2. Efficiency Constant κ

By (1.73), $\kappa \equiv I/J$, so that κ measures the information transfer efficiency of the system. That $\kappa \leq 1$ also ties in with classical ideas of Plato and Kant [15], according to which observations are generally imperfect versions of absolute truth. In the language of Kant, J is the level of information in the or absolute truth, *noumenon*, while I is that in its imperfect observation, called the *phenomenon*. Thus, man observes phenomena, never noumena. In classical cases (mechanics, electrodynamics), $\kappa = 1/2$, and in quantum effects (quantum mechanics, quantum gravity) $\kappa = 1$. These two values of κ together imply that the 50% loss of information in classical effects results from ignoring their quantum aspects.

Interestingly, in cancer growth (see Chapter 3 or [3]) there is a net $\kappa = 4.27$, seeming to violate inequality (1.73). Rather, this indicates a "ganging up" effect, whereby the many cells in a cancer mass *independently* send messages to a nearby healthy cell. The messengers are ions of lactic acid, and the messages are "die," since the lactic acid poisons the healthy cell. Under these conditions of independence the individual amounts of information I from cancer cells add, forming a net information that much exceeds that indicated by (1.73) for a *single* message event (1.72). Since each individual $\kappa \leq 1$, the result 4.27 predicts that, on average, at least 4.27 cancer cells contribute to the poisoning process.

1.3.3. Fisher I-Theorem

The loss of information (1.73) occurs during the measurement transition $J \to I$. This takes place over the measurement time interval $(t, t + dt)$. This loss is consistent with the Second law of statistical physics, as follows.

On a qualitative level, the Second law states that *disorder increases with time*. From Eq. (1.55), the information I due to a one-dimensional fluctuation x, at a fixed time t, obeys

$$I = I(t) = 4 \int dx q'^2, \quad q \equiv q(x|t), \quad q' \equiv \partial q / \partial x. \tag{1.74}$$

This shows that I in fact *measures* the level of disorder of the system. Directly, I is small if and only if the gradients q' are small in magnitude. By *normalization* of $q^2(x|t)$, this implies a broadened function $q(x|t)$ which, by $p(x|t) = q^2(x|t)$, implies a broadened $p(x|t)$. This, in turn, describes a more random scenario, where the frequency of occurrence of small events and large events x tend toward equality.

This increase of randomness means an increase of disorder with the time. Thus the loss of information (1.73).

1.3.3.1. I as a Montonic Measure of Time

Again, by the Second law of statistical physics, the level of disorder must decrease (or remain constant) with time. Then, by the preceding, this suggests that a classical statistical system obeys the *I-theorem*,

$$\frac{dI}{dt} \le 0. \tag{1.75}$$

This suggestion was, in fact, the first *prediction* of the modern Fisher approach [7]. It was proven rigorously [3, 16] for $p(x|t)$ that obey a Fokker–Planck (generalized diffusion) equation. The same F–P condition also gives rise to the Boltzmann H-theorem, which is the usual statement of the Second law (using entropy instead as its measure of disorder). The "Fisher I-theorem" (1.75) is taken to be an alternative expression of the Second law.

One interesting consequence of (1.75) is that *it predicts the direction of an "arrow of time"* [3]. That is, if the Fisher information level of a system is observed to decrease, $dI < 0$, its history is necessarily moving forward, $dt > 0$.

A second consequence is its implication as to the state of the system immediately *prior to* its observation. If the system is already in a *maximized information state* then it can only suffer a loss of information after the measurement, obeying the requirement (1.75). This condition of a maximized prior state of information is therefore a premise of EPI.

The premise of a maximum is generalized to a case of *extremized* information in the nuclear strong-force scenario (see material above Eq. (1.63)).

1.3.3.2. Re-Expressing the Second Law

We showed in the preceding that the evolution of disorder is not uniquely measured by entropy. Fisher information is an alternative measure that likewise changes monotonically with the time. In fact in Section 4.4.6 (Chapter 4) an *entire family* of such measures is defined. Thus, the Second law as stated in standard textbooks is overly restrictive. The correct statement is that *disorder by any appropriate measure* monotonically changes with the time. By the I-theorem (1.75), this includes the Fisher measure.

Indeed, the Fisher measure is sufficiently powerful to provide *by itself* a distinct development of thermodynamics (Chapter 4). This ignores any use of Boltzmann entropy, except for purposes of comparison.

1.4. Extreme Physical Information (EPI)

Many of the EDA approaches in this book are based upon use of the EPI principle. A brief introduction is as follows (see [3] for full details). We return to the *information channel* (1.72). It is known, since Heisenberg, that any measurement perturbs the thing being measured: Because any datum is formed out of particles that illuminate and, hence, interact with the object, these messenger particles perturb the object. Thus, information level J is perturbed, by some amount δJ. Then the flow Eq. (1.72) implies that I is perturbed as well. This is by an amount

$$\delta I = \delta J. \tag{1.76}$$

Depending upon the measurement scenario, this is either a postulate or holds identically. It holds identically if the space of J is unitary to that of I (see Eq. (1.21)), and if J space has a physical reality. Examples are momentum values for quantum particles [3], or *investment momentum* in a program of optimum investment (Chapter 2). It follows from (1.76) that $\delta I - \delta J = 0$ or, equivalently, $\delta(I - J) = 0$. This implies that the value of $I - J$ is some extreme value,

$$I - J = \text{extremum}. \tag{1.77}$$

On the other hand, Eq. (1.73) states that the information change $(I - J) = J \cdot (\kappa - 1) < 0$ is *a loss*. Thus, (1.77) expresses a principle of "extreme loss of information," or to put it in a more positive way, "extreme *physical* information" owing to its many physical applications.

As with I, information J is in practice a functional, i.e., an integral involving the likelihood law and the input–output law. The principle (1.77) is implemented by regarding the integrand of functional $(I - J)$ as the Lagrangian $\mathcal{L}$ of the problem. The solution is then generally a *differential equation*. This is found by the use of the Euler–Lagrange Eqs. (1.6), (1.9), or (1.65), depending upon dimensionality.

1.4.1. Relation to Anthropic Principle

It is interesting to interpret the foregoing in light of the *anthropic principle* (AP) [17] in its "weak" form. Basically this states that observed effects are the way they are because, if they were not, our carbon-based, intelligent life form could not exist to observe them. This idea has been used to rule out ranges of the physical constants that would not permit the existence of such life [17].

We show next that the anthropic principle is consistent with EPI.

Consider, in this regard, the EPI property that $I \approx J$ (Sections 1.3.2 and 1.4.5). *Without* this property, e.g., with $I \ll J$, observation would not mirror reality. Let us define as "intelligent" any carbon-based life form that can formulate and utilize many of the laws of our universe. Consider, under these conditions, a scenario of amoebae searching for food. With $I \ll J$, the amoebae would be basing their search upon randomly erroneous data, so that even those with an increased ability to learn would *not* be rewarded with food. Consequently, *any tendency toward intelligence would not confer an evolutionary advantage.* Hence, such life forms would not tend to evolve toward intelligent life forms.

Conversely, because our universe does in fact obey $I \approx J$, *evolution rewards life with increased intelligence.* (This is not to say that all animals evolve toward intelligence. Many use other strategies to increase their fitness. The condition $I \approx J$ *allows* intelligence to develop, but does not demand it.) Thus, man is an intelligent life form, i.e., capable of formulating and utilizing the laws of the universe. At this point in time, apes and dolphins, e.g., are not. If and when they become so, it will be but an example of convergent evolution.

Hence, EPI is consistent with increased information, therefore the existence of man, and, consequently, the anthropic principle. We emphasize, however, that AP and EPI are not, on this basis, equivalent. AP is not an approach for the discovery of physical laws, as EPI is, but rather, a tautology leading to some useful considerations on values of the physical constants (as above).

1.4.2. Varieties of EPI Solution

A basic premise of EPI is that both the zero-principle (1.73) and the extremum principle (1.77) are to hold for *any* observed phenomenon. For some effects both principles have a common solution (examples [3] are the Boltzmann energy distribution and the in situ cancer growth law). For other effects there are two distinct solutions (an example [3] is in quantum mechanics, where the Klein–Gordon equation arises out of (1.77) while the Dirac equation arises out of (1.73)). By the overall approach either the system likelihood law $p(\mathbf{y}|a)$ or the system input-output law $\mathbf{y} = \mathbf{y}(a, \mathbf{x})$ (see (1.29)), or both, can be sought. Operationally, this is simply a matter of choice as to which is varied to attain the extremum. Most applications of EPI have been to estimate likelihood laws [3], although system input-output laws may also be derived. An example of the latter is the famous biological law expressing attributes, such as the metabolism rate, in terms of the mass of the organism. This is as the mass to a multiple of a quarter power (Chapter 8). Another such example is the Lorentz transformation, which follows from the EPI zero-principle (1.73) *alone* [3].

See also the "exhaustivity" property of EPI below (Section 1.5.2).

1.4.3. Data Information is Generic

I is the level of information from the standpoint of the observer. On this basis all *data are regarded generically* by EPI. That is, data are simply data irrespective of their source. Hence, in all uses of EPI, regardless of the nature of the source effect, I always has one of the *equivalent forms* derived in Section 1.2. By comparison, J has a form that is *specific to* each application. This is because J is the level of information that is supplied *by the source*, before it is transmitted to the observer. The operational result is that J acts as the driving force in each differential equation (1.6) or (1.9) that results.

How does one know the particular J for a given problem?

1.4.4. Underpinnings: A "Participatory" Universe

The well-known physicist John Wheeler [13, 14] has posited the following world-view:

All things physical are information-theoretic in origin and this is a participatory universe . . . Observer participancy gives rise to information; and information gives rise to physics.

Clearly, EPI fits well into this worldview. The "observer participancy" is the observer's measurement, the information is Fisher's, and it "gives rise to physics" through the EPI principle (1.73), (1.77).

EPI further develops this idea, in recognizing that there is not one but two informations to consider in any measurement. The received information I is always about "something," and this is the source effect at information level J.

In most applications, the extremum attained by EPI principle (1.77) is a mini-mum. Thus, the acquired information I is as close to the true level J as is possible, that is, $I \approx J$. In fact, in quantum effects, it is rigorously true that $I = J$ [3]. The full level of information comes through. This also more generally holds for all effects that physically admit of a *unitary space* (e.g., Fourier or momentum space in quantum mechanics) to measurement space. This has an elegant interpre-tation: Although the observer can never *directly* view a noumenon (Section 1.3), his observation of its phenomenon can carry as much information about it as is available. This allows him to infer the nature of the noumenon as well as possible. Philosophically, this is as well a comforting thought. Otherwise, what would be the point of observing?

The EPI approach may also remind students of philosophy of the famous adage of "the cave" of the Greek philosopher Plato. A person born and raised in isolation in a cave sees shadows on the wall cast by people outside the cave. From these he concludes that the shadows *are* the people. Here is an example where the acquired information level I is much less than the intrinsic level J defining the people outside. That is, the information efficiency constant (1.73) obeys $\kappa \ll 1$.

This has a parallel in the philosophy of the German philosopher Immanuel Kant (see also Section 1.3.2):

Man observes a phenomenon that is only a sensory version of the "true" effect, the latter called a *noumenon*. Hence the noumenon is some unknown, perhaps unknowable, absolute statement of the nature of the effect. Man cannot know the absolute noumenon, and so contents himself with merely observing it as a *phe-nomenon*, within the limited dimensionality of some sensory framework. Various frameworks have been used for this purpose through the ages: witchcraft, astrology, religion or (in our era) differential equations! How does EPI fit within this scheme, and can it perhaps provide an absolute framework for defining the noumenon?

The EPI framework is in fact provided by the notion of an observation. Here the noumenon is an unknown physical process. To be identified as a noumenon, it must first be observed as a phenomenon. (Parenthetically, we do not consider here *unobservable* noumena, such as possible parallel universes or the strings of string theory.) The observation is an absolute truth about the noumenon. This is in the sense that it is known to exist (whether accurately or not). It is important to emphasize at this point that the use of EPI allows us to learn properties of *the noumenon*. In this respect it departs from Kant's model, which regards the noumenon as generally unknowable. The EPI approach consists of the following program: observation, measurement, the use of *a priori* system information in the form of an invariance principle (see below), and, finally, some mathematics (typically a variational problem). In this manner, EPI codifies, organizes, explains, systematizes, gives numerical solutions, and predicts properties of the noumenon. All without directly viewing it, as per Kant.

1.4.5. A "Cooperative" Universe and Its Implications

Some followers of Kant, notably Schopenhauer, were struck by the largely negative nature of his philosophy. They became despaired of the fact that we never directly "see" the noumenon. Furthermore, as discussed above, we often do not see the full information level of the noumenon, since generally $I \leq J$. In fact, Kant thought that we could never attain full information $I = J$ about the noumenon (see above). However, EPI (and, equivalently, the philosopher B. Spinoza) indicates that Kant was incorrect in this regard. Cases $I = J$ exist. For example, $I = J$ holds for all unitary phenomena, including quantum mechanics (Table 1.1). Thus, EPI turns the apparent disadvantage around: It shows that "the truth" (the mathematical law of physics at the source) can, anyhow, be *inferred*. This is by using EPI, i.e., either by *extremizing* the difference $|I - J|$ of information or by inferring that $I = J$ in quantum, and other unitary, cases.

As we see below in Section 1.6, the extremum is often a *minimum*. For these noumena EPI accomplishes a *minimizing of the loss $I - J$* of the absolute truth in observations of them. Thus, it achieves a level of information I in the data that obeys $I \approx J$, i.e., contains *a maximal level of truth* about the source. In fact most fundamentally, on the quantum level, EPI accomplishes $I = J$, i.e., the maximum *possible* level of truth in the observations. This is comforting, since it satisfies what a measurement is meant to accomplish, and might likewise have lifted Schopenhauer's spirits.

It should be noted that to accomplish a quantum measurement for which $I = J$, the object must be in a *coherent state*. In practice this can only be accomplished when viewing microscopic objects, since a loss of coherence, i.e., *decoherence*, occurs for macroscopic objects. Therefore, in principle, the *potential* exists for observing an absolute maximum level $I = J$ of truth.

In the preceding, Wheeler called our universe "participatory." However, it is really more than that. By the participatory use of EPI, we infer its laws. However, since in fact the resulting information levels I are close to being their maximum possible, or noumenal, level J, *this universe cooperates maximally with our objective of understanding it through observation*. Hence, it can be further described as *cooperative*.

The degree of cooperation is actually of *three* types: (1) Since the data contain a maximum of information I about the parameter under measurement, this allows

TABLE 1.1. Summary of some Fisher applications.

Source effect	I(a), coord.	Unitary trnsf.	J(a)	Type solu., κ
Lorentz transf. [3] from $\mathbf{q}, a \to \mathbf{q}', a'$	Eq. (1.53) $\mathbf{x} \equiv$ sp-tm	Rotation by imaginary angle	$J(\mathbf{a}) = I'(\mathbf{a}')$ in moving frame (info. invar. to ref. frame)	EPI zero (1.73) $\kappa = 1$
Relativistic quant. mech. [Chapter 4 of 1.3]	(58) $\mathbf{x} \equiv$ sp-tm	Fourier	$\iint d\boldsymbol{\mu}\,dE\,P(\boldsymbol{\mu}, E)(-\mu^2 + E^2/c^2)$ (μ = momentum, E = energy)	EPI type (A) $\kappa = 1$
Quantum gravity [3]	(1.59) $\mathbf{x} \equiv$ sp-tm of univ.	Functl Fourier	$\int d\mathbf{x} \sum_{ijkl} D\mu\, P(\mu(\mathbf{x}))\mu^{ij}\mu^{kl}$	EPI type (A) $\kappa = 1$
Higgs mass effect [23]	(1.54b) $\mathbf{x} \equiv$ sp-tm	Rotat., Weinberg angle θ_W	$J^{Higgs} = I^Z$ (1.58), $Z \equiv Z^0$ boson	EPI type (A) $\kappa = 1$
Collapse of wavefn. at measurement, Chapter 10 of [3]	(1.58) $\mathbf{x} \equiv$ sp-tm	Fourier	As preceding	EPI Eq. (1.77) + log-likelihood. $\kappa = 1$
Heisenberg unc. princ. Chapter 4 of [3]	(1.58), $N = 2$ $\mathbf{x} \equiv$ position	1D Fourier	Not used	Cramer–Rao ineq. (1.36)
Pop. biol. unc. princ., Chapter 3 Decision rule for pop. cataclysm., Chapter 3	(1.51b), $a \equiv t$, the time	Not used	Not used	Cramer–Rao ineq. (1.36)
Classical electrodynamics	(1.54) $\mathbf{x} \equiv$ sp-tm	Not used	$\iint d\mathbf{r}\,dt \sum_n j_n q_n$ (j_n = four-current density)	EPI type (B) $\kappa = 1/2$
Classical gravitation	(1.54b)	Not used	$\iint d^4x\, T^{\mu\nu} q_{\mu\nu}$ ($T^{\mu\nu}$ = stress-energy tensor)	EPI type (B) $\kappa = 1/2$
Boltzmann energy E distrib.	(1.54) with $\mathbf{x} \equiv x = iE$	Not used	$\int dE\, q^2(E)$ (E = energy)	EPI type (B) $\kappa = 1$
Maxwell–Boltzmann law	(1.53) with $\mathbf{x} \equiv (c\boldsymbol{\mu})$	Not used	$\int d\mathbf{x}\, x^2 \sum_n A_n q_n^2$, $A_n =$ const. ($\mathbf{x} \equiv c \cdot$ momenta)	EPI type (B) $\kappa = 1$

Phenomenon	Data/measurement model	Prior/transform	EPI measure	EPI type
Quasi-incompressible turbulence	(1.54) with $\mathbf{v}$ = vel., $\mathbf{x} = (\rho\mathbf{v}, \rho c)$, ρ = density	Not used	$\iint d\mathbf{w}\, d\rho\, q^2[\lambda_1 w^2/(2\rho) + H(\rho - \rho_1)(\lambda_2 + \lambda_3 \epsilon(\rho))]$ ($\mathbf{w} = \rho\mathbf{v}$, H = step funct., $\epsilon(\rho)$ = internal energy)	EPI type (C)
EPR-spin entanglement	(1.54a) with $(y_n|a) \equiv (S_a|x)$	Not used	$\sum_{ab} \int dx\, q_{ab}^2$ with x = angle between 2 analyzer orientns $\mathbf{a, b}$	EPI type (B)
Optimum investment schedule for prodn. fn. F(K(t)) (Chapter 2)	(1.58) with $N = 2$, $\mathbf{x} \equiv K$, capital	Fourier	$\int dt\, F(K(t)) \exp(-\rho t) \equiv (F(K(t)))_t$, t = time	EPI type (C)
(1) Price fluctuations in stocks, via "technical approach" [1]	(1.55), $\mathbf{x} \equiv x \equiv t$	Not used	$\sum_m \lambda_m \int dx\, f_m(x)q^2(x)$	EPI type (C)
(2) Fluctuations in extrinsic thermodyn. parameters [24]	(1.55), $\mathbf{x} \equiv$ vel. or E_{kin}			
Population growth	(1.51b) with $a \equiv t$, n = species type	Not used	$\sum_n p_n(g_n + d_n)^2$ (g_n, d_n = growth, death comps. of fitness coeffs.)	EPI type (B), $\kappa = 1/2$
Molecular replication, RNA molecules	(1.51b) with $a \equiv t$, n = RNA strand	Not used	$\sum_n p_n \left(\frac{b_n A}{A_n + A} - d_n \right)^2$	EPI type (B), $\kappa = 1/2$
Cancer growth [3], in situ	(1.55) with $x \equiv t$	Not used	$\int dt(q^2/t^2)$	EPI type (B), $\kappa = 4.27$
Quarter-power laws biology [25], chapter 8	(1.55) with $\mathbf{y} \equiv y$	Not used	$\sum_n A_{nj} \cos(4\pi na)$, $j = 0, 1, 2, \ldots$	EPI type (C)

us to *optimally estimate it*. (2) By using EPI, we can infer the laws of physics *that gave rise to* the data. That is, *we optimally learn the laws that govern the universe*. (3) Finally, by telling us that $I = J$ nature lets us know that *there is nothing more to be learned* about the observed effect (at the given scale of observation). Thus the EPI-based observer (at least) is saved from embarking on wild-goose chases. Indeed, the level of cooperation provided by the universe is so strong that it ought not only be regarded as our school, but also our playground.

In fact this cooperative property is a very special one. Before it was tried (first by Maupertuis and Lagrange as an "action" principle—see Section 1.4.7), there was no guarantee that EPI would give rise to correct laws. Certainly the weak form of the anthropic principle does not imply it. That is, our mere *existence* as a carbon-based life form does not imply an optimum ability to understand the universe. Certainly, other life forms, such as apes, do not have this ability. Rather, it is our existence as *obsessively curious and thinking creatures* that permits it, as discussed next.

The cosmologist E.R. Harrison [17] has proposed that universes such as ours evolve specifically to nurture *intelligent* life forms. Presumably, these life forms play an active role in the evolution of their universe. And information-oriented creatures like us can only do this if we first *learn* how it works. Of course, the learning process is most effective if $I \approx J$, and there we have the EPI principle.

Thus, Harrison's is a kind of "strong interpretation" of the anthropic principle that leads to a cooperative universe and to EPI. In summary, any Harrison-type universe that forms and nurtures intelligent life must be *cooperative* in imparting information to such life.

This begs the question of what active role intelligent life *will play* in the evolution of such a universe. The EPI approach in fact logically supplies us with one. As we saw, this is an information-dominated universe that cooperates by allowing us *maximum* information gain at each observation. Therefore, assuming that we want to maintain this universe, our ultimate role is *to preserve these maximal gains of information*. Certainly our invention of libraries and the Internet, not to mention secure encoding devices (Chapter 6), are concrete steps in this direction.

However, of course natural effects exist that tend to thwart such activities. The Second law of thermodynamics is one well-known example. Its end, after all, is a "heat death" for the universe, when asymptotically *zero* information would be gained at each observation. On this basis, then, *our goal is straight-forwardly to effect a reversal of the Second law within this universe*. A first thought is to use other universes or dimensions as depositories for our excess entropy, analogous to the way life processes locally gain order (lose entropy) by dumping waste entropy into their surroundings (Chapter 3).

A clarification, however, is in order: This ultimate goal of reversing the Second law is not something that we *ought to* do but, rather, is something that we will *automatically* do, as a natural effect of living in a Harrison universe. As precursors, does not our activity in the arts and sciences already have as its effective goal the creation of order out of chaos?

1.4.6. EPI as an Autopoietic Process

We have found that the probe particles of the channel fill a dual role. They both carry information about the subject into the observations *and*, by their perturbing effects, give rise to an extremum principle, EPI. This principle defines the very physical law that gave rise to the observations. This dual process is self-perpetuating or autopoietic in nature. It describes an Escher-like cycle that intimately connects the human observer with the source effect he/she wants knowledge of. In this respect the process is a good example of the "participatory" nature of Wheeler's universe (Section 1.4.4), and also of the "cooperative" nature of the universe (Section 1.4.5).

1.4.7. Drawbacks of Classical "Action Principle"

Of course, historically, the *action* principle, not EPI, is the original Lagrangian approach to physics. Conceived of by Maupertuis, developed by Lagrange and generalized by Helmholtz, the action principle has had many successes in analyzing physical problems. However, it has notable drawbacks. Most fundamental are its whys and wherefores [18]: Why should the action be *extremized*? Why not some other condition? What does "action" really mean?

By an action-based approach, information I is replaced with a generalized kinetic energy functional T, and information J is replaced with a generalized potential energy functional V. Thus, the action principle is phenomenological in nature. This is, in fact, at the root of its problems of practical use: With a new phenomenon, how does one know what to use for the "generalized" kinetic energy and potential? This question has led to useful extensions of these concepts in various scenarios, such as electromagnetics. But what about scenarios where there is no ready association of T with a kinetic energy or V with a potential energy? For example, what about "open system" scenarios, which do not obey conservation of particle number, energy, etc.? Examples are in population dynamics [19], economics [20], biological growth [21], sociology [22], and other fields that are outside the traditional mainstream of physics. Undoubtedly kinetic and potential energies "exist" in these fields, but it is unclear how they contribute to the T and V needed in the Lagrangian.

By comparison, Lagrangians have been formed in these fields, on the basis of informations I and J (see preceding references) and not energies T and V, thus enabling a Lagrangian approach—EPI—to be utilized. The solutions have proven to be meaningful and useful.

In attacking a new problem, then, which should be used—the action approach or EPI? From the preceding, EPI has a wider scope of application since it includes nonconservative systems. It is, in fact, *inclusive* of the action approach to systems problems, becoming mathematically equivalent to it in many cases (e.g., nonrelativistic quantum mechanics). EPI also provides a class of solutions that action approaches entirely ignore—the EPI zero solutions (1.73). It also provides Lagrangians for the solution of others ([19]–[22]) for which no action Lagrangian was known to exist. A principal aim of this book is, in fact, to solve such problems. Finally, EPI follows a well-defined procedure, wherein I always obeys (1.47) or one of its specialized forms in Section 1.2.2, and J is constructed as follows.

1.5. Getting the Source Information J for a Given Scenario

There are three fundamentally different ways of getting the functional J for a given problem. These depend upon the level of prior knowledge at hand, and can be termed (A) abduction, (B) deduction, and (C) induction (American philosopher C.S. Peirce; see also Sec. 1.6.5). Each gives rise to a corresponding level of *accuracy* in the EPI output, as follows.

1.5.1. Exact, Unitary Scenarios: Type (A) Abduction

The various expressions for Fisher I found among Eqs. (1.35) to (1.71) all have basically the same form, a sum-of-squares or L^2 measure. We found at Eq. (1.26) that such a measure is invariant under a unitary transformation. This has a very important ramification.

In a wide class of problems, the observer has prior knowledge of a conjugate space to observation space, such that a unitary transformation U connects the probability amplitudes of the two spaces. Also, the conjugate space coordinate is *physically meaningful*. An example is momentum-coordinate space in quantum mechanics, where $U \equiv F$, the Fourier transform.

In these unitary scenarios, J is found very simply. The EPI principle (1.77) has been shown to hold [3] if, in such unitary scenarios, J is set equal to I as *evaluated in conjugate space* (for example, by the use of Parseval's theorem). Note that the zero-principle (1.73) now holds trivially, as well, with $\kappa = 1$. Hence, in a unitary scenario both EPI principles (1.77) and (1.73) hold. Indeed, they hold independently, i.e., give generally different (but consistent) solutions $\psi(\mathbf{x})$. Examples [3] are, respectively, the Klein–Gordon and Dirac equations in quantum mechanics, and both Newton's Second law $f = ma$ and the Virial theorem in classical mechanics.

Using EPI, unitary scenarios give rise to theories that are as *exact* as are currently known, i.e., on the level of relativistic quantum mechanics. These were (historically) called type (b) approaches in [3]. Such exact theories are called *abductions*, or first principles, by Peirce.

1.5.2. Exhaustivity Property of EPI

Experience with EPI indicates that every unitary transformation defines a different physical effect. For example, the Fourier unitary transformation (1.27) gives rise to the effect called *quantum mechanics* [3]. See Table 1.1 for this and other examples. This property of *exhaustivity* is a vital property of the approach.

As a corollary, most effects are that EPI gives in the *absence* of a known unitary transformation are *inexact*. Experience likewise bears this out, as summarized in Table 1.1.

Another corollary is that new physical effects can be mathematically defined and anticipated, even before they are observed, by the activity of discovering new unitary transformations and applying them to EPI. Of course, for this approach to be useful, the observer must also be capable of interpreting the (at first) abstract co-ordinates of the unitary transformation as physical observables. This takes physical insight.

A final, if bizarre, corollary is the possibility that new physical effects could be *invented* or *formed* by the hand of man. This would be by devising unitary transformation spaces that currently do not naturally exist. (An analogous situation is the recent past invention, in high-energy nuclear physics, of the so-called manmade elements.) Their use in EPI would lead to *laws by design*. This is to be compared with current laws of physics, which could be termed *laws of default*.

1.5.3. Inexact, Classical Scenarios: Type (B) Deduction

A second class of problems gives *inexact* solutions, as indicated by $\kappa < 1$. Effects of this kind are classical in nature—gravitation and electromagnetics [3], for example. Here, J is not trivially known as the transform-space version of I as in the preceding. Rather, J *must be solved for,* by simultaneous solution of Eqs. (1.73) and (1.77). The supposition is that the lower level of information now disallows the knowledge of two distinct solutions to (1.73) and (1.77), so that they are effectively "blurred together" into one composite law that approximates both. For brevity, this is called the *self-consistent* EPI solution.

In order to obtain such a common solution, some prior knowledge must be at hand about $\psi(\mathbf{x})$. This is generally in the form of an *invariance principle,* such as continuity of flow. (In principle an equivalent symmetry given by Noether's theorem could also be used.)

Such a state of knowledge is called *deduction* by Peirce. It does not give an exact solution. Past examples are continuity of flow and gauge invariance [3], used in derivation of the laws of classical gravitation and classical electromagnetics. Although these laws are approximations, they are of course very good ones. The subsidiary variable κ is found to be exactly $1/2$ in these cases, rather small considering how good the approximations are. Table 1.1 shows a number of examples of these type (B) solutions (historically called type (A) in [3]).

In many of these scenarios, although the *input* form of the functional J is not in inner-product form (since it is not a type (A) scenario), its *solution* must, in fact, *have* inner-product form (see *classical electrodynamics* item in Table 1.1). The explanation is simple. There is a distinction between J in its general form for a problem and J in its *solution* form. By (1.73), J is always proportional to I. But I is an L^2 measure and hence, as we saw, invariant to unitary transformation at solution. Therefore, so is J at solution, meaning, it must likewise be in inner-product form.

1.5.4. Empirical Scenarios: Type (C) Induction

The next level down in accuracy ensues are scenarios where the prior knowledge is so limited that κ and J *cannot* now be solved for. This occurs, e.g., in econophysics ([3] and Chapter 2) where the "technical viewpoint" of investment assumes price data as *effective* invariants. Or in statistical mechanics (Chapter 4) the invariants are extrinsic measurements of means. Here the information J is replaced by empirical constraints, so that the EPI approach devolves into the MFI approach. See Eq. (2.3), Chapter 2, and examples in Table 1.1. This level of knowledge is called *induction* by Peirce.

1.5.5. Summary

In approximate cases (C), we found that the form of J is fixed by *empirical* data. Since these have random components, the resulting EPI outputs cannot have universal validity. These economic and statistical mechanical outputs are mathematically equivalent to those of MFI, which minimizes I plus a pure constraint term as in the Lagrange-undetermined multipliers approach (1.15). Nevertheless, indications are that these approaches are at least moderately accurate (see Chapters 2 and 4). In one notable case [2], where the exact answer is known to be the Boltzmann (exponential) law on energy x, the MFI answer for $p(x)$ is instead a squared Airy $Ai(x)$ function, which differs from the Boltzmann curve by only the order of a few percent over the useful (significantly high) part of $p(x)$. However, the level of accuracy is not as high as in a classical, type (B) answer.

In summary, EPI solutions are of types (A), (B), or (C), depending upon the level of prior knowledge. Type (A) solutions are of the highest accuracy and require knowledge of a physically meaningful unitary transformation. Type (B) solutions are down one notch in accuracy, and require knowledge of one, or more, invariance principles. These are not associated with a unitary transformation of type (A), as mentioned previously. Finally, type (C) solutions are least accurate, dependent upon empirical data as prior knowledge. The implication is that the highest form of knowledge is of a unitary transformation, next down is an invariance principle, and the lowest is empirical data. This makes sense and, as we saw, agrees with Peirce's classes of prior knowledge.

Table 1.1 summarizes many past applications of EPI, classifying them as to whether of type (A), (B), or (C). It also lists the particular unitary transformations and functionals J that have, so far, been found to give rise to the various source effects. Also listed are the corresponding forms of $I(a)$, as referenced by equations in this chapter.

1.6. Information Game

Equation (1.10) shows that during variation of I in principle (1.77), I is actually *minimized*. That is, using $\mathcal{L} = 4q'^2$ from representation (1.55) for I, the Legendre condition (1.10) gives $\partial^2 \mathcal{L}/\partial q'^2 = +8 > 0$, the condition for a minimum. (This assumes that coordinate x is real, in particular not an imaginary time ict.) This has two important ramifications.

Note that the information term J seldom depends upon gradient terms such as q'. Then, by the preceding paragraph, the output value of $I - J$ in a single-coordinate EPI solution (1.77) only achieves *a minimum* as its extremum. This means that the data information I is maximally close to the source information J. Or, although one usually cannot directly observe the full level of information J about an effect, he can infer it optimally as acquired information I. Hence, nature tends to cooperate with the observer in disclosing its effects. See further discussion of the effectively "cooperative universe" in Section 1.4.5.

Second, by the form $\mathcal{L} = 4q'^2$, the minimum results from low-gradient values $q'(x)$. By normalization of $q^2(x) \equiv p(x)$ this is equivalent to a maximally spread-out or *blurred* $q(x)$ or $p(x)$ curve. Hence the information about the unknown parameter value a (Section 1.2.1.2) tends to be minimal. This is to be compared with the "cooperative" effect of the preceding paragraph, whereby information *about the effect* is maximized.

This blurring effect also implies that an act of observing is effectively a move in a mathematical game with nature. This *game aspect* of EPI is discussed in the next few sections.

1.6.1. Minimax Nature of Solution

In comparison with the minimizing tendency in I due to blurring, we found that the general form (1.47) of I that is used represents a state of the system that *maximizes* the information, as if it arose out of independent data. These would have correlation coefficient $\rho = 0$. Or conversely, for any fixed state of blur, I decreases with increased correlation of the data. Consider in this light values of I in a two-dimensional space of correlation value ρ (vertical axis) and degree of blur (horizontal axis). Coordinate ρ increases vertically and the blur coordinate increases to the right. *The EPI solution $\hat{q}(x)$ occupies a point in this space on the horizontal axis, since this is where $\rho = 0$.* The finite degree of blur then places the solution point on the horizontal axis and somewhere to the right of the origin (defining the state of blur). *The exact position of this solution point is determined by the nature of the invariance principle that is used in EPI.*

1.6.2. Saddlepoint Property

Let us test the kind of extreme value in I that exists at the solution point in the above space. In particular we are interested in the possibility of a *saddlepoint*, which would be where I is locally maximized in one direction but minimized in the other. Consider, in this light, excursions toward the solution point from a nearby point to its left and above. As the test point moves to the right (toward the solution), the blur increases and, so, I *decreases*. On the other hand, as the test point moves downward (toward the solution), correlation decreases so that I *increases.* Therefore, the EPI solution point is a maximum value in one direction (vertical) and a minimum in the other (horizontal). Hence it is a saddlepoint solution. And as we saw in the preceding paragraph, that particular saddlepoint is fixed by the nature of the EPI invariance principle.

1.6.3. Game Aspect of EPI Solution

A saddlepoint solution also characterizes the solution to a mathematical, fixed-point, zero-sum game [26] between two protagonists. Here one player selects a horizontal coordinate value as its "move," and the other a vertical coordinate. The game has information I as the prize. Hence, each player tries to maximize his level of I, and this is at the expense of the other since it is a zero-sum game.

Owing to the fixed nature of the EPI invariance principle in use, the game is "fixed point"; i.e., the solution point is always the same, independent of which player makes the first "move."

Since information is the prize, this is called an "information game" or "knowledge game." Such a game can also be used to optimize an encryption procedure; see Chapter 6.

1.6.4. Information Demon

A useful mnemonic device is to regard the enforcer of correlation as the observer, and the enforcer of the variational principle—which blurs the amplitude function $q(x)$—as nature in the guise of an "information demon." (Note: This is not the Maxwell demon who, by contrast, tries to be purely useful. It is more helpful to instead think of the demon as aiding and abetting "Murphy's laws.")

By this picture, the observer initiates "play" by putting a question to the demon (nature), in the form of a request for data. Since the observer wants maximum information in his data, he collects the data independently. Since the data are provided by nature, any information that is gained by the observer is at its expense. Hence the demon (nature) wants to minimize his expenditure of information. For this purpose, the demon's move is to inject an amount of blur into the output law. As we saw, the payoff point of the game is the EPI solution.

Note that the independence of the data justifies the assertion following Eq. (1.45), that the Fisher I used in EPI should be in a maximized state, *as if* it arose out of independent data. This is a concrete result of the game. Hence the game is not merely a mnemonic device. Another practical result of the game is the "game corollary" described below.

1.6.5. Peirce Graphs

The American logician Charles S. Peirce was seen to contribute ideas, in Section 1.5, that underly the three types of EPI solution. Peirce also predates this idea of a game played with nature, in his "existential graphs or diagrams" (1897).

This graphical approach to logic can also be regarded as a forerunner of today's familiar *gedanken experiments*. An extension of the familiar Venn diagram of probability theory, graphical reasoning was regarded by Peirce as "the only really fertile reasoning," from which not only logic but every science could benefit. As in a gedanken experiment, Peirce regarded the use of such diagrams in logic as analogous to the use of experiments in chemistry. Moreover, as in the EPI game, which puts questions to nature, the experiments upon diagrams may be understood as "questions put to the nature of the relations concerned." In fact, Peirce regarded the system of Existential Graphs as inseparable from a game-like activity, to be carried out by two fictitious persons, the Graphist and the Grapheus. The two protagonists are very different. The Graphist is an ordinary logician, and the Grapheus is the creator of "the universe of discourse." These have obvious counterparts in the observer and demon, respectively, of the EPI knowledge game. (By the way, this connection to Peirce came as a very recent, and pleasant, surprise to the author[3].)

The knowledge game provides another practical tool of analysis, as discussed next.

1.6.6. Game Corollary

In many EPI derivations the output amplitudes contain unknown constants, usually arising out of integration of given Euler–Lagrange equations. We found by experience with many such derivations [3] that such a constant—call it b—can, in fact, usually be inferred, and by a simple approach. This traces from the knowledge game. Assume that the game has been played (an EPI solution formed), but the demon has an additional, final move up his sleeve. Since his aim is to minimize the level of I that the observer acquires, the demon adjusts the value of b such that the absolute level $|I|$ of the observer's information is further minimized.

The mathematical approach this suggests is to form the EPI solution $\hat{q}(x, b)$ containing the unknown constant b, and then back substitute this into the expression for I (e.g., (1.55)). After integrating out x, the result is a function $I(b)$ of the unknown constant. This function is then minimized through choice of b, i.e., by solving

$$|I| = \min \tag{1.78}$$

in b. Usually the minimum is at an interior point to the range of values of b, so that it can be obtained by elementary calculus,

$$\frac{\partial |I|}{\partial b} = 0. \tag{1.79}$$

Thus, b is found as the net result of a *double minimization*: first, through the action of EPI or the knowledge game (which allows a general b) and second, by further adjustment of b to satisfy (1.78) or (1.79). An interesting use of this approach is in

---

[3] Thanks to a personal communication from philosopher V. Romanini (2005).

Chapter 8, where it is used to get the particular powers $a = n/4, n = 0, \pm 1, \pm 2, \ldots$ of the quarter-power laws of cosmology and biology.

Because it is an outgrowth or corollary of the knowledge game, this approach to finding b is called the *game corollary*. There are many applications of the game corollary in [3]. Some are as follows:

(a) The definition (1.54) or (1.58) of the channel capacity information indicates that I grows with N. Therefore, by (1.78), if the value of N is not known, it should be estimated as the value that minimizes $|I|$. For example, $N = 8$ is the smallest value of N that obeys the Dirac commutation rules [3] for the EPI solution for an electron. By (1.56) this defines a four-component wave function $\psi_n, n = 1\text{-}4$.
(b) EPI predicts that in situ cancer grows with time as a simple power law ([3, 21] and Chapter 3) with, at first, an unknown value b of the power. Use of the rule (1.79) gives the specific value $b = (1/2)(1 + \sqrt{5}) = 1.618 \cdots \equiv \Phi$, the Fibonacci golden mean.
(c) The universal physical constants are by definition empirically defined. The preceding indicates that any such constants that can be incorporated within an EPI problem might be computable by the use of either Eq. (1.78) or (1.79).

See also the revised form of the game corollary defined in Chapter 6.

1.6.7. Science Education

These uses of Fisher information essentially regard all of science as a single effect—the reaction of nature to its observation. Of course the act of observation has long been regarded as the *sine qua non* of physics. This is also an intuitive notion—observation *ought* to somehow define physics. On these grounds, the concept of observation should be useful in providing a route to *teaching* physics and, in fact, *all science*. A course that unifies science in this way is, in fact, a current aim of educators. This book indicates that it can be accomplished, through the use of information as the unifying concept. In fact such a course already exists. A version has been taught at the University of Arizona by this author for many years. Some prior knowledge of probability and statistics is, of course, required. A further benefit is that the approach is flexible: The level of physics that is taught can be varied according to the level of competence in statistics of the students.

1.7. Predictions of the EPI Approach

Predictions are the hallmarks of a physical theory. Any theory that does not make predictions cannot be falsified and is, therefore, not physical [27]. (String theory is, e.g., often criticized in this respect.) We end this introduction with a summary of those predictions that have, so far, been made by EPI:

(1) Free quarks must exist. This was verified—Free-roaming quarks and gluons have apparently been created in a plasma at CERN (2000) [28].

(2) The Higgs particle, currently unmeasured, should have a mass of no more than 207 GeV [23].

(3) The Higgs particle has not been found, because it is equally probable to be anywhere over all of space and time. The particle has a flat PDF in space-time. Therefore it may never be detected [23].

(4) The Higgs mass effect follows completely from considerations of information exchange, and is in no way dependent upon ad hoc models such as the usual Cooper-pair one (although the latter can of course be alternatively used to derive the effect as in textbooks) [23].

(5) The allowed quark combinations in formation of hadrons are much broader than as given by the "standard model." For example, the combination qqqqq is allowed, whereas this is forbidden by the standard model [28].

(6) Electrons combine in the same combinations as do quarks in formation of electron clusters and composite fermions [28]. Composite electron clusters have been verified experimentally [29].

(7) The Weinberg- and Cabibbo angles θ_W, θ_C of particle theory obey simple analytical expressions in terms of the Fibonacci golden mean $\Phi = 1.618\ldots$ These are, respectively, $\tan^2 \theta_W = (1/2)(\Phi - 1) = 0.30902\ldots$, and $\tan^2(4\theta_C) = \Phi$, giving $\theta_C = 12.95682°$. These two results θ_W, θ_C are close to current midrange experimental values of the angles, exceeding them by about $0.3°$ [3].

(8) Any untreated, in situ cancer mass has a *minimum level* of Fisher information. This represents a reversal of evolution from normally functioning cells to cells that do not function, instead rapidly reproducing as a cancer mass ([3, 21], Chapter 3).

(9) The so-called "technical approach" to investment, when used in conjunction with EPI, gives excellent estimates of price fluctuation curves for financial issues. Empirical uses of the approach agree well with the log-normal nature of the Black-Scholes valuation model. This approach also regards market dynamics as following generally *non*equilibrium processes ([3, 20], Chapter 2).

(10) In EPRB (Einstein-Podolsky-Rosen-Bohm) experiment of two-particle entanglement, information J obeys the theoretical requirements of an "active" or "interactive" information as defined by Bohm and Hiley [3, 30].

(11) EPI provides a common basis for many physical and biological growth processes. This includes all transport and population growth phenomena, including the Boltzmann transport equation, the equation of genetic change, the neutron reactor growth law, and mixtures of these [3].

(12) The speed of light c is, aside from its definition as a speed, a simple measure of *the ability to acquire knowledge*. It obeys $c \geq (dH/dt)/\sqrt{I}$. Thus, c is an upper bound to the rate dH/dt at which information H about the position of an electron can be learned, relative to the amount $\sqrt{I}$ that is already known [3].

(13) Equilibrium- and non-equilibrium statistical mechanics ([24], or Chapter 4), as well as economic valuation, can be derived using an EPI output in the mathematical form of a constrained Schrodinger wave equation ([3, 20], or Chapter 2).

(14) The universal physical constants should be analytically determinable [3], as those that minimize I via the "game corollary" Eqs. (1.78) or (1.79). Also, certain

physical and biological effects must exist to fill missing entries in Table 5.2, as discussed in Section 5.1.7.

(15) The game corollary can be used to imply that neutrinos have finite mass, as is now known to be the case [3].

(16) It had long been thought that black holes transmit zero information. By comparison, EPI predicts that a black hole transmits Shannon information at a *maximum* bit rate [31]. The physicist S.W. Hawking has lately come to agree [32] that black holes do, in fact, transmit information.

(17) EPI is consistent with the existence of three space dimensions and one time dimension. Essentially, only in such a universe can EPI *estimates* be made and, therefore, can man exist in his current form! (These results are based upon physical analyses due to Tegmark [33].)

First, consider the time dimension. Suppose an observer wants to predict the near-future position of a particle based upon observation of its past positions. With more, or less, than one time dimension the problem becomes ill-posed. Therefore the mean-squared error in the predicted position approaches infinity, which is equivalent to a Fisher information $I = 0$. It results that the estimated position has no validity. Hence, there must be one time dimension.

Next, if there are more than three space dimensions then neither classical atoms nor planetary orbits can be stable. Hence, the observer cannot make measurements on these, again incurring effectively $I = 0$ or infinite error in any estimates of trajectories. Given such inabilities to estimate, man could not exist in his present form. Hence, there must be three or less space dimensions.

Finally, what if there are less than three space dimensions? This has two untenable results: (a) There can be no gravitational force of general relativity. Hence the attempted measurement of the position of a graviton would give infinite error or $I = 0$ again. (b) Because of severe topological limitations, all known organisms, including the EPI observer, could not biologically exist (e.g., their arteries could not cross). Hence, there cannot be less than three space dimensions. Combining this with the previous paragraph, there are exactly three space dimensions.

In summary, man's existence requires the collection of finite amounts of information $I > 0$, and this requires $(3 + 1)$ dimensions.

(18) EPI [3] gives rise to the Schrodinger wave equation (SWE). In the presence of a special, purely imaginary scattering potential, the Hartree–Fock approximation of the SWE for a system of particles gives rise to a system of particles obeying Lotka–Volterra growth equations (Chapter 5). L–V equations describe the growth of living (and nonliving) systems. This suggests that life could have originated out of the scattering of a special particle by an initially lifeless system.

(19) EPI can be used to derive the basic *dynamics of human groups* (companies, societies, countries, etc). These are dynamical equations of growth over time t describing the group's relative subpopulations and/or resources. Here, informations J and I respectively represent a group's levels of "ideational" and "sensate" complexity. The ideational system is the group's body of ideals and principles, e.g., the US Constitution. The sensate system is the group's body of reactions to the principles in the form of its activities and interactions. Then the EPI principle that

functionally $I - J = $ min. or, equivalently, that *numerically* $I \rightarrow J$, amounts to a prediction that any group tends to, but does not perfectly, live by its ideals. This is empirically obeyed by most groups (indeed, otherwise they dissolve). In the special case where the EPI minimum is zero, that is when $I - J = 0$ is attained, this represents a *Hegelian alliance* between group ideals and practice. Application of the EPI principle to general human groups gives a prediction that groups whose constituencies are locked into constant, unregulated growth eventually degenerate into monosocieties, where one constituency completely dominates over the other ([22] and Chapter 9).

(20) In biology, time and fitness are complementary variables, obeying a new uncertainty effect (Chapter 8). That is, the mean-squared uncertainty in the age of an ecological system, as estimated by random observation of one of its creatures, obeys Heisenberg-type complementarity with the mean-squared fitness over all creatures of the system. Thus, small mean-squared fitness tends to be associated with large uncertainty in age (old species tend to lack variability).

(21) A flow of investment capital represents a flow of information. The Tobin approach (so-called "q-theory") to obtaining an optimum schedule of investment is, for a certain class of cases, equivalent to attaining an optimized flow of information via EPI (Chapter 2). EPI also has, in this application, other output solutions. These demark alternative schedules of investment, and also imply a new uncertainty principle of economics.

Acknowledgments. It is a pleasure to thank Bernard H. Soffer for help in clarifying some fundamental philosophical issues that are addressed in this chapter.

2
Financial Economics from Fisher Information

B. Roy Frieden, Raymond J. Hawkins,
and Joseph L. D'Anna

In this chapter we will cover three general applications of Fisher information in the analysis of financial economics. The first two applications (Sections 2.1 and 2.2) demonstrate how constraints based on knowledge of system data can be used *to construct probability laws*. This is by the use of a form of extreme physical information (EPI) known as minimum Fisher information (MFI). The third application (Section 2.3) shows how optimum *investment strategies* can arise out of the application of EPI to a financial system. That is, a dynamical investment program that enforces an *optimization of information flow*, achieving $I - J = extremum$, can also, in certain cases, achieve a program of *optimal capital investment*.

2.1. Constructing Probability Density Functions on Price Fluctuation

2.1.1. Summary

Information flow is both central to economic activity [1–3] and a primary causal factor in the emergence, stability, and efficiency of capital markets [4–8]. The well-known interaction between information and economic agents in the price discovery process suggests that information flow may also play a central role in determining the *dynamical laws* of an economic system. Our view is that this determination arises out of perturbation of both the perceived and intrinsic information levels of the economic system. Specifically, an economic agent perturbs *both* (i) the Fisher information I about system value that is based solely on *observations* of the system and (ii) the Fisher information J about system value that is *fundamental to* the system. Thus, $I - J$ represents the difference between the perceived and intrinsic values of the system, and it is this difference that (a) implies an opportunity for investment and (b) is perturbed by the economic agent. If the perturbations δI and δJ are equal, then $\delta(I - J) = 0$, $I - J$ is an extremum, and we can employ the variational approach known as extreme physical information [9] to determine the dynamical laws of the economic system. In this way the dynamical laws arise out of an analysis of the flow of information about investment value and the opportunities for investment that these flows engender.

We begin this section with the construction of *equilibrium* distributions, and illustrate this in the construction of yield curves. The framework that we develop for understanding the structure of equilibrium probability laws will form the basis for the second general application: the construction of *dynamics* (Section 2.2). In this section we will see that the equilibrium distributions are but the lowest order modes in a multimode system. The higher order modes give rise to relaxation dynamics: A financial realization of this structure is the model structure of yield curves as below.

2.1.2. Background

In this section, the focus of the research is upon the structure of financial economics, as defined by its probability density functions (PDFs) $p(x)$ on price valuation x. Historically, much knowledge of economic structure and dynamics has arisen by taking a physical perspective. The resulting "econophysics" has emerged as a distinct field of physics. Its basic premise is that the same phenomenological and mathematical insights that are used to provide unification for complex *physical* systems can also apply to problems of economics and finance. A now-classic example of this cross-fertilization is *Brownian motion*, first analyzed by Louis Bachelier in his Ph.D. thesis [10]. Rediscovered in the mid-twentieth century by workers such as Osborne [11][1], econophysics research achieved a high point with the discovery of the celebrated option pricing equation by Black, Scholes, and Merton [13,14], which created the profession of financial engineering. The current state of this field is well represented by a number of substantial texts [15–17]. These, together with the uniquely informed perspective on econometrics [18], form what might be considered the current basis of this field.

2.1.3. Variational Approaches to the Determination of Price Valuation Fluctuation

Variational approaches have been found useful for determining price fluctuation curves $p(x)$. As with much of the work applying statistical mechanics to finance, initial forays in this direction were based on the application of Shannon entropy [19], as popularized by Jaynes [20], to problems in financial economics [21,22]. Also, a principle of maximum entropy has found application as a useful and practical computational approach to financial economics [22–26].

However, an alternative approach arises out of the *participatory*, or human, component of any financial or economic system. Recall Wheeler's principle of a "participatory universe" ([27,28], Section 1.4.4):

---

[1] A remarkable collection of papers from that time, including Osborne's, can be found in Cootner [12].

All things physical are information-theoretic in origin and this is a participatory universe.... Observer participancy gives rise to information; and information gives rise to physics.

By the basic premise of econophysics, financial economics is an eminent candidate for obeying Wheeler's principle, and EPI. In this chapter we try it out on economic problems, in this section deriving probability laws $p(x)$ on price fluctuation x. In one mode of application, EPI provides an operational calculus for the incorporation of *new knowledge*. This is via Lagrange constraint terms discussed in Chapter 1, an approach long exploited in the physical sciences. While the criterion of *smoothness* has often been used to motivate the use of maximum entropy [29], other variational approaches provide similar—and potentially superior—degrees of smoothness to probability laws [30, 31]. It is the purpose of this section to show that EPI—which provides just such a variational approach—can be used to reconstruct probability densities of interest in financial economics. Note in this regard that "smoothness" is an automatic result of using EPI, and not its motivation.

How, then, does the phenomenon of price fluctuation fit within the EPI framework of measurement?

2.1.4. Trade as a Measurement in EPI Process

The trade (purchase, sales, etc.) price y of a security is a direct measure of its "valuation." (The simpler word "value" would be preferred, but is too general in this context.) Consider the trade, at a time t, of a security (stock, bond, option, etc.) called A. Denote by θ the "true" or ideal valuation θ of A at the time t. There is no single, agreed-upon way of computing θ, but for our purposes it can be taken to be simply the arithmetic average value of y over all trades of A worldwide at that time.

Regarding time dependence: All quantities θ, $p(x)$, $q(x)$ are assumed to be evaluated at the time t of the trade. However, for simplicity of notation, t is suppressed. Note that since t has a general value, we are not limiting attention to cases where t is large and distribution functions approach equilibrium distributions. The analysis will hold for general t, i.e., for generally *non-equilibrium* distributions, although we first apply it to equilibrium cases.

Basically, trading is a means of price discovery. Therefore a trade valuation y generally differs from its "true valuation" θ: The disparity is due to a wide range of causes, among them imperfect communication of all information relevant to valuation of the security, and the famous emotional effects of greed and fear in the minds of buyers and sellers. (This is a case where the phenomenon—the valuation—is affected not only by the *act* of observation, i.e., the trade, but also by the *emotions behind* the trade.) Even in cases of perfect communication, there is inherent heterogeneity in the mapping of such information to prices. The result is differing expectations as to whether the price of a security will increase or will decrease. As Bachelier aptly noted [10] "Contradictory opinions concerning these changes diverge so much that at the same instant buyers believe in a price increase and sellers in a price decrease."

Indeed, if this were not so the trade would not take place.

Let x denote the difference between the traded price y and ideal valuation θ, so that

$$y = \theta + x. \tag{2.1}$$

This is the basic data Eq. (1.29) of this EPI process. In effect, a traded price is a sample, or *measurement*, of the valuation of the security. The trader is also therefore an observer, in the sense used in Chapter 1. Likewise the difference x between the ideal and observed valuation is generally unpredictable and, hence, regarded as random. Hence, we often call x the "noise" in the price.

By definition, the PDF $p(x)$ on the valuation noise governs the ensemble of such trades of A by the observer under repeated initial conditions. Thus, $p(x)$ is a well-defined PDF. Given the efficiency of today's communication networks, to a good approximation $p(x)$ should also represent the PDF on valuation noise over all worldwide trades of A at that time. That is, $p(x)$ should also obey ergodicity. How can $p(x)$ be estimated?

EPI is the outgrowth of a flow of Fisher information, as discussed in Section 1.3 of Chapter 1. The carriers of the Fisher information must be known (e.g., photons in an optical microscope [9]). Here it is valuation y in, say, *dollars*. Therefore we apply EPI to the given problem of estimating $p(x)$.

The approach also turns out to apply to the problem of estimating a PDF $p(x)$ of *classical statistical mechanics*, where x is now some macroscopic observable such as energy, pressure, velocity, etc. (see also Chapter 4 of this text, and Chapter 7 of [9]). The following development is in the specific context of the economic valuation problem. However, we point out along the way corresponding results for the statistical mechanics problem.

We emphasize that *estimates* of PDFs, rather than exact answers, are sought in these problems. As discussed in Chapter 1, Sections 1.5.1–1.5.4, the answer to a variational problem depends intimately upon the nature of the prior knowledge. Here it is type (C) *induction*, in the form of data. Hence the outputs $p(x)$ will only be approximate. By comparison, the bound information quantities J utilized in our later discussion of the investment process act as the constraints imposed by *nature*. They therefore give exact ("nature's") results for the PDFs.

2.1.5. Intrinsic vs Actual Data Values

In Eq. (2.1), y denotes an *intrinsic* datum, i.e., serves to define the PDF $p(x)$ on intrinsic fluctuations x, rather than representing actual data. The actual data are here valuations, denoted as d_m. These depend upon $p(x)$ through relations

$$d_m = \int_a^b dx f_m(x) \, p(x), \quad m = 1, \ldots M, \tag{2.2}$$

as given below. Thus the data are expectations or mean values. The limits a, b and kernel functions $f_m(x)$ are known for each set of data.

In all applications of EPI the activity of taking data plays a vital role in initiating the EPI process. This is the case here as well. The execution of the trade (and of all other trades) during the time interval $(t, t + dt)$ necessarily affects the PDF $p(x)$ on valuation of the security. Because the time interval is very short, the result is a perturbation of $p(x)$. Therefore informations I and J are perturbed, initiating the EPI extremum principle described in Section 2.1.1 and Chapter 1.

Returning to the classical statistical mechanics problem, the observer ordinarily knows certain macroscopic data values d_m, $m = 1, \ldots, M$ such as temperature, pressure, etc. These data likewise obey known equality relations of the form Eq. (2.2). How should the taking of the data effect an EPI solution $p(x)$ to the problem? As discussed in Sections 1.5.1–1.5.4, the answer depends upon the quality of the prior knowledge. If this is type (A) or (B), any macroscopic data at hand are *ignored*, in favor of acquiring *exact* distribution laws (e.g., the Boltzmann and Maxwell–Boltzmann laws). Interestingly, so ignoring the data and relying instead on fundamental EPI analysis to establish the distribution shapes corresponds to a "fundamentalist" approach to investment. Once these laws are so established, only its undetermined *constants* are to be fixed by the data.

However, in the given economic problem we do not have type (A) or (B) prior knowledge and, so, cannot get an exact solution. Instead, we settle for an approximate solution (type (C) as described in Section 1.5.4). In this kind of problem, the empirical data are used, not only to fix the constants of the distributions but also to fix the *forms* of the distributions by the variational procedure.

How the data can be built in is discussed next.

2.1.6. Incorporating Data Values into EPI

As discussed in Chapter 1, in a type (C) EPI solution, constraints are imposed upon the variational solution in place of the use of the information term J. Here the constraints are the data Eq. (2.2). Data give information about the particular member of an ensemble $\psi(\mathbf{r}, t)$ of possible amplitude functions that are present. Thus, knowledge of, for example, a particular electron position gives information about the particular Feynman path it has taken. This should, for example, cause $\psi(\mathbf{r}, t)$ to have large values along that path $\mathbf{r}(t)$ and nearby ones, but small values elsewhere. How, then, can the variational EPI approach be modified to accommodate data?

As discussed in Chapter 1 (see Eqs. (1.12) and (1.13)), equality constraints may be imposed upon a Lagrangian extremum problem by simply adding them to the objective functional via undetermined multipliers λ_m. Here the objective functional is the physical information $I - J$. Thus, for our problem the EPI extremum principle becomes a constrained problem:

$$I - J + \sum_{m=1}^{M} \lambda_m \left[\int_a^b dx f_m(x) \, p(x) - d_m \right] = \text{extrem.} \qquad (2.3)$$

(The EPI zero condition will not be used.)

Continuing the parallel development for statistical mechanics, by analogous reasoning the distribution laws $p(x)$ for mechanics problems obey the same general principle (see, e.g., Chapter 4).

The use of data as in Eq. (2.3) corresponds to a "technical" approach to investment, as discussed next.

2.1.7. Information J and the "Technical" Approach to Valuation

As mentioned above, two extreme strategies for valuating a security are termed (a) fundamental and (b) technical. A pure fundamentalist arrives at a valuation decision y only after detailed examination of all attributes of the business—its yearly sales, debts, profits, etc. By comparison, a pure technician bases the pricing decision purely upon knowledge of the sales history of prices y. (This is an extreme form of empiricism.) Most investors use some combination of the two approaches. Both approaches (a) and (b) seem to be equally successful.

In taking a purely technical stance, the observer avoids any knowledge of the fundamentals of the security. Thus, the observer presumes that there exists zero prior information about *valuation*, the quantity θ. In EPI, prior information equates to bound information J. Thus,

$$J = 0. \tag{2.4}$$

Since the information efficiency κ occurs in EPI only in multiplication of J, the result Eq. (2.4) means that for these valuation problems the value of κ is irrelevant.

This kind of prior knowledge has a powerful effect upon the statistical mechanics problem. In limiting his or her observation to *macroscopic* variables, the observer rules out any prior knowledge about $p(x)$ that might follow from theoretical knowledge of *microscopic* effects. Hence, in effect the "technical approach" of financial economics corresponds to the "macroscopic observables approach" of classical mechanics. Both approaches take the stance of purposely ignoring the microscopic details of the unknown processes. Thus, Eq. (2.4) is obeyed in the macroscopic statistical mechanics problem as well. Hence, so far, both problems have the EPI solution of Eqs. (2.3) and (2.4).

However, despite Eq. (2.4), a *microscopic* approach can be taken in statistical mechanics, with the result that J is *not* zero (see, e.g., Eq. (7.47) of [9]). Thus, Eq. (2.4) is actually an approximation, so that when it is used in the Lagangian for these problems, the resulting PDFs $p(x)$ must likewise be approximations.

The same kind of reasoning holds for the economic valuation problem: To ignore microscopic (fundamental) effects is to ignore vital information. This is effectively why the PDFs of this chapter are approximate (albeit smooth).

2.1.8. Net EPI Principle

Representing I by functional (1.55), and using Eq. (2.4), Eq. (2.3) becomes

$$4 \int_a^b dx q'^2(x) + \sum_{m=1}^M \lambda_m \left[\int_a^b dx f_m(x) q^2(x) - d_m \right] = \text{extrem.},$$

$$p(x) \equiv q^2(x), \quad q' \equiv dq/dx. \qquad (2.5)$$

The extremum solution is always a minimum (Section 1.6, Chapter 1). Thus, for this problem the EPI solution is also the constrained MFI solution.

As we shall see, minimizing the Fisher information has the benefit of yielding a probability law where smoothness is ensured across the full range of support. This is in contrast to maximum entropy, where smoothness tends only to be concentrated in regions where the probability density is very small.

A further interesting aspect of the problem Eq. (2.5) is that its solution will obey a Schrodinger wave equation (SWE). This is used in Chapter 4 as the foundation for a new approach to statistical mechanics.

Since maximum entropy is comparatively well known in financial circles, and shares with the Fisher information a common variational structure, we shall use it as a point of comparison. Maximum entropy solutions $p(x)$ from data d_m obey

$$-\int_a^b dx p(x) \ln p(x) + \sum_{m=1}^M \lambda_m \left[\int_a^b dx f_m(x) p(x) - d_m \right] = \text{extrem.} \qquad (2.6)$$

The two principles Eqs. (2.5) and (2.6) will be applied to corresponding data d_m in three valuation problems below. To distinguish the resulting two solutions to each problem, Fisher information-based solutions to Eq. (2.5) will be denoted as q_{FI} or p_{FI}, and maximum entropy solutions to Eq. (2.6) will be denoted as p_{ME}.

Returning to the statistical mechanics problem, principle Eq. (2.6) is precisely that of E.T. Jaynes [20] for estimating an unknown PDF $p(x)$.

2.1.9. SWE Solutions

The two estimation principles Eq. (2.5) and Eq. (2.6) are readily solved in general. The net Lagrangian of Eq. (2.5) is

$$\mathcal{L}_{FI} = 4q'^2(x) + \sum_{m=1}^M \lambda_m f_m(x) q^2(x). \qquad (2.7)$$

From this we see that

$$\frac{\partial \mathcal{L}_{FI}}{\partial q'} = 8q'(x), \quad \frac{\partial \mathcal{L}_{FI}}{\partial q} = 2 \sum_{m=1}^M \lambda_m f_m(x) q(x). \qquad (2.8)$$

Using these in the Euler–Lagrange Eq. (1.6) gives

$$q''(x) = \frac{q(x)}{4} \sum_{m=1}^M \lambda_m f_m(x). \qquad (2.9)$$

This is the wave equation on valuation fluctuation x we sought at the outset. It holds at a general trade time t (Section 2.1.4), so that in effect $q(x) = q(x|t)$ with the t-dependence suppressed for brevity. Particular solutions $q_{FI}(x)$ of the wave equation are defined by particular values of the multipliers λ_m, in applications below.

As usual for EPI outputs, the solution Eq. (2.9) has the form of a differential equation. In fact it has the form of the energy eigenvalue SWE (cf. Eq. (D9) of [9]). However, since Planck's constant $\hbar$ does not explicitly appear, it is not descriptive of quantum mechanics in particular, although, depending upon application, $\hbar$ can enter *implicitly* through an appropriate constraint term.

Since the economic problem and the statistical mechanics problem have so far shared a common EPI approach, the statistical mechanics problem must likewise have the solution Eq. (2.9)[2]. In fact they do: See Eq. (4.99) of Chapter 4. Moreover, as in these statistical mechanics applications, the time t is arbitrary. That is, the output $q(x)$ of the EPI solution Eq. (2.9) represents generally a *nonequilibrium* solution to the valuation problem. In summary, EPI predicts the following:

Economic valuation problems and nonequilibrium statistical mechanics problems have the same general solution as does stationary quantum mechanics: the energy eigenvalue SWE[3]. A portion of this chapter is devoted to applying the general solution Eq. (2.9) to specific economic problems[4]. In practice, the form of the differential Eq. (2.9) as a SWE is fortuitous, since it permits past decades of development of computation SWE solutions to now be applied to economic problems. In particular, a number of analytic solutions are known, for which a substantial collection of numerical solutions exist (see below).

We next turn to the maximum entropy solution. The Lagrangian of Eq. (2.6) is

$$\mathcal{L}_{ME} = -p(x)\ln p(x) + \sum_{m=1}^{M} \lambda_m f_m(x)\, p(x). \qquad (2.10)$$

From this we see that

$$\frac{\partial \mathcal{L}_{ME}}{\partial p'} = 0, \quad \frac{\partial \mathcal{L}_{ME}}{\partial p} = -1 - \ln p(x). \qquad (2.11)$$

Using these in the Euler–Lagrange Eq. (1.6) gives directly

$$p_{ME}(x) = \exp\left[-1 - \sum_{m=1}^{M} \lambda_m f_m(x)\right]. \qquad (2.12)$$

---

[2] See also in this regard Eq. (7.48) of [9] and [32, 33].
[3] *Note on terminology: The SWE Eq. (2.9) is also called in physics the "stationary SWE." Unfortunately, the term "stationary" when applied to economic problems suggests relaxation of the system to an equilibrium state over some large time t. But in fact we are permitting the time t to be arbitrary (see above). Hence use of the word "stationary" would be misleading, and it is replaced with the alternative terminology "energy eigenvalue."*
[4] Its application to particular statistical mechanics problems may be found in [32, 33] and Chapter 4.

Since $p_{ME}(x)$ must obey normalization Eq. (2.12) is more conveniently cast in the prenormalized form [20]

$$p_{ME}(x) = \frac{1}{Z(\lambda_1, \ldots, \lambda_M)} \exp\left[-\sum_{m=1}^{M} \lambda_m f_m(x)\right], \qquad (2.13)$$

where

$$Z(\lambda_1, \ldots, \lambda_M) \equiv \int_a^b dx \exp\left[-\sum_{m=1}^{M} \lambda_m f_m(x)\right]. \qquad (2.14)$$

The reader can easily verify that $p_{ME}(x)$ defined by Eq. (2.13) explicitly obeys normalization. Given this, it should not be reimposed by a multiplier λ_m corresponding to a normalization datum $d_m = 1$.

From Eqs. (2.5) and (2.6), both the EPI and maximum entropy estimates depend upon empirical information in the form of equality constraints. In the following sections, increasingly complex forms of empirical information are given. The efficient generation of yield curves from observed bond prices, and probability densities from observed option prices, will result. The calculations are based upon the mathematical correspondence between (i) the resulting differential equations for probability laws in these examples from financial economics and (ii) past work with the Schroedinger equation. In addition, we shall see that the probability densities generated using Fisher information are, in general, smoother than those obtained using maximum entropy.

2.2. Yield Curve Statics and Dynamics

2.2.1. Summary

We use the EPI approach of the previous section to motivate practical computational approaches to the extraction of probability densities and related quantities from fixed-income observables. In the first two cases we contrast the results obtained by this method with those of maximum entropy. We begin with observables expressed in terms of partial integrals of a probability density and of first moments of the density. We conclude the applications by examining the origin of yield curve modes. The analysis predicts a new measure of the volatility of value. The analysis also indicates that yield curve dynamics obey a Fokker–Planck differential equation, verifying the common ansatz that *a diffusion process underlies interest-rate dynamics.* Finally, the analysis provides an information-theoretic derivation of the famous *Nelson–Siegel approach* to yield curves.

2.2.2. Background

An ability to accurately extract probability densities from a limited number of fixed-income observables is central to a knowledge of economic processes. The overall aim of this section is to attempt to estimate such probability densities by the use of EPI. To judge the efficacy of such approaches, test-case comparisons will be made with existent estimation approaches such as the use of splines or the use of maximum entropy.

Past research has indicated that a useful description of economic dynamics is provided by equations like the Chapman–Kolmogorov and Fokker–Planck equations. These are equations of non-equilibrium statistical mechanics. Therefore it would be useful to find whether these likewise follow from the use of EPI.

2.2.3. PDF in the Term Structure of Interest Rates, and Yield Curve Construction

A yield curve is a representation of the rate of interest paid by a fixed-income investment, such as a bond, as a function of the duration of that investment. The yield curve is often called a "term structure of interest rates" curve. The interest rate over a given time interval between, say, today and a point of time in the future, determines the *value* at the beginning of the time interval of a cash flow to be paid at the end of the interval. This is also known as the present value of the cash flow. Since the value of any security is the present value of all future cash flows, yield curve fluctuations give rise to fluctuations in present value and, thus, play an important role in the variation of security prices.

We begin with a more formal description of the notion of present value mentioned above. The yield curve is closely related to the function $D(t, T)$ known as the discount function. This represents the fraction of (say) a dollar such that, were it invested in some risk-free financial today, it would appreciate to the value of one dollar at a given future date. Algebraically, if t is the present time and T is the prescribed future time, the discount function has the value

$$D(t, T) = e^{-r_s(t,T)(T-t)} = e^{-\int_t^{t+T} ds\, r_f(t,s)}. \tag{2.15}$$

Here $r_s(t, T)$ is the "spot rate" yield curve at time t, specifying the continuously compounded interest rate for borrowing over the period $[t, T]$, and $r_f(t, T)$ is known as the "forward rate" yield curve at time t, specifying the interest rate for instantaneous borrowing at the future time $[T, T + \delta T]$. The exponential falloff discounts the anticipated compound interest.

Our approach, based on deriving yield curves that extremize Fisher information, is facilitated by associating such yield curves with complementary PDFs where the time to maturity T is taken to be an abstract random variable [34, 35]. We assume the associated probability density $p(t, T)$ satisfies $p(t, T) > 0$ and is related to the discount factor $D(t, T)$ via

$$D(t, T) = \int_T^\infty ds\, p(t, s), \tag{2.16}$$

$$= \int_0^\infty ds\, \Theta(s - T)\, p(t, s). \tag{2.17}$$

$\Theta(x)$ is the Heaviside step function.

Discount factors are not always observable. On the other hand, consider a coupon bond. This is a debt obligation that makes periodic interest payments during the life of the obligation, and pays off the principal amount on the maturity date. Such bonds are commonly available with prices $B(t, N)$ related to the discount factor by

$$B(t, N) = \sum_{i=1}^{N} C(T_i) D(t, T_i), \qquad (2.18)$$

where N indicates the number of remaining coupon payments and $C(T_i)$ is the cash flow at time T_i in the future. For a typical coupon bond $C(T_i)$ is equal to the ith coupon payment for $i < N$ and equal to the final coupon payment plus the principal payment for $i = N$.

In these expressions one can see that discount factors and bond prices share a common structure as averages of known functions. Discount factors are the average of $\Theta(s - T)$ and coupon bond prices are the average of $\sum_{i=1}^{N} C(T_i) \Theta(s - T_i)$. Generally, where observed data $d_1, \ldots, d_M = \{d_m\}$ such as discount factors and the prices of bonds can be expressed as averages of known functions $\{f_m\}$ at a static point in time, we may write

$$\int dT f_m(T) p(T) = d_m, \quad m = 1, \ldots, M. \qquad (2.19)$$

The PDF $p(T)$ implicit in the observed data can be obtained by forming a Lagrangian using Fisher information [36] in its shift-invariant form

$$I = \int dT \frac{(dp(T)/dT)^2}{p(T)}. \qquad (2.20)$$

Employing the usual variational approach Eqs. (1.6), (2.7), we obtain a solution (2.9)

$$p(T) = q^2(T), \qquad (2.21)$$

$$\frac{d^2 q(T)}{dT^2} = \frac{q(T)}{4} \left[\lambda_0 + \sum_{m=1}^{M} \lambda_m f_m(T) \right]. \qquad (2.22)$$

The λ's are Lagrange multipliers that enter by incorporating a normalization constraint on $p(T)$ (using Lagrange multiplier λ_0) and observed data (using multipliers λ_m) into the Fisher information Lagrangian.

This is an approximate, type (C) use of the EPI approach (Section 2.1.4) wherein the constraint equations (Eq. (2.19)) are used in place of the fundamental source information J. To review, this replacement amounts to a technical (as compared with fundamental) approach to valuation.

To illustrate the differences between the Fisher information and maximum entropy approaches to term structure estimation, we consider the case of a single zero-coupon bond of "tenor" $T = 10$ years. This describes a bond that pays only the principal amount at the maturity date, with no preceding interest payments, and for which there are 10 remaining years to maturity. Also, let it have a price of 28 cents on the dollar or, equivalently, a discount factor of 0.28. Comparing Eqs. (2.2) and (2.17), this is a case $M = 1$ with $f_1(t) = \Theta(t - T)$. Hence the maximum entropy solution Eq. (2.13) is

$$p_{ME}(t) = \frac{e^{-\lambda\Theta(t-T)}}{T + 1/\lambda}. \tag{2.23}$$

The corresponding discount factor is readily found from Eq. (2.17) to be $D(T) = 1/(\lambda T + 1)$, from which the Lagrange multiplier λ is 0.2571.

The Fisher information equation is (2.9) with the constraints of normalization and one discount factor:

$$\frac{d^2q(t)}{dt^2} = \frac{q(t)}{4}\left[\lambda_0 + \lambda_1\Theta(t - 10)\right], \tag{2.24}$$

This problem is known to have the solution

$$q(t) = \begin{cases} A\cos(\alpha t) & \text{if } t \leq T, \\ B\exp(-\beta t) & \text{if } t > T, \end{cases} \tag{2.25}$$

where the coefficients A and B are determined by requiring that $q(t)$ and $q'(t)$ (or the logarithmic derivative $q'(t)/q(t)$) match at $t = T$. Carrying through yields

$$\tan(\alpha T) = \frac{\beta}{\alpha}. \tag{2.26}$$

Choosing $\alpha T = \pi/4$ and matching amplitudes, we find that $B = A\exp(\pi/4)/\sqrt{2}$, so that

$$p_{FI}(t) = \begin{cases} 8(\pi + 4)^{-1}\alpha\cos^2(\alpha t) & \text{if } t \leq T, \\ 4(\pi + 4)^{-1}\alpha\exp(\pi/2 - 2\beta t) & \text{if } t > T. \end{cases} \tag{2.27}$$

Also $D(T) = 2/(\pi + 4) = 0.28$.

The maximum entropy PDF Eq. (2.23), Fisher information PDF Eq. (2.27), and the corresponding term structure of interest rates are plotted in Figure 2.1. In the uppermost panel we see both PDFs as a function of tenor with the Fisher information result denoted in all panels by the solid line. The maximum entropy estimate is uniform until the first (and in this case only) observation is encountered; beyond which a decaying exponential is observed. The discontinuity in slope traces from the *global* nature of the entropy measure (Section 1.2.2.12, Chapter 1). The Fisher information estimate is smoother, reflecting the need to match both the amplitude and derivative at the data point. This traces from the *local* nature of the Fisher measure (Section 1.2.2.12).

It is difficult to make a real aesthetic choice between the two curves in the upper panel. To aid in making a choice we turn to two important derived quantities: the

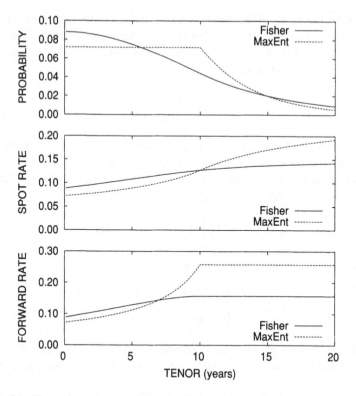

FIGURE 2.1. The estimated probability densities and interest rate curves for a single 10-year discount factor of 0.28. (Reprinted from [39], Copyright 2004, with permission of Elsevier.)

spot rate $r(t)$ and the forward rate $f(t)$. These are related to the discount factor by

$$D(T) = e^{-r(T)T}, \qquad (2.28)$$

$$= e^{-\int_0^T dt f(t)}. \qquad (2.29)$$

The spot rates are shown in the middle panel of Figure 2.1. Both methods yield a smooth result with the Fisher information solution showing less structure than the maximum entropy solution. A greater difference between the two methods is seen in the lowermost panel of Figure 2.1, where the forward rate is shown. It is the structure of this function that is often looked to when assessing the relative merits of a particular representation of the discount factor. The forward rate reflects the structure of the probability density as expected from the relationship

$$f(t) = p(t)/D(t). \qquad (2.30)$$

The maximum entropy result shows more structure than the Fisher information result, again, due to the continuity at the level of the first derivative imposed on $p(t)$ by the Fisher information approach.

It is a comparatively straightforward matter to extend this approach to the construction of a term structure of interest rates that is consistent with *any number* of arbitrarily spaced zero-coupon bonds. A particularly convenient computational approach based on transfer matrices is given in Appendix A of Chapter 13 in [9].

Previous work on inferring the term structure of interest rates from observed bond prices has often focused on the somewhat ad hoc application of splines to the spot or forward rates [40–43].

By comparison, the work of Frishling and Yamamura [44] is similar in spirit to the EPI approach, in that it minimizes $df(t)/dt$. Their paper deals with the often unacceptable results that straightforward applications of splines to this problem of inference can produce. In a sense, EPI can be seen as an information-theoretic approach to imposing the structure sought by Frishling and Yamamura on the term structure of interest rates. A similar pairing of approaches is the minimization of $d^2f(t)/dt^2$ by Adams and Van Deventer [45] and the cross-entropy work of Edelman [31].

2.2.4. PDF for a Perpetual Annuity

Material differences between the probability densities generated by Fisher information and maximum entropy are also seen when the observed data are moments of the density—a common situation in financial applications. Consider a perpetual annuity, i.e. an annuity that pays interest forever. The value ξ of a perpetual annuity due to all possible values t of the tenor is given by Brody and Hughston [34,35]

$$\xi = \int_0^{+\infty} dt\, t p(t). \tag{2.31}$$

This first-moment constraint provides an interesting point of comparison for the maximum entropy approach employed by Brody and Hughston and our Fisher information approach. Comparing Eqs. (2.2) and (2.31), here $M = 1$ and $f_1(t) = t$. Then the maximum entropy solution Eq. (2.13) is directly

$$p_{ME}(t) = \frac{1}{\xi} \exp(-t/\xi). \tag{2.32}$$

The EPI solution $q(t)$ to Eq. (2.9) was, equivalently, solved by Frieden [30] as

$$p_{FI}(t) = c_1 \text{Ai}^2(c_2 t). \tag{2.33}$$

Ai(x) is Airy's function, and the constants c_i are determined uniquely by normalization and the constraint of Eq. (2.31).

We show these two probability densities as a function of tenor t for a perpetual annuity with a price of $1.00 in Figure 2.2. The two solutions are qualitatively similar in appearance: Both monotonically decrease with tenor and are quite smooth. However, since the Fisher information solution starts out at zero tenor with a lower value than the maximum entropy solution and then crosses the maximum entropy

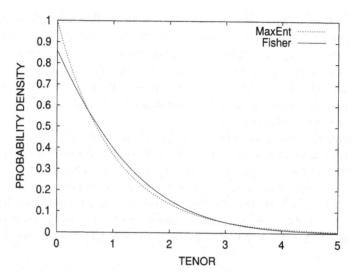

FIGURE 2.2. The two PDFs associated with a perpetual annuity with a price of 1.0. (Reprinted from [39], Copyright 2004, with permission of Elsevier.)

solution so as to fall off more slowly than the maximum entropy solution in the mid-range region, the Fisher information solution is in appearance even smoother. This can be quantified as follows.

A conventional measure of smoothness of a PDF is the size H of its entropy: The smoother the PDF, the larger is H. As we found in Chapter 1, a smooth PDF also has a *small* value of the Fisher information I. Therefore, to compare the smoothness of PDFs using alternative measures H and I, it is useful to form out of I a quantity that becomes *larger* as the PDF becomes smoother. Such a quantity is of course $1/I$ [30]. By Eq. (1.37), $1/I$ has the further significance of defining the Cramer–Rao error bound, although that property is not used here.

The $1/I$ value of the Fisher information solution Eq. (2.33) in Figure 2.2 is found to be 1.308, by integrating Eq. (1.55) over $0 \leq t \leq \infty$. By comparison, $1/I$ for the maximum entropy solution Eq. (2.32) in Figure 2.2 is found to be 1.000. Hence, the Fisher information solution is significantly smoother on the basis of $1/I$ as a criterion.

We also compared the relative smoothness of these solutions using Shannon's entropy as a measure. The maximum entropy solution is found to have an H of 1.0, while the Fisher information solution has an H of 0.993. Hence the maximum entropy solution does indeed have a larger Shannon entropy than does the Fisher information solution, but it is certainly not much larger. Since the two solutions differ much more in their $1/I$ values than in their Shannon H values, it appears that $1/I$ is a more sensitive measure of smoothness, and hence biasedness, than is H.

An interesting result emerges from an examination of the smoothness as a function of *range*. Since over the range $0 \leq t \leq 5$ shown in Figure 2.2 it appears that the Fisher information is smoother than the maximum entropy solution by *any* quantitative criterion, we also computed the Shannon H values over this limited interval. The results are $H = 0.959$ for the maximum entropy solution and $H = 0.978$ for the Fisher information solution. Thus the Fisher information solution has the larger entropy! This shows that the Fisher information solution is smoother than the maximum entropy solution, *even by the maximum entropy criterion*, over this *finite* interval. Of course the maximum entropy solution must (by definition) have the larger entropy over the *infinite* interval. It follows that the larger entropy value results from values of the PDF in the long *tail* region $t > 5$.

However, in this tail region the maximum entropy PDF has negligibly *small* values for most purposes. Hence, the indications are that the criterion of maximum entropy places an unduly strong weight on the behavior of the PDF in tail regions of the curve.

Note by comparison that over the interval $0 \leq t \leq 5$ the $1/I$ values for the two solutions still strongly differ, with 1.317 for the Fisher information solution and 1.007 for the maximum entropy. Again, the Fisher $1/I$ measure is the more sensitive measure. Moreover, these are close to their values as previously computed over the entire range $0 \leq t \leq \infty$. Hence, Fisher information gives comparatively lower weight to the tail regions of the probability densities.

2.2.5. Yield Curve Dynamics

Yield curves are remarkable in that the fluctuations of these structures can be explained largely by a few modes, and the shape of these modes is largely independent of the market of origin: a combination of parsimony and explanatory power rarely seen in financial economics. While these modes play a fundamental role in fixed-income analysis and risk management, both the origin of this modal structure and its relation to a formal description of yield curve dynamics remain unclear. The aim here is to show that this modal structure is a natural consequence of the information structure of the yield curve, and that this information structure, in turn, implies an *equation of motion* for yield curve dynamics.

Our application of Fisher information to yield curve dynamics is an extension of prior work [9,37,39] to derive equations of motion in physics and static PDFs in asset pricing theory[5]. Though less well known as a measure of information in physics and mathematical finance than Shannon entropy, the concept of Fisher information predates Shannon's and other information statistics, and remains central to the field of statistical measurement theory [46]. Fundamentally, it provides a representation of the amount of information in the results of experimental measurements of an

---

[5] The relationship between Fisher information and related approaches such as maximum entropy [20] and minimum local cross entropy [31] in the context of financial economics is discussed in [39].

unknown parameter of a stochastic system (Section 1.2.1.4). Fisher information appears most famously in the Cramer–Rao inequality (1.36) that defines the lower bound on variance of a parameter estimate for a parameter dependent stochastic process. It also provides the basis for a comprehensive alternative approach to the derivation of probability laws in physics and other sciences [9, 37, 38].

The aim of the present work is to derive a differential equation for yield curve dynamics, ab initio, with the minimal imposition of prior assumptions, save that bond price observations exist and that a stochastic process underlies the dynamics. In a sense our approach is *an inversion* of the perspective of a maximum likelihood estimate, where one solves for the most likely parameter values given observations within the context of a presumed model. Here we instead derive the stochastic model that is implied by minimizing Fisher information, given known parameter measurements (bond prices).

The notion that a modal structure underlies yield curve dynamics comes from common empirical experience with two related yield curve measurements—the construction of static yield curves from observed market prices and the analysis of correlations between corresponding points in the time evolution of successive static yield curves. Yield curves are inferred from observed fixed-income security prices, and since the prices of these securities change over time so does the yield curve. Yield curves are usually generated after the close of each trading session, and this process can therefore be viewed as a *stroboscopic measurement* of the yield curve. Yield curves can assume a variety of shapes, and many methods have been proposed for their construction[6]. Of these methods, the Nelson Siegel approach [48] of representing yield curves as solutions of differential equations has gained wide acceptance in the finance industry [49] and in academic research [50–54]. In using a second-order differential equation to represent the yield curve, the Nelson Siegel approach is essentially a proposal that yield curves can be represented effectively by a modal expansion, and the practical success of this approach to yield curve fitting in markets around the world is a measure of the correctness of this assertion.

The modal structure of the yield curve is also implied in the eigenstructure of the two-point correlation function constructed from yield curves. Specifically, diagonalization of the covariance matrix of yield changes along the yield curve produces an eigenstructure where most of the variance—as measured by summing the normalized eigenvalues—is explained by the first few eigenmodes [55–57]. The consistency of the general shape of the eigenmodes derived from *empirical* yield curve data, and the explanatory power of the truncated expansions in those eigenmodes, is surprisingly robust over time, and largely independent of the country in which the interest rates are set [58–60]. This analysis motivated the use of yield curve modes by fixed-income strategists and risk managers some time ago. However, an explicit link between *yield curve modes and dynamics* only appeared in comparatively recent research. Taking the viewpoint of econophysics, it demonstrated the eigenstructure to be consistent with both the existence of a *line tension*

---

[6] See, for example, [47] and references therein.

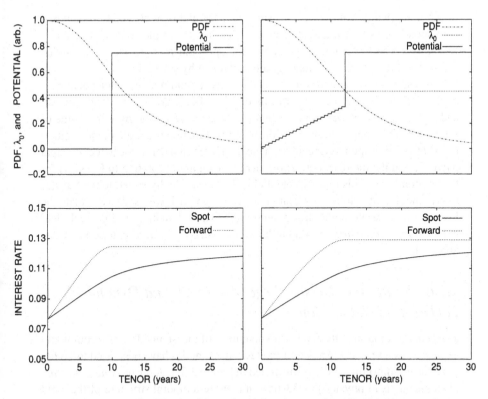

FIGURE 2.3. The equilibrium densities, related functions, and implied yield curves for a discount factor (left) and a coupon bond (right). (Reprinted from [63], Copyright 2005, with permission of Elsevier.)

along the yield curve and a *diffusion-like* equation of motion for yield curves [61]. This notion of a line tension along the yield curve has found further expression in descriptions of the yield curve as a vibrating string [62]. While the yield curve phenomenology just described *can* be described well by modal expansions, there has been little to motivate *why* this should be the case, and it is to this question that we now turn.

Our explanation for the existence of dynamic yield curve modes builds on a recent application of Fisher information methods [9,39]. This is to the construction of well-mannered, static yield curves from a finite set of discount functions or observed bond prices. In [39] we exploited the formal equivalence of Eq. (2.9) and the time-independent SWE to calculate the equilibrium densities $p(T)$ implicit in security prices, as shown in Figure 2.3.

In the graphs on the left-hand side of Figure 2.3 we see the results for a single discount factor with a price of 35% of par and a tenor of 10 years. The upper-left graph illustrates the elements of Eq. (2.22) with the potential term of the SWE $(\sum_{i=1}^{M} \lambda_m f_m(T)/4)$ being a single-step function. The amplitude of the potential is well described by the step function, λ_1, the level of λ_0, and the PDF $p(T)$. These follow numerically from a self-consistent field calculation with the $\{\lambda_0, p(T)\}$ pair corresponding to the ground state of Eq. (2.9) subject to the constraint given by Eq. (2.16). The lower-left graph shows the spot and forward yield curves that follow from the PDF in the upper graph as defined by Eq. (2.15).

The results of this analysis for a 6.75% coupon bond making semiannual payments with a maturity date of November 15, 2006, a price of 77.055% of par, and a pricing date of October 31, 1994 [44] are shown on the right-hand side of Figure 2.3. The stepped structure of the potential function is a result of the cumulative sum of the coupon payments, with the final large step being due to the principal payment. Unlike the discount factor, there is no analytic solution to Eq. (2.9) for the coupon bond. This type of potential is, however, ideally suited to the transfer matrix method of solution [64], and that is the approach we used to calculate the PDF solution shown in the upper-right graph. The calculation of a general yield curve from *a collection* of coupon bonds is a straightforward extension of this approach.

2.2.6. *Relation to Nelson Siegel Approach and Dynamical Fokker–Planck Solution*

The general solution to the SWE with potentials of this stepped form is commonly expressed as a series expansion of modes. These modes have been used to go beyond the equilibrium solutions illustrated in Figure 2.3 to describe *nonequilibrium* phenomena, as in physics [32, 33, 65]. In fact the temporal structure of the yield curve follows directly from its modal structure (Eq. 2.34). This result also provides an information-theoretic derivation of the *Nelson Siegel approach*. Of interest as well is the behavior of the solutions illustrated in Figure 2.3 in the range of tenor where there are no observed security prices. The solution to Eq. (2.9) is known to be an exponential decay that leads to a constant interest rate: a result consistent with most priors concerning long-term interest rates.

The temporal evolution of yield curves now follows directly from the known relationship between solutions of Eq. (2.9) and those of the Fokker–Planck equation[7][67, 68]. Specifically, the solutions of Eq. (2.9) $\{\lambda_0^{(m)}, q_m(T)\}$ can be used to construct a general solution [67]

$$p(T, t) = \sum_{m=0}^{\infty} c_m q_0(T) q_m(T) e^{-\vartheta\left(\lambda_0^{(m)} - \lambda_0^{(0)}\right)t/4} \tag{2.34}$$

to the Fokker–Planck equation

$$\frac{\partial p(T, t)}{\partial t} = \frac{\partial}{\partial T}\left[\frac{\partial U(T)}{\partial T} + \vartheta \frac{\partial}{\partial T}\right] p(T, t), \tag{2.35}$$

---

[7] Formally, the Fokker–Planck equation can be obtained from our Fisher information-based variational approach by incorporating a Lagrangian term enforcing the constraint that total probability density is conserved under time evolution [66].

where the potential function $U(T)$ is related to the ground state $q_0(T)$ as

$$U(T) = -2\vartheta \log q_0(T). \tag{2.36}$$

Taken together, Eqs. (2.9) through (2.36) explain the existence of a modal structure of yield curves and *provide a theoretical basis for the common ansatz that a diffusion process underlies interest-rate dynamics [47, 57, 69, 70]*.

There are a variety of ways to solve Eq. (2.34), but the observation that the the eigenstructure of the two-point correlation function is dominated by a few modes suggests that this infinite series can be reduced to a few terms using the Karhunen–Lòeve expansion[8] together with the Galerkin approximation [71–74]. Specifically, writing Eq. (2.34) in the slightly more general form $p(T, t) = \sum_{m=0}^{\infty} a_m(t)\phi_m(T)$, where $\phi_m(T) \equiv q_0(T)q_m(T)$, substituting this into the Fokker–Planck equation written suggestively as $\dot{p} = \mathcal{L}_{FP}(p)$, and projecting along the eigenfunctions $\phi_m(T)$, gives

$$\dot{a}_i(t) = \int_0^{\infty} \mathcal{L}_{FP} \left(\sum_{m=0}^{\infty} a_m(t)\phi_m(T) \right) \phi_i(T) \, dT, \tag{2.37}$$

$$a_i(0) = \int_0^{\infty} p(T, 0)\phi_i(T) \, dT, \, \dot{a}_i(t) \equiv da_i/dt. \tag{2.38}$$

Truncating the series expansion for $p(T, t)$ at $m = N$ gives a Galerkin approximation of order N [73], and this truncation is justified in our case because of the dominance to the two-point correlation function by a few modes.

As an example, this approach is applied to the dynamics of the Eurodollar yield curve in Figure 2.4. The PDF implicit in the Eurodollar futures market from the beginning of 1994 to the end of 1996 is shown in the upper-left panel of the figure, and the average of these density functions is shown in the lower-left panel[9]. Using the method of snapshots [71] we obtained the eigenstructure shown partially in the two right-hand panels of Figure 2.4. As the normalized eigenvalues indicate, more than 99% of the variance of this system is contained in the first two modes. Thus a Galerkin approximation of order 2 would be expected to provide an adequate representation of the temporal yield curve.

2.2.7. A Measure of Volatility

A measure of risk that is based on the standard deviation of an asset return is typically called its "volatility." We showed that Fisher information, via the EPI principle, provides a way of calculating the densities implied by security prices.

---

[8] This approach appears under a variety of names including factor analysis, principal-component analysis, and proper orthogonal decomposition.
[9] The PDF was obtained from constant-maturity Eurodollar futures prices as discussed in [61]

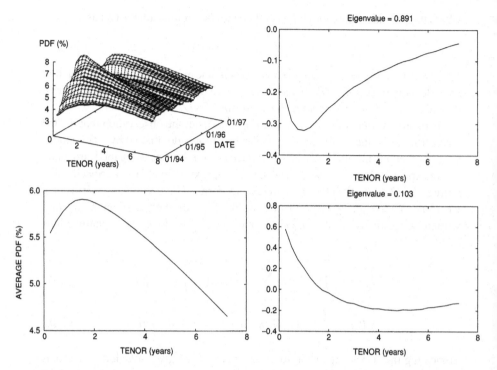

FIGURE 2.4. The PDF as a function of time for Eurodollar futures (upper left) together with the average PDF during this period (lower left) and the empirically determined eigenfunctions with corresponding normalized eigenvalues (upper and lower right). (Reprinted from [63], Copyright 2005, with permission of Elsevier.)

Fisher information also generally provides, via the Cramer–Rao inequality (1.37) of Chapter 1, a value

$$e_{min}^2 = \frac{1}{I} \tag{2.39}$$

of the intrinsic uncertainty e_{min}^2 in knowledge of an unknown ideal parameter θ. In valuation problems, θ is in fact the unknown valuation of the security (Section 2.1.4). Then the uncertainty e_{min}^2 is that in knowledge of the valuation of the security. Therefore, if we want the term volatility to measure the smallest expected level of uncertainty in knowledge of the valuation, then e_{min}^2 is the volatility and it relates to I by Eq. (2.39). In this application the value of I to be used in Eq. (2.39) would be formed out of the EPI solution $q(x)$ to the valuation problem.

Equation (2.39) is also a *conservative* measure of the volatility, since it expresses the minimum expected level of uncertainty in the observer about the valuation of the security. The actual uncertainty is the volatility figure or higher.

As a trivial example, if $p(x)$ is a Gaussian, the information $I = 1/\sigma^2$ with σ^2 the variance. Then by Eq. (2.39) the volatility $e_{min}^2 = \sigma^2$. Of course $p(x)$ does not have to be Gaussian. Equation (2.39) provides a natural generalization of the concept of implied volatility to *non*-Gaussian underlying-asset distributions.

2.2.8. Equilibrium Distributions

We have shown how, in the presence of limited data, EPI can be used to solve some canonical inverse problems in financial economics. The general solution is the differential Eq. (2.9). The resulting probability densities are generally smoother than those obtained by other methods such as maximum entropy. They also often well approximate the underlying PDF that gave rise to the data. EPI also has the virtue of providing a natural measure of, and computational approach for determining, implied volatility via the Cramer–Rao inequality (Section 2.2.7).

EPI (or, equivalently in this problem, MFI) provides a variational approach to the estimation of probability laws that are consistent with financial observables. Using this approach, one can employ well-developed computational approaches from formally identical problems in the physical sciences. A key benefit of the use of the Fisher measure is that the output PDFs are forced to have continuous first derivatives. This imparts to the estimated PDFs and resulting observables (integrals of these PDFs) a degree of smoothness that one expects in good estimates.

As a parallel development, we also attacked the corresponding problem of PDF estimation in non-equilibrium classical mechanics. There again, the lack of microscopic analysis and the limited amount of data do not allow for an exact answer. The approximate answer to both the economic and mechanics problems is found to obey the differential Eq. (2.9). This has the mathematical form of the SWE. In fact, the entire Legendre-transform structure of thermodynamics can be expressed using the Fisher information in place of the Shannon entropy—an essential ingredient for constructing a statistical mechanics [75]. (See also Chapter 4.)

Comparing the maximum entropy approach to that of EPI, it was seen that the former has the virtue of simplicity, always giving an exponential answer (2.12) for the PDF. By comparison the EPI solution is always in the form of a differential equation (2.9). Although somewhat more complicated, this is, in a sense, a virtue as the probability densities of physics, and presumably of finance and economics, generally obey differential equations. Indeed, it is in these applications that we see perhaps the greatest potential for future applications of EPI.

2.2.9. Aoki Theory

The term "macroeconomics" describes the overall state of an economy, as to its total output, employment, consumption, investment, price and wage levels, interest and growth rates, etc. A theory of macroeconomics that is related to EPI is that of Aoki [76,77]. The aim of this approach is to place all or nearly all heterogeneous microeconomic agents (individual commodities and securities) within one stochastic and dynamic framework. The result is a description of economic dynamics that is provided by equations like the Chapman–Kolmogorov and Fokker–Planck equations—equations of non-equilibrium statistical mechanics. Since the Fokker–Planck equation and the SWE Eq. (2.9) share eigensolutions (cf. Risken [68]), Aoki's theory might often share solutions with those of EPI.

2.2.10. Non-equilibrium Distributions

We have derived an equation of motion for yield curves that is consistent with observed statics and dynamics starting from the Fisher information of the PDF that underlies the discount function. Our derivation leads to an SWE for the probability amplitude of the density function underlying the discount factor and, thus, explains why solutions of equations of mathematical physics involving second-order tenor derivatives work so well as a representation of yield curves. This result also provides an explanation for the existence of a line-tension term in the equations of motion found in string models of yield curves. Using the well-known relationship between solutions of the SWE and the Fokker–Planck equation, we obtained an equation of motion (2.34) for the yield curve consistent with the common ansatz that diffusion processes underlie yield curve dynamics. Finally, since the eigenstructure of the yield curve two-point correlation function is dominated by a few modes, a practical numerical solution to this equation of motion can be had by using the Karhunen–Lòeve expansion together with Galerkin's method.

2.3. Information and Investment

2.3.1. Summary

We next show that Tobin's q-theory of investment can be expressed as a direct consequence of optimized information flow; specifically, the EPI notion that $I - J$ is an extremum where I is the Fisher information in the observed data and J is the Fisher information fundamental to the system. Tobin's q-theory is one of a class of investment strategies that result from the EPI viewpoint (see Eq. (2.57)). This solution maximizes the present value of the net earnings of this firm. Hence, a program of investment that achieves an optimum flow of information about value also achieve turns out to an optimum gain in *capital*. Knowledge equates to gain of capital. Interestingly, this is as firm a result as any law of physics that, likewise, follows from the EPI principle.

The solution also follows a physical Bohmian [78] view of investment. That is, it posits a well-defined trajectory of investment rate $\dot{K}(t)$ even in the presence of inherently random uncertainties in capital K.

The results indicate that effective investment is only possible in a scenario of incomplete knowledge. Uncertainties in capital give opportunities for investment, and optimal knowledge of uncertainties defines optimal investment. Conversely, with complete *certainty* there is no real investment opportunity. These follow from the fact that the very uncertainty K in the observation (see Eq. (2.40)) or, more precisely, its time rate of change $\dot{K}(t)$, *defines* the optimum investment program $\mathcal{I}(t)$ that is derived.

Finally, we return to Wheeler's principle of the participatory universe (Section 2.1.3). This perspective is gaining currency in physics. Its counterpart in economics is the following: All things economic are information-theoretic in origin; economies are participatory; economic agent participancy gives rise to information; and information gives rise to economics. In fact this economic version of the idea is already relatively well established (cf. [1, 2] and references therein). And a goal of this chapter is to emphasize the common structure shared by these fields at the level of information.

In addition we show find that market efficiency follows naturally from this approach, and find two apparently new effects common to all of these strategies: (i) an investment uncertainty principle stating that *it is impossible to instantaneously know both the level of capital stock and the investment flow with arbitrary precision* (see Eq. (2.73)); and (ii) a prediction (Eq. (2.74)) that *the mean-squared investment (plus interest) can never exceed the mean production function.*

2.3.2. Background

A canonical problem of finance is that of forming an optimal program for the investment of capital in the stock of a firm. If $K(t)$ represents the amount of capital K to be so invested at a time t, what form should the function $K(t)$ have if it is to result in maximum expected gain? One classic answer to this question is the Tobin q-theory of investment [79].

From the point of view of information, a program of capital investment represents as well a system with inputs $K(t)$ and resulting outputs $F(K(t))$ as defined by a production function F. These inputs and outputs impart information about the ideal level $K_0(t)$ of capital that is present at the given time. Then, is there a program that optimizes the flow of Fisher information $I - J$ about the level of $K_0(t)$? And if so, are there conditions under which this program achieves, as well, maximum expected gain of capital for the investor?

Hence, we consider the canonical problem of investing an amount of capital $K(t)$ in the stock of a firm characterized by a production function $F(K)$. Let it also have a quadratic adjustment cost function $\phi(\mathcal{I}) = (\alpha/2)\mathcal{I}^2$. Here α is a constant and $\mathcal{I}(t) \equiv dK/dt \equiv \dot{K}(t)$ is the flow of investment capital into the firm at the observed value t of the present time t_0 [80, 81], where $t = t_0 + \Delta t$. The production function is the net income as a function of capital stock [81], and is assumed to be known. The adjustment cost function $\phi(\mathcal{I})$ represents the cost of adjusting capital stock, i.e., installation costs, and captures the notion that it is more costly to do this quickly [80]. The quadratic cost function case cited is also the most common, since it represents a scenario of constant cost/time.

Using the usual Fisher information form I and a source information J that is simply related to the production function $F(K)$, we apply EPI to this investment scenario, seeking the investment program $\mathcal{I}(t)$ that extremizes the information flow $I - J$.

2.3.3. Information I

In this scenario the observer is an investor who invests by increasing the capital stock, which for a given time instant t_0 we write as K_0. At each investment time t the investor measures the level of the capital stock as K_{obs}, but this measurement is necessarily imperfect due to a fluctuation K, or

$$K_{obs} = K_0 + K. \tag{2.40}$$

(cf. Eq. (1.29)) All quantities in the analysis are functions of t, which is suppressed for simplicity. The fluctuation arises out of imperfect knowledge of the present time t_0 and the fact that the economy is a *complex* system, i.e., one undergoing persistent random change. The greater the fluctuation K, the greater is the investor's ignorance about the level of capital stock.

To represent this fluctuation we introduce a generally complex probability amplitude $\psi(K)$ for the fluctuation K. Its associated probability $p(K)$ is defined as usual by

$$p(K) \equiv \psi(K)\psi^*(K) \equiv |\psi(K)|^2 \tag{2.41}$$

(cf. Eq. (1.57)) where the asterisk denotes the complex conjugate. If we assume that the same law of fluctuation holds regardless of the level of the capital stock, i.e., shift invariance holds, then the law $p(K)$ governing the fluctuations follows from the likelihood law $p_{12}(K_{obs}|K_0)$ as

$$p_{12}(K_{obs}|K_0) \equiv p(K_{obs} - K_0) = p(K) \tag{2.42}$$

(cf. Eq. (1.48)) by Eq. (2.40).

In general, classical Fisher information I has one defining form that, for our one-dimensional problem, is

$$I \equiv \left\langle \left[\frac{\partial \ln p(K_{obs}|K_0)}{\partial K_0} \right]^2 \right\rangle, \tag{2.43}$$

(cf. Eq. (1.35)) where the angle brackets denote an expectation over all possible data K_{obs}. This universal information form applies to all problems of data acquisition and measures the level of information in the data irrespective of the physical nature of that data.

In our shift-invariant scenario the information simplifies further, to Eq. (1.58), which for our single-coordinate, single-component case is

$$I = 4 \int_{-\infty}^{+\infty} dK |\psi'(K)|^2, \tag{2.44}$$

where $|\psi'(K)|^2 = \psi'\psi'^*$ and $\psi' \equiv d\psi/dK$. This is also the form of information used in the EPI derivation of the wave equations of quantum mechanics [9].

This definition of information corresponds well to our expectation, given the nature of the observation given in Eq. (2.40): Intuitively, if the fluctuation K is confined to small values, then the level of information in the observation ought to be high; whereas if K is large, the information should be low. This is expressed by Eq. (2.44) since if K is restricted to small values, then the PDF $p(K)$ must be very peaked and narrow about $K = 0$, implying a high gradient dp/dK and, consequently, a high gradient $d\psi/dK$ and a large value for I. Conversely, if K has many large values, then $p(K)$ will be broad, will have low gradients, and a low value for I.

2.3.4. Information J

In general, functional J represents the information in the source. As with any source, it drives the problem. In contrast with the universal nature of information I, a generally *different* functional $J(K)$ drives each problem of observation: I is a general property of data, while J is specific to each source scenario (Section 1.4.3).

The fundamental driving force of this problem is the production function $F(K)$. This drives the problem by providing the net instantaneous *output* of capital due to each fluctuation (input) K of capital stock. Hence J must be constructed from the known $F(K)$. As I is defined as an expectation, so too must be J, namely that of the production function

$$J \equiv 4 \langle F(K(t)) \rangle \equiv 4 \int_0^\infty dt\, F(K(t)) P(t). \tag{2.45}$$

The factor 4 is merely chosen for later convenience, and $P(t)$ is the term-structure density,

$$P(t) = f(t) D(t), \tag{2.46}$$

with $f(t)$ the instantaneous forward rate and $D(t)$ the discount factor [9, 34, 35, 39, 63]. In our example the term structure of interest rates is taken to be a constant cost of capital ρ, from which it follows that $f(t) = \rho$, $D(t) = \exp(-\rho t)$, and $P(t) = \rho \exp(-\rho t)$. Prior knowledge (2.45) is of type (B) (Section 1.5.3).

2.3.5. Phase Space

As defined above, informations I and J are represented in different spaces: I in K-space and J in t-space. To extremize the difference between these functions, they must be placed in the *same* space. Toward this end we create a phase space by defining a canonical momentum μ as

$$\mu \equiv m\dot{K} \equiv m\mathcal{I}, \tag{2.47}$$

were m is a positive constant indicating an a priori tendency to invest at a high or low interest rate ρ; a sort of intertial mass of investment as we shall demonstrate below. The particular value chosen for m represents prior knowledge. Together with the capital stock K, this momentum μ defines a joint phase space (K, μ) of

TABLE 2.1. Correspondence between the phase spaces of statistical and investment mechanics. These relationships are discussed in the text; $\mathcal{F}$ represents the Fourier transform.

Item	Statistical mechanics	Investment mechanics
Direct space	Position x	Capital stock K
Conjugate space	Momentum $\mu = mv = m\dot{x}$	Momentum $\mu = m\mathcal{I}$
Time dependence	$x(t),\ \mu(t)$	$K(t),\ \mu(t)$
Conjugate relations	$\psi(t) = \mathcal{F}\{\Theta(\mu)\}$	$\psi(t) = \mathcal{F}\{\theta(\mu)\}$
Uncertainty principle	$\Delta x \Delta \mu \geq \hbar/2$	$\Delta K \Delta \mu \geq b/2$

investment events. The momentum μ is a fluctuation, regarded as random, from a "true" momentum value μ_0. It obeys a probability law $p(\mu)$ assumed to be shift invariant. As with the definition (2.41) of $p(K)$, $p(\mu)$ is assumed to be related to a probability amplitude function $\theta(\mu)$ as

$$p(\mu) \equiv |\theta(\mu)|^2. \tag{2.48}$$

The new amplitude function $\theta(\mu)$ is assumed to be the Fourier spectrum of $\psi(K)$,

$$\psi(K) = \frac{1}{\sqrt{2\pi b}} e^{i\rho K/b} \int_{-\infty}^{\infty} d\mu\ \theta(\mu) e^{i\mu K/b}, \tag{2.49}$$

where $i \equiv \sqrt{-1}$ and $b = \text{const.}$

The motivation for this relationship between $\psi(K)$ and $\theta(t)$ follows from either an econometric or an econophysical perspective. From the econometric perspective, the Fourier relationship can be seen as a statement that any well-behaved function $\psi(K)$ has a spectrum $\theta(t)$ in the abstract sense. From the econophysical perspective—developed further in Table 2.1.—the Fourier relationship draws from the literature on phase spaces (cf. [9] and references therein), particularly that of quantum mechanics. It also provides an economic interpretation of the coefficient b that, as we shall now see, gives new insight into the nature of the investment process.

A consequence of Eq. (2.49) is that K and μ simultaneously suffer ultimate uncertainties whose product obeys $\Delta K \Delta \mu = b/2$ (cf. Eq. (2.73)). This is a resolution limit for joint knowledge of capital stock and investment flow. Thus, investment phase space (K, μ) is effectively quantized, consisting of contiguous boxes of constant size $b/2$, where b represents the investor's level of ignorance about the actual state (K, μ) of the investment space. The intuition behind this joint uncertainty is that μ is a rate that, in practice, can only be computed as a finite difference $[K(t_0 + \Delta t) - K(t_0)]/\Delta t$ over time. This has arbitrary accuracy only as an *instantaneous* rate in the limit as $\Delta t \to 0$. But in fact Δt must be finite in order to accurately observe the value of K as a sum of capital stock values. Therefore, both K and $\dot{K}$ cannot be determined with arbitrary accuracy, in qualitative agreement with the finite size $b/2$ of $\Delta K \Delta \mu$.

With this phase space construction, we can now transform our expression for I to t-space. Beginning with Eqs. (2.49) and (2.44), we obtain

$$I = \frac{4}{b^2} \int_{-\infty}^{\infty} d\mu \, |\theta(\mu)|^2 (\mu + \rho)^2 , \tag{2.50}$$

which, by Eq. (2.48), is an expectation

$$I = \frac{4}{b^2} \langle (\mu + \rho)^2 \rangle. \tag{2.51}$$

This is an ensemble average over values of μ at any one time t. We can evaluate this by invoking *ergodicity*: that any one trajectory $\mu(t)$ "sees," over a sufficiently long time interval, all characteristic fluctuations in μ that occur at the *one* time t. By ergodicity, the ensemble average equals an equivalent time average

$$I = \frac{4}{b^2} \int_0^{\infty} dt (\mu + \rho)^2 P(t), \tag{2.52}$$

$$= \frac{4\rho}{b^2} \int_0^{\infty} dt (\mu + \rho)^2 e^{-\rho t}, \tag{2.53}$$

using the term-structure density $P(t)$ (Section 2.3.4).

2.3.6. Optimized Information and q-Theory

With the information I now in the same space as the information J, the problem of extremizing $I - J$ becomes solvable. By Eqs. (2.45) and (2.53), and ignoring an irrelevant constant multiplier -4,

$$I - J = \rho \int_0^{\infty} dt \left[F(K(t)) - \frac{(\mu + \rho)^2}{b^2} \right] e^{-\rho t} \equiv \text{extrem.} \tag{2.54}$$

As only the investment space trajectories $(K(t), \mu(t))$ are to be varied to obtain the extremum, we can ignore the additive term in ρ^3/b^2 after squaring out Eq. (2.54). Also using Eq. (2.47), we obtain a principle

$$\rho \int_0^{\infty} dt \left[F(K(t)) - \frac{m^2 \dot{K}^2 + 2\rho m \dot{K}}{b^2} \right] e^{-\rho t} = \text{extrem.} \tag{2.55}$$

The Euler–Lagrange equation solution (1.6) to this extremization problem is

$$\frac{\partial}{\partial t} \left(\frac{\partial \mathcal{L}}{\partial \dot{K}} \right) - \frac{\partial \mathcal{L}}{\partial K} = 0 \tag{2.56}$$

with $\mathcal{L}$ the integrand of Eq. (2.55). This gives

$$\ddot{K} - \rho \dot{K} + \frac{b^2}{2m^2} F'(K) - \frac{\rho^2}{m} = 0, \tag{2.57}$$

a differential equation that may be solved for the trajectory $K(t)$ and, therefore, the investment program $\mathcal{I}(t)$.

Since

$$\frac{\partial^2 \mathcal{L}}{\partial K^2} = -\frac{2m^2}{b^2} e^{-\rho t} < 0, \tag{2.58}$$

for any real m and b, Legendre's condition for a maximum is satisfied. The extremum in principle (2.54) is therefore a *maximum*. As we shall see below, a particular case of this solution is equivalent to Tobin's q-theory of investment, whose aim is likewise to maximize a functional.

The dependence of the uncertainty parameter b in Eq. (2.57) is of interest because more weight is given to the production function when b is large; i.e., when the investor has more ignorance about the total investment space. Conversely, when there is complete *certainty*, $b = 0$ and there is no dependence on the production function. Since the production function obviously matters, it must be that $b = 0$ is not a realistic choice. That is, the investor following an optimal investment program must admit of finite uncertainty about investment space. In fact he must quantify it through the stated size of b.

The significance of parameter m can be further clarified. It can be related to the uncertainty parameter b through the notion of a canonical momentum $\partial \mathcal{L}/\partial \mu \equiv \dot{K}_c$. Differentiating in this way, the Lagrangian $\mathcal{L}$ of Eq. (2.54) gives

$$\frac{\partial \mathcal{L}}{\partial \mu} \equiv \dot{K}_c = -\frac{2\rho}{b^2} (\mu + \rho) e^{-\rho t}, \tag{2.59}$$

where $\dot{K}_c$ is a canonical velocity, not quite equal to the investment rate $\dot{K}$, as indicated by the subscript c. This velocity therefore has two components where, by Eq. (2.47), $\mu \propto \dot{K}$. This suggests that $\dot{K}_c$ be rewritten as

$$\dot{K}_c = -(\dot{K} + \dot{K}_0) e^{-\rho t} \tag{2.60}$$

with identically $\dot{K} = 2\rho\mu/b^2$. The latter, together with Eq. (2.47), implies that $m = b^2/2\rho$, and with this result and the identification that the adjustment cost constant $\alpha = 2(m/b)^2$ the information difference $I - J$ to be maximized in Eq. (2.55) becomes identically

$$\int_0^\infty dt\, [F(K(t)) - \phi(\mathcal{I}) - \mathcal{I}] e^{-\rho t} = \max. \tag{2.61}$$

after ignoring the irrelevant multiplier ρ. This is the integral representation of Tobin's q-theory of investment [81].

2.3.7. Other Optimized Strategies

In addition to extremizing the difference $I - J$, EPI is supplemented by the zero-principle $I - \kappa J = 0$, where κ represents the efficiency with which the information J is transferred to I [9]. Substituting the expressions from above, Eq. (2.54) may be used to get as the zero principle

$$\langle (\mu + \rho)^2 \rangle - \kappa b^2 \langle F(K(t)) \rangle = 0. \tag{2.62}$$

Recasting these expectations explicitly we obtain

$$\int_0^\infty dt \left[(\mu + \rho)^2 - \kappa b^2 F(K(t)) \right] e^{-\rho t} = 0, \tag{2.63}$$

or

$$\int_0^\infty dt \left[\left(\frac{b^2 \dot{K}}{2\rho} + \rho \right)^2 - \kappa b^2 F(K(t)) \right] e^{-\rho t} = 0. \tag{2.64}$$

The local nature of Fisher information requires that the zero-principal be satisfied *at each coordinate value* of the integrand, or

$$\left(\frac{b^2 \dot{K}}{2\rho} + \rho \right)^2 - \kappa b^2 F(K(t)) = 0, \tag{2.65}$$

for each t. This can be solved immediately as

$$\dot{K} = \frac{2\rho}{b^2} \left[b\sqrt{\kappa F(K(t))} - \rho \right], \tag{2.66}$$

and integrated for any production function $F(K(t))$. The result is a transcendental equation in $K(t)$, which can be solved numerically if not analytically. This obviously represents an alternative investment program to that of Tobin (cf. Eq. (2.61)). It is not known how well the program works for purposes of investment.

2.3.8. Investment Parameters

In our discussion of optimal investment we introduced a number of concepts and parameters—inertial investment mass, fundamental uncertainty in investment, market efficiency, and the role of participancy in the formulation of dynamics—to which we now return.

The coefficient m was introduced in Eq. (2.47) as an inertial mass of investment. Given the expressions for m and α immediately preceding Eq. (2.61) we see that

$$m = \alpha \rho \tag{2.67}$$

and, using Eq. (2.47),

$$\mathcal{I} = \frac{\mu}{m} = \frac{\mu}{\alpha \rho}. \tag{2.68}$$

With this expression we see that the investment rate is inversely proportional to both the cost of adjustment and the discount rate. As the adjustment cost decreases, the rate of investment in capital stock increases. Similarly, as the discount rate decreases, the rate of investment in capital stock increases. Both these trends make sense. Also, from this relationship an interpretation of m as the inertial mass of investment follows naturally.

2.3.9. Uncertainty Principle on Capital and Investment Flow

The intrinsic uncertainty associated with the investment process can also be seen to be a direct consequence of the Cramer–Rao inequality. Specifically, the Cramer–Rao inequality (1.36) for estimating the level of capital stock K_0 can be expressed in terms of the information I by

$$\langle \Delta K^2 \rangle I \geq 1, \tag{2.69}$$

where $\langle \Delta K^2 \rangle$ is the mean-squared error in any unbiased estimate of the capital stock. With I given by Eq. (2.51), this inequality becomes

$$\langle \Delta K^2 \rangle \frac{4}{b^2} \langle (\mu + \rho)^2 \rangle \geq 1, \tag{2.70}$$

or

$$\langle \Delta K^2 \rangle \langle (\mu + \rho)^2 \rangle \geq \frac{b^2}{4}. \tag{2.71}$$

A particularly interesting scenario is that where investment dominates, or $\rho \ll \mu$ at all or most times. Then our inequality becomes

$$\langle \Delta K^2 \rangle \langle \mu^2 \rangle \geq \frac{b^2}{4}, \tag{2.72}$$

and taking the square root of both sides gives

$$\Delta K \Delta \mu \geq \frac{b}{2}, \tag{2.73}$$

where $\Delta K = \sqrt{\langle \Delta K^2 \rangle}$ and $\Delta \mu = \sqrt{\langle \mu^2 \rangle}$ are now root-mean-square errors. This shows that investment space has a finite resolution size of $b/2$. The inequality means that it is impossible for both uncertainties ΔK and $\Delta \mu$ to be arbitrarily small. However, the smaller is b, the smaller both ΔK and $\Delta \mu$ can be. Therefore, the finite size of b defines the level of uncertainty the investor has about the investment scenario. Moreover, by the ergodicity assumption, these uncertainty averages also hold over the ensemble of such systems *at a given* time value t. Hence, this inequality states that one cannot know *simultaneously both* the amount of capital stock in the investment system and the the investment flow with arbitrary accuracy. If one is known very finely, the other can be known only crudely: This principal of complementarity is a direct prediction of the approach.

2.3.10. Market Efficiency

Market efficiency has a natural definition in this approach: The efficiency in transforming information intrinsic to the phenomenon (specified by J) to that of the observation (specified by I) is represented by the coefficient $\kappa \equiv I/J, 0 \leq \kappa \leq 1$.

From Eq. (2.62) we find the efficiency to be

$$\kappa = \frac{\langle(\mu + \rho)^2\rangle}{b^2 \langle F(K(t))\rangle}.$$ (2.74)

Since $\kappa \leq 1$, this indicates that the mean-squared investment (plus interest) can never exceed uncertainty b^2 times the mean production function. This quantifies the intuitive notion that growth of capital needs a foundation in mean production.

3
Growth Characteristics of Organisms

ROBERT A. GATENBY AND B. ROY FRIEDEN

In this chapter a systems viewpoint is taken of the growth characteristics of normal and malignant tissue. We find that such growth is well analyzed by the concepts of Shannon and Fisher information. In Section 3.1 conventional mechanisms of information transmission via DNA, RNA, and proteins are identified, as well as *unconventional* structures such as lipids and ion gradients. Information storage, flow, and utilization are analyzed, both within cells and over a system of cells. In Section 3.2, malignant tissue growth is found to be accurately described by the use of Fisher information in particular. Cancer growth is seen to occur as a disease of information, in fact an *information catastrophe* due to the regression of cells to a minimally ordered state consistent with life. The analysis yields many predictions about the growth of healthy tissue and cancerous tissue, some of which are nonintuitive and have a strong bearing on cancer diagnosis and treatment.

3.1. Information in Living Systems: A Survey

3.1.1. Summary

Defining the properties that distinguish "living" from "non-living" systems is surprisingly difficult. A relatively simplistic description of a living organism is that of an open system which must maintain a stable, ordered state far from equilibrium. However, unlike other non-equilibrium systems in nature, living systems maintain a stable, low-entropy state by the continual use of various forms of information. The information theory (IT) of biology traditionally focused on genes and proteins as providing these inputs. Newer methods, by the use of bioinformatics, dynamical systems, and molecular machine models, emphasize the role played by intracellular information *networks*. The topological and dynamical properties of these networks arise in novel ways out of the informations of Fisher, Shannon, and Kullback–Leibler. This analysis is allowing us to understand the complex, multiscalar dynamics of health and disease as, respectively, *normal* and *disordered* processes of information; in particular, of genomic information storage, transmission, and transcription. In addition, recent IT applications suggest that critical information may be stored and transmitted by ordered cellular structures *other* than the usually assumed mechanisms of DNA, RNA, and proteins. Such nontraditional information storage structures include lipids and ion gradients. Also, information may be transmitted by molecular flux through cell membranes. Gradients in the cell membrane potentials are regarded as imparting the information to the ionized molecules as they pass through. Many fascinating challenges remain, including defining the intercellular information dynamics of multicellular organisms and the role of the storage/flow of disordered information in controlling disease.

3.1.2. Introduction

A biological organism is of course alive-a state that requires remarkable system properties. This is from both the standpoints of function and mathematical analysis. However, a living system also has many properties in common with some nonliving systems, such as being both highly complex and in a general state of non-equilibrium. This makes such a system amenable to analysis by much of the same mathematics as for nonliving systems. In particular, the concept of information applies quite successfully, and gives many predictions that are often insightful and, yet, can be nonintuitive. We begin this chapter on applications of information to biology by briefly surveying the various forms of information that have in the past been used to analyze living systems.

Living organisms are both stable, highly ordered structures and dynamical, semiopen systems. Remarkably, these operate far from thermodynamic equilibrium [1]. (Note the resemblance of these properties to those of the economic systems of Chapter 2.) The task of maintaining this non-equilibrium state is formidable. Disorder tends to continuously increase as large molecules are broken down, ions diffuse along concentration gradients, and the metabolism of glucose and other substrates (neighboring resources) produces metabolites and heat. Thus, a stable intracellular thermodynamic state seems to require a steep cellular transmembrane entropy gradient that is maintained through appropriate energy-dependent exchange mechanisms. Although information storage, transmission, reception, and utilization must clearly play central roles in maintaining cellular order, the dynamics governing these processes remain incompletely understood.

In a 1970 review [2], Johnson described IT as a "general calculus for biology." It was long known that life without energy is impossible. Johnson's article emphasized that *life without information is likewise impossible*. Since the article, remarkable progress has been made toward understanding the informational basis for life. Automated techniques now allow a complete characterization of the mRNA-and protein contents of cellular populations and, even, individuals. The human genome has been cataloged along with those of increasing numbers of other organisms. Correlations between specific gene configurations and various human diseases are reported almost daily.

In parallel with these important advances, IT has grown far beyond the methods and concepts that dominated investigation of information dynamics in living systems in 1970. Some of these approaches use new applications of traditional mathematical models developed by Fisher, Shannon, and Kullback [3–6]. Others address limitations of the IT methodology by applying new statistical and modeling approaches to information dynamics, including bioinformatics, dynamical systems, game theory, graph theory, and measurement theory. This research has provided insights, through statistical inference and fundamental analysis, into biological processes over many orders of magnitude. These processes include ion distribution and flow, molecular structures such as DNA and proteins, and the organization of cells, multicellular organisms, and ecological communities.

On the systems level, these theories are inclusive of the activities of information storage, transmission, and utilization in living systems. Our focus in this introduction will be upon new, and sometimes controversial, IT approaches that have provided novel insights into the general principles that govern these activities. Results of these IT approaches will be discussed, along with future goals and challenges.

As a matter of nomenclature, we use the acronym IT to stand for both information theories in general and for those informations that are special cases of a general information measure called Kullback–Leibler (K–L) information, or cross-entropy. The context will define the particular sense in use.

3.1.3. Ideal Requirements of Biological Information

It seems clear that information storage and processing differ markedly from living to nonliving systems. Also, there is much uncertainty in the choice of an appropriate, and unique, measure for the information. Even settling upon a definite information measure for biological systems, in particular, is surprisingly difficult, and often contentious. To start with, information can be variably expressed in terms of its relationship to a variety of other system traits, including [7–10]

- randomness,
- uncertainty,
- predictability, and
- thermodynamic entropy.

As next discussed, other traits exist as well.

In the particular context of biology, information is probably best viewed as a quantity deeply imbedded in a dynamical *process,* so that the information content of a message cannot be separated from the underlying system, which includes the

- intent of the source and
- ability of a receiver to use that information to perform work or other function.

Thus, a complete system approach is mandated. Perhaps more than in most fields of science, biological information must convey both order and meaning. For example, a protein synthesized by *random* attachment of amino acids is an ordered structure that decreases local entropy but carries no information, just as a word consisting of randomly placed letters is meaningless. Similarly, the information

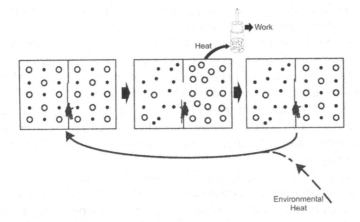

FIGURE 3.1. A modified version of Maxwell's famous demon *gedanken* experiment. Circles and dots represent fast and slow gas molecules, respectively. In the initial state, they pass freely through an open channel so that the particles are distributed equally in both containers which are, therefore, at equal temperature. Here we will assume that the same distribution is present in the environment around the box. Suppose that a demon opens and closes a gate across the channel. It allows only circles to move left to right, and dots right to left, producing a new state in which the right box contains mostly circles and the left mostly dots. The increased temperature in the right box can then be used to do work as shown. Subsequently, the internal state of the boxes is at a lower energy (temperature) than the surroundings. Heat will then flow into the box with a return to the initial state.

encoded in a protein can be used to perform work/function only by the specific ligand to which it can bind. Thus, the information encoded in a growth factor is conveyed only if a growth factor receptor is present, just as a message in English conveys no information to a receiver who speaks only Russian. This critical role of context and meaning in information is a significant part of the problem of trying to develop a comprehensive mathematical formalism to describe information dynamics in living systems.

Although there is currently no information metric that measures all these traits, Shannon information and extreme physical information (EPI-Fisher) seem to come closest. These information measures are, respectively, used in system models that regard a biological organism as a communication channel or a classical estimation channel.

IT is typically defined as the study of information storage, communication, and utilization. Its origin is often traced to Shannon's pioneering article [5] in 1948, which described a theory of communication. Interestingly, he generally eschewed the term "information theory," probably because the term "information" was already in use (from the early 1920s onward) in describing Fisher's information [3, 4]. In fact the role of information in physical processes, particularly thermodynamics, had been investigated by Szilard and others [11] since the era of Maxwell's famous nineteenth century [12] *gedanken* demon (Figure 3.1).

This simple experiment illustrates the dynamical and complex interactions of biological information and thermodynamics. The apparent change in molecular distribution due to the demon's efforts is the result of its use of information to distinguish between the two types of molecules and to judge the direction of their motion. This utilization of information is manifested thermodynamically by heat flow between the boxes and the addition of information to the system. That is, the nonrandom distribution of the molecules, by Eq. (3.1) below, now confers a nonzero value of H in the containers. This information/energy is dissipated by performing work. Energy flow from the environment will then return the containers to their original state.

Note that if the work is exported, the total system loses energy with each cycle. If the work is performed in the environment in contact with the boxes under ideal conditions, the process would be adiabatic. In reality, of course, energy would be inevitably lost in the process of performing work and by the metabolism of the demon in making measurements and moving the gate. Thus a hidden cost of the *gedanken* experiment is the energy the demon expends to gain the information necessary to allow it to perform its functions.

3.1.4. Some Alternative Information Types

Through the years, many other forms of information have been invented and utilized. Aside from those due to *Fisher* and *Shannon*, a partial list includes informations [13] due to *Renyi, Wootters, Hellinger, Gini–Simpson, Tsallis, and Kullback–Leibler*. In general, each type of information is tailored to satisfy the needs of a particular application. For example, the Gini–Simpson information is sensitive to the degree of diversity that is present in the population mix of a given ecology. Also, certain of these informations are so general as to include others within their scope.

3.1.5. Kullback–Leibler Information

Chief among these is the Kullback–Leibler (or K–L) information, denoted as $K_{KL}(p/r)$ and defined as

$$K_{KL}(p/r) \equiv - \sum_i p(y_i) \log_b \frac{p(y_i)}{r(y_i)}. \tag{3.1}$$

It is assumed that the basic events y_i occur independently. (Otherwise the probabilities would be of joint events.) All sums over i are from $i = 1$ to $i = N$. The logarithmic base b is at the discretion of the user. The base choice imparts an artificial unit to K_{KL}. For $b = 2$, the unit is called the "bit." Although this term was not actually coined by Shannon [5, 6], it came into popular use through his work. The curve $p(y_i)$ is the subject probability law (usually regarded as unknown and to be found) for independent events y_i, $i = 1, \ldots, N$, and $r(y_i)$ is a known or *input* probability law. It is often called a *reference* probability law, as discussed below.

Information (3.1) is inclusive of Fisher information (Eq. (1.51a), Chapter 1), Renyi information [13], and both Shannon transinformation and Shannon information (shown below).

The form (3.1) is that of a "distance measure," specifically, the distance between the two curves $p(y_i)$ and $r(y_i)$. As a check, note that if $p(y_i) = r(y_i)$ at all y_i, i.e., the two curves are the *same* curve, any legitimate measure of the distance between them should be zero. In fact, the K–L distance, or information, (3.1) is identically zero under these circumstances ($\log_b 1 = 0$). Conversely, the distance (3.1) grows in magnitude as $p(y_i)$ and $r(y_i)$ increasingly differ from one another. As a fine point, the distance is, however, a "directed" one, in that the distance from p to r is not equal to that from r to p, i.e., from definition (3.1) $K_{KL}(p/r) \neq K_{KL}(r/p)$.

Another interesting case is where r is "a differential away" from p, i.e.,

$$r(y_i) = p(y_i + \Delta y), \quad \Delta y \to 0. \tag{3.2}$$

Then if also the subdivision $y_i \equiv i\Delta y$ in (3.1), the discrete sum in (3.1) becomes a continuous integral over y, and Eq. (1.51a) shows that

$$K_{KL}(p/r) \to - \lim_{\Delta y \to 0} \frac{\Delta y^2}{2} I. \tag{3.3}$$

The amplitude of the K–L information becomes essentially the Fisher information.

The probability density function (PDF) $r(y_i)$ is often called a "reference" law, for the following reason.

3.1.6. Principle of Extreme K–L Information

Suppose that, for a given PDF $r(y_i)$, the $p(y_i)$ are sought that extremize $K_{KL}(p/r)$ subject to constraints, as in a problem (Section 1.1.2.4)

$$-\sum_i p(y_i) \log_b \frac{p(y_i)}{r(y_i)} + \sum_j \lambda_j \sum_i p(y_i) f_j(y_i) = \text{extrem.} \tag{3.4}$$

This is called the "principle of extreme K–L information." Here the $f_j(y_i)$, $j = 1, \ldots, J$ are known constraint kernels. (An example might be $f_j(y_i) = y_i^j$, where the constraints are moments of order j.) This is characteristically a Bayesian approach to estimating the $p(y_i)$. That is, the constraints are any the observer thinks "reasonably characterize" the given process. They are not necessarily the most important or most fundamental in some sense. Obviously, the resulting solution for $p(y_i)$ cannot, then, be regarded as fundamental either. Rather, it is considered to be as "reasonable" as were the constraints.

3.1.7. Bias Property

The solution to (3.4) is obtained simply by differentiating $\partial/\partial p(y_k)$ and equating this to zero. The result is a solution

$$p(y_k) = r(y_k)e^{-1}b^{\sum_j \lambda_j f_j(y_k)}, \quad k = 1, \ldots, N. \tag{3.5}$$

This shows a proportionality

$$p(y_k) \propto r(y_k) \tag{3.6}$$

between the estimate $p(y_k)$ and the input law $r(y_k)$. That is, the estimate $p(y_k)$ is biased toward $r(y_k)$. Or, $r(y_k)$ acts as a *reference function* for $p(y_k)$. Also, from (3.6), $p/r \sim 1$. This is the reason for the division notation $K_{KL}(p/r)$, and also why this information is often called the "cross entropy" between p and r.

3.1.8. A Transition to the EPI Principle

Principle (3.4) of extreme K–L information actually includes the EPI principle within its scope. In the continuous limit $y_i \equiv i\Delta y$, $\Delta y \to 0$, and with the special reference function choice (3.2), correspondence (3.3) holds, so that principle (3.4) becomes

$$-\frac{\Delta y^2}{2} I + \sum_j \lambda_j \sum_i p(y_i) f_j(y_i) = \text{extrem}. \tag{3.7}$$

If the constraints j are chosen such that their sum in (3.7) is effectively the Fisher source function $J = J[f_1(y), \ldots, f_J(y)]$ for the problem, then (3.7) takes the form of the EPI principle (1.77). This means that for any problem that is attackable by EPI, the problem could also have been attacked using the extreme K–L principle (3.4), provided that the reference choice (3.2) was used and that the user (exercising good intuition) constructed the special set of constraints that are identical with the theoretical source functional J for the problem.

3.1.9. Biological Interplay of System and Reference Probabilities

This interplay (3.5) between p and r seems vital to biological systems. In fact, biological systems extensively apply K–L cross entropy, since living organisms continuously generate probability distribution functions that measure their internal and external environments. In turn, the fitness of each organism is dependent on the accuracy of these functions. That is, the distance between the curve $p(y_i)$ internally generated by the organism and the reference curve $r(y_i)$ representing the external environment must be minimized, if the organism is to maximize its fitness. The reference curve $r(y_i)$ can, for example, model possible external threats from predators or natural phenomenon, as well as available food sources. The internal picture $p(y_i)$ is generated through energy-dependent utilization of the senses.

In living systems, the energy cost of measuring and reconstructing the external probability distribution function requires optimization strategies that trade off accuracy for cost. That is, the most accurate measurement of the environment would require the senses to be maximally "tuned." In fact, most organisms scan their environment with a single highly tuned sensory system (sight, or smell, hearing etc.) while others are used much less. Presumably these strategies have evolved to give the organism a sufficiently accurate picture of the environment to maximize its survival probability while minimizing the cost in terms of energy output. These are both for the use of the sensory system and for storing, transmitting, and transcribing the greater heritable information content required to build more complex, high-resolution systems.

3.1.10. Application of K–L Principle to Developmental Biology

An interesting long-term problem for Kullback–Leibler information may be seen in *developmental biology*. Here the information content of a single fertilized egg is translated into a large, complex multicellular organism, through the process of growth. Then, for efficient development, the probability distribution function that describes the *mature* organism—call it $p(y_i)$—must be nearly identical in information content to the probability distribution function within the *egg*—call it $r(y_i)$. That is, the bias property Eq. (3.6) must hold again. Developmental biology can thus be modeled as a K–L process in which the probability function of the fertilized egg acts as a *reference function* for future development, and is replicated stepwise with maximal accuracy. This process should, then, obey the extremization principle (3.4), with appropriate constraint kernels $f_j(y_i)$. *This relationship should define the first principles that govern developmental biology.* To our knowledge, this idea has not yet been carried through. We predict that its implementation will lead to new insights into developmental biology, both regarding normal and abnormal development.

As a working alternative, extremization principle (3.4) can also go over into the *EPI principle*, as we showed in Section 3.1.8. The EPI principle is mathematically different enough from the more general principle (3.4) to give different results from it. Hence, it might well be worth trying. It certainly works in application to abnormal, i.e., cancerous, growth as taken up in Sections 3.2.7 to 3.2.11.

We next examine special cases of K–L information that are of the Shannon form.

3.1.11. Shannon Information Types

Shannon information and transinformation follow as special cases $r(y_i)$ of the K–L information (3.1). In the case of a constant, unit reference function $r(y_i) = 1$,

Eq. (3.1) becomes

$$K_{KL}(p/r) \equiv H(Y) = -\sum_i p(y_i) \log_b p(y_i). \qquad (3.8)$$

This is called the "Shannon self-information" or simply "Shannon information." Note that it is in the form of a thermodynamic entropy. Definition (3.8) regards H as the amount of "information, choice, and uncertainty" [5, 6] in a signal Y that possesses possible configurations y_i with respective probabilities $p(y_i)$. Note that this information is also identically $-\langle\log_b p(y_i)\rangle$, the expectation $\langle S_i \rangle$ of a more elementary information $S_i \equiv -\log_b p(y_i) \equiv \log_b(1/p(y_i))$. This information shows an important property: *Improbable* events y_i carry high elementary information, and conversely. Hence S_i is also called the *surprisal* [14]. On this basis, by (3.8) $H(Y)$ is the *mean* surprisal of the system.

An example of self-information (3.8) is that supplied by a *codon*. A codon is a triplet message—a sequence of three of the nucleotides A, T, C, and G that encodes for an amino acid. The number of different possible codon sequences is then 4^3, so that the probability $p(y_i)$ of any one sequence is 4^{-3}. Then the elementary information in a codon is $S_i \equiv \log_2(1/p(y_i)) = 6$ bits, so that by (3.8) the self-information is 6 bits as well. However, a limitation to this approach arises out of an unknown amount of degeneracy in codon structure, in particular at the third position of the triplet code. There is resultingly an unknown level of redundancy in the genome. In turn, this will tend to overestimate the level of Shannon information, so that it is actually *less than* 6 bits.

In the more general case of joint or *two*-dimensional probabilities $\mathbf{y} \equiv (y_i, y_j)$, $i, j = 1, \ldots, N$, definition (3.1) is replaced by

$$K_{KL}(p/r) \equiv -\sum_{i,j} p(y_i, y_j) \log_b \frac{p(y_i, y_j)}{r(y_i, y_j)}. \qquad (3.9)$$

If the reference function is constructed as $r(y_i, y_j) = p_1(y_i)p_2(y_j)$, where the latter are the marginal laws for $p(y_i, y_j)$, the K–L form (3.9) becomes

$$K_{KL}(p/r) \equiv -S = -\sum_{i,j} p(y_i, y_j) \log_b \frac{p(y_i, y_j)}{p_1(y_i)p_2(y_j)}. \qquad (3.10)$$

This is exactly the negative of the Shannon trans- (or mutual) information S. Information S represents the information in a signal y_i about *another* signal y_j (thus the "trans"); or vice versa. A good introduction to properties of Shannon information in general is [15].

In summary, the K–L information becomes either form (3.8) or (3.10) of Shannon information, or Renyi information, or Fisher information (1.51a), depending upon the particular choice of the reference function $r(y)$.

3.1.12. *Information as an Expenditure of Energy*

Shannon defined H as an "entropy" [5, 6], because of its formal similarity to H in Boltzmann's theorem. This has led to the concept of information as "negative" entropy [16] (sometime called "negentropy"). There is clearly a qualitative relationship present as well, in that the information content of a system increases only if entropy (randomness) decreases and inflow of information increases system order (Figure 3.1). A quantitative relationship, although subject to disagreement, between information and energy has been proposed [17]:

$$H_{bit} \geq kT \ln 2. \tag{3.11}$$

Here H_{bit} is the amount of energy required to acquire 1 bit of information, k is the Boltzman constant, and T is the absolute temperature.

The main point of Eq. (3.11) is that *information must be purchased by expending energy and, therefore, resources.* This places limits on biological information dynamics, because living systems must balance their need for information with their ability to pay the energy cost of its storage and transmission. Optimization of this process by evolution will favor organisms that maintain enough information to encode a complex and robust system, but no more than *suffices.*

It is interesting that the dependence of information upon energy continues at finer and finer levels, even existing at the quantum level. Here it takes the form [18]

$$I = \frac{8E^2}{c^2 \hbar^2}, \tag{3.12}$$

with $E = mc^2$ the relativistic energy of a particle of mass m, c the speed of light, and $\hbar$ Planck's constant divided by 2π. Information I is the Fisher information about the space-time location of the particle.

It may be noted that Shannon information (3.11) is linear in the energy kT, whereas the Fisher information (3.12) is quadratic in the energy E. Thus, the Fisher measure increases more rapidly with the energy. This is indicative of a trend—Fisher information is usually a faster, more sensitive function of its parameters than is the Shannon variety. Ultimately, this is because Fisher information (1.50) is a *local* measure of the system, explicitly dependent upon the *slope* of the PDF, whereas Shannon information is a *global* measure, effectively going as the PDF itself (times its logarithm, a slow function).

3.1.13. *Some Problems with Biological Uses of Shannon Information*

Perhaps the first comprehensive analysis of information dynamics in living systems was performed by Schrodinger in his book "What is Life" [19]. He pointed out that cells are continuously buffeted by thermodynamic perturbations, so that atomic level intracellular structures are not sufficiently stable to encode heritable information. Rather, he predicted that transgenerational information must be encoded in molecules with an *aperiodic* structure. Subsequent discoveries of the DNA structure, consisting of sequences of the four nucleotides A, T, C, and G of course confirmed his prediction. As a sequence of symbols, polynucleotides and polypeptides are ideally suited to analysis by Shannon's methods [20–24]. A variety of authors have used information theory to quantify the information content of genes, to examine the conservation of information during evolution [25–28], and to analyze the origin and evolution of biological complexity [29, 30]. Application of Shannon methods to ecology have met with some success, particularly in describing the interactions among multiple species [31, 32]. (Fisher information has likewise been so used; see Chapter 7.) While quite interesting, this type of analysis was well described in the previous review of Johnson and will not be further examined here.

Rather, the focus will be on new methods, since it seems clear that, in the 36 years since Johnson's original article, IT using traditional Shannon methods has *not* become, as predicted, the "general calculus" of biology. Although the lack of widespread application is probably the result of multiple limitations, several issues stand out:

First, Shannon methods that quantify the content of information in a biological structure say nothing about its *meaning*, *cost*, or *function*.

Second, such traditional IT models typically did not address the precise mechanisms by which information is used to perform work and maintain cellular stability.

Third, IT tended to view information as a quantity that is simply exchanged between individuals. Rather, it has become increasingly clear that biological information flows along complex *pathways*, with positive and negative *feedback* loops and substantial temporal and spatial plasticity.

These limitations have been addressed by a number of investigators who have applied Shannon methods in novel ways or developed new mathematical approaches to *information dynamics*. Some are as follows.

3.1.14. Intracellular Information Dynamics

3.1.14.1. Bioinformatics and Network Analysis

The aim of intracellular information dynamics is to describe the formation, flow, and storage of information for a *system* of cells. Some of the most important advances in our theoretical understanding of these dynamics have stemmed from applications of methods from dynamical systems and bioinformatics. The former grew out of graph theory and is used to characterize the structure and interconnections of complex systems such as electrical grids and the Internet [33]. Initial biological applications examined metabolic pathways [34, 35]. More recent focus has been on information pathways—the *interactome* of Figure 3.2—in which a node represents an information carrier (typically protein or RNA) and the flow of information is defined by the interactions of these carriers [36, 37]. Most of this has focused on the interactome of *C. elegans* [38], but it seems likely that the organizational principles will be similar in eukaryotic cells.

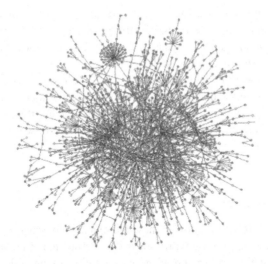

FIGURE 3.2. The organizational principles of a scale-free network of information flow in cells called an *interactome*. Each node of the interactome (shown as a dot) is a class of proteins, whose connections are based on direction of interactions of the protein molecules or flow of substrate from one protein to another (these could be metabolites from an enzyme-catalyzed reaction or ions from a protein pump within the membrane). The number of connections for each node is described by a *power law distribution*, where most nodes have few connections but a few nodes have many connections; the latter are called *hubs*. The visible result is a network pattern showing typical self-similarity at all scales. The information flow dynamics of hubs is bimodal with some ("party hubs") demonstrating temporally synchronous connections, while others ("date hubs") exhibit greater heterogeneity and typically connect with nodes that are more physically separated.

This work has been made possible by rapid advances in genomics and proteomics that measure the intracellular expression of the information in each gene based on the number of RNA or proteins present. A necessary parallel development included new methods in bioinformatics [38], which encompass diverse statistical methods that are used to extract information from large data sets. The latter typify the new molecular biology technologies.

Information networks contain junctions, or nodes, which typically represent a distinct protein species or connection based on the interaction of one node or protein with another. This is either by direct contact or indirectly, as the products of one interaction become the substrate for another enzyme. It appears that intracellular information networks are, at least to a first approximation, organized in a scale-free topology, so that nodal connections can be described by a distribution function

$$P(k) \sim k^{-\gamma}. \tag{3.13}$$

Here $P(k)$ is the probability that a node will have k connections, and γ is the scaling exponent. This is a simple power law. Thus, a few nodes have many connections while most have only one or a small number. Highly connected nodes (arbitrarily defined as $\geq$ 5 connections [39]) are called *hubs*. This topology has the advantages of allowing high connectivity and small diameter (defined as the average length of the shortest path between any two nodes).

Interestingly, power law behavior such as (3.13) follows as well from EPI, as shown below. This means that power law behavior follows not only as an efficient allocation of energy within a system but also as an efficient flow of information to its observer. In effect, living systems are "meant" to be observed optimally by other living systems (Section 1.4.5).

3.1.14.2. Hub Dynamics

The dynamics of information flow in these biological networks is beginning to be understood. Han et al. [39] used microarray data before and after various cellular perturbations to define the synchronization of activity among nodes. They found consistent evidence for two distinctly different hub dynamics. Some of the hubs are highly connected with others, so that the groups react synchronously—a pattern they describ as "party hubs." In other hubs, the connected nodes demonstrat greater temporal heterogeneity in their responses and are termed "date hubs." This bimodal behavior is highly dependent of spatial distribution of the proteins—date hubs tend to involve information flow over greater distances within the cell, while party hubs connect proteins that are "near neighbors." Interestingly, they also demonstrate that removal of date hubs result in much greater disruption of the network than that of party hubs. These results demonstrate that information flow within the interactome is complex and probably nonlinear. In fact, maps of cell signaling typically demonstrate multiple parallel, but interacting, pathways. These suggest that the initial message may, as in the Internet, be broken into components and sent along different paths to insure rapid and maximally accurate transmission.

It is important to recognize that the interactome represents a map of a *potential* information network—the actual network will vary between individuals and in the same individual over time. For example, Luscombe et al. [40] demonstrated that only a few hubs are completely stable over time, while most exhibit *transient* variations in connections in response to a variety of intracellular and environmental changes. In other words, information flows through highly plastic networks that are *continuously rewired* in response to stimuli that carry information from outside the network. These results suggest that future work will also need to focus on the dynamics within nodes, so that the flow and connectivity become explicit functions of other system parameters, such as the concentration of the node protein that is present. For example, a node and its connections will be lost when the protein is not expressed, or if its concentration falls below some threshold. Similarly, the number of functioning links to a node, or the speed of flow through the links may be dependent on protein concentration.

Many additional challenges remain for this exciting, and highly productive, avenue of investigation. Significant variations among the data sets on which the models are based [41] will need to be resolved, and other analytic methods such as the network motif approach should be examined [36].

3.1.14.3. Dynamics of System Failures

In addition, much may be learned from understanding the topology and dynamics that lead to failures in the intracellular information system. Topological analysis of scale-free networks demonstrates that their heterogeneity confers resistance to failure caused by random loss of nodes. For example, it has been demonstrated that network function is maintained even when up to 8% of the nodes are randomly removed [42, 43]. On the other hand, scale-free networks are vulnerable to selective attack. For example, removal of 8% of the *most-connected* hubs resulted in a 500% decline in network function as measured by average path length [44].

The dynamical aspects of system failure are suggested by Zhao et al. [45]. He proposed that in each time step of normal system function the relevant quantity (in this case information) is exchanged between every pair of nodes, and travels along the shortest possible pathway. A node i will ordinarily have some number of shortest pathways running through it—defined as its load L_i. The capacity C_i of the node is the maximum load it can carry, with

$$C_i = (1 + \alpha)L_i, \tag{3.14}$$

where α is a tolerance parameter. They found that the loss of a node results in a redistribution of flow through other nodes. When this exceeds the node capacity, cascading failure will result in *system disintegration*. Consequently, the vulnerability of the system to attack is dependent on the value of α.

Another way of stating this is that the robustness of intracellular information networks comes at a cost—the expenditure of resources necessary to increase the number of connections and/or the nodal capacity [45]. Thus, both topological and dynamical analysis of system failure illustrates the central biological cost–benefit calculation apparent in Eq. (3.14).

Living systems must balance their need for robust information dynamics with cost. Therefore, eventually the *energy costs* of forming and maintaining large information networks will need to be included in the analyses. These constraints place boundary conditions on biological information networks, such that each cell can be expected to evolve system topology and dynamics that approach the *minimum necessary* to achieve sufficient robustness. A level of robustness is needed that enables the cell to resist the range of typical environmental disturbances. Chief among these are perturbations in physical parameters such as temperature and biological disturbances due to parameters such as infectious agents. This optimization process may well prove useful in modeling evolutionary dynamics.

3.1.15. Information and Cellular Fitness

One approach to addressing the value and cost of information assigns to the information content (3.8) of each gene a fitness value defined by its contribution to the survival minus its cost [46, 47]. In the context of the dynamical system models, date hubs, for example, will have a higher fitness value than party hubs, which typically have higher values than individual nodes. This potential variation in the fitness value of each component of the information network may play a role in its observed plasticity. For example, the nodes and connections of the four proteins that constitute decarboxylase under normal conditions have no function and, therefore, low fitness values. Not surprisingly, normally these proteins are not detectable. However, these components of the system assume great value when a cell is compelled to use glycine as a carbon source. Since the benefit now exceeds the cost, these hubs now become highly active.

In addition the "fitness" of cells in multicellular organisms is highly contextual. When it is defined as survival and function of the multicellular organism, the value of cellular differentiated genes is high because they maintain the stability of the entire system. However, if fitness is defined as simply individual cellular proliferation, then gene segments that encode differentiated cell function have a *negative* value (i.e., their contribution to proliferation is zero, so subtraction of the energy cost of the information yields a negative fitness value). This binary fitness effect is not merely conceptual. It may be observed during *carcinogenesis*, as the transformed cells act increasingly as individuals rather than as components of a multicellular society. Thus, random mutations that disable differentiated functions will be favored during somatic evolution—a process that manifests itself as de-differentiation or loss of function. This is typical of invasive cancers.

3.1.16. Cellular Information Utilization

The preceding methods have provided many insights into the general organization and dynamics of information flow within a cell, but do not address the nitty-gritty question of how this information is actually *used* by the cell to perform work. Schneider and colleagues [17, 48] have examined biological utilization of information through a novel approach in which macromolecules (typically polypeptides or polynucleotides) or macromolecular complexes function as *isothermal machines* operating in a cycle. Each such machine contains information that is originally encoded in the genome, then transmitted through the sequence of amino acids and, finally, manifest in a tertiary structure that permits highly specific allosteric binding. This allows a cycle (Figure 3.3) which begins when the machine is primed to a higher energy state so that its binding sites are available and, therefore, ready to do work. During the cycle, the machine expands its information content by "choosing" a specific ligand over all other possibilities—a process manifested by the energy of binding. In the *operation* phase of the cycle, the machine dissipates this energy by performing work that benefits the cell. This approach explicitly integrates information *with* cellular thermodynamic processes, and allows the value and meaning of the encoded information to be defined by the specific cellular function performed.

More recent studies suggest that this general approach can be broadly applied to mechanical and enzymatic protein function [49]. In fact a theoretical model of

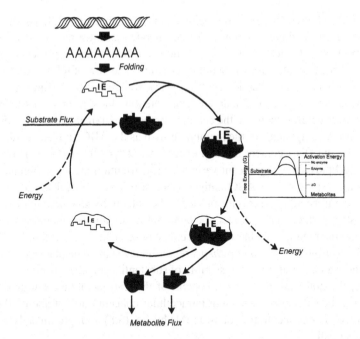

FIGURE 3.3. This sequence illustrates the molecular machine model that translates information into useful work, as developed by Schneider. Information stored in DNA flows to proteins through the amino acid sequence which, after protein folding, results in a specific shape. On the top left, the protein is primed with a baseline level of information and energy (designated by the size of the **I** and **E**). The protein chooses to bind with a specific substrate and rejects all other possibilities. This creates a molecular complex in which the information content and energy (through binding) are maximized. This is dissipated by reducing the activation energy of the reaction, allowing the substrate to be split into two metabolites with a release of energy (ΔG). This reduces the energy and informational content of the enzyme, which is then primed by the addition of new energy from the environment (e.g., through thermodynamic fluctuations or ATP). The net cycle allows information to be used to generate work. Note that the reaction produces molecular flux in which substrate is consumed and metabolites are generated.

cells as a collection of *molecular machines* has been proposed [50], and it seems likely that a combination of molecular machines and network analysis will yield substantial insights into the coupling of cellular information dynamics with cellular function.

3.1.17. Potential Controversies—Is All Cellular Information Stored in the Genome and Transmitted by Proteins?

Implicit in much of the work on cellular information is that the genome and its RNA and protein products define a complete representation of information dynamics. However, the results of Luscombe et al. [37] suggest that the interactome, in fact, adapts to information flow from some *external* sources.

Gatenby and Frieden [51] have pointed out, e.g., that Eq. (3.8) permits the structure of any cellular *component* to be encoded as information. The linear distribution of nucleotides A, T, C, and G along a DNA "string" represents a clear example of such an information encoding structure. The flow of information to the comparable linear string of amino acids is by now well recognized. However, there are many structures in the cell that have this potential. In addition to DNA, RNA, and protein structures, there are the cell membranes and transmembrane gradients of ions. Some components of the membrane such as the ABO antigen are directly encoded by the DNA content through expression of specific enzymes. However, membrane lipids show significant variation in distribution among different cells [47]. Using Eq. (3.8), the ABO antigen—because it is identical on all cells—encodes no information. This is in contrast to the potentially *abundant information* in the distribution of lipids. In fact, the potential information in membrane lipids of a typical mammalian cell has been calculated to be on the order of 5×10^{10} bits. Similarly, *transmembrane ion gradients* represent substantial displacements from randomness and, therefore, the possibility of substantial encoded information. For example, the potential information content of the transmembrane ion gradients of Na^+ can be calculated (assuming intracellular concentration of about 20 mM and extracellular concentration of 140 mM) by Eq. (3.8) as approximately 1.2×10^8 bits per cell.

What role might these information carriers play (cf. Section 1.3)? Clearly the genome acts as the long-term storage mechanism for heritable information. However, the dynamics over short time periods require rapid interactions, and these could very well utilize nongenomic, nonprotein information carriers. For example, information encoded in molecular gradients could relate to the constantly changing metabolic status of the cell (Figures 3.3 and 3.4), or to the extracellular concentrations of substrate or signals. Finally, long-term information storage and transgenerational transmission is costly, but can be augmented if heritable information is expanded by encoding proteins that can, in turn, generate additional information. Thus, proteins that are transmembrane ion pumps can act as Maxwell demons generating additional information (through the transmembrane ion gradients as outlined above) in each new cell.

Independently, Keener [53, 54] recently demonstrated that information stored in gradients can be transmitted by molecular and ionic flux (Figure 3.4). The general concept is evident in the equation for diffusion of particles across membranes M:

$$J_M = \frac{D_M A_M}{L_M}(C_O - C_I) \qquad (3.15)$$

where J_M is the flux across the membrane, D_M is the diffusion coefficient, A_M is the surface area of the membrane, L_M is the diffusion length across the membrane, and C_O and C_I are the outside and inside concentrations. Keener points out that the flux may actually encode information about the parameters of Eq. (3.15). For example, if the system is subject to perturbations in the extracellular concentration of some nutrient C, a measurement of flux will also allow measurement of the extracellular concentration C_O. He has recently demonstrated that bacteria can

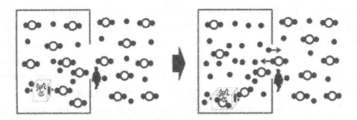

FIGURE 3.4. An alternative demon *gedanken* in which a container is initially at equilibrium with an environment, so that two different molecules are equally distributed (shown in the left box). Within the container is a box housing an ogre who at some time t_0 begins to "eat" the central portion of the larger molecule and releases the two peripheral metabolites back into the environment (similar to the enzyme in Figure 3.3). As a result of the ogre's appetite, the concentration of particles within the box changes, producing a gradient with the environment, and molecular flux through the door. A demon stationed at the door can, by measuring the concentration gradients or the flux, gain information regarding the consumption rate of the ogre. In other words, *concentration gradients and fluxes of small molecules represent potential sources of information storage and transmission* that are distinct from the protein interactome in Figure 3.2.

measure flux through positive and negative feedback loops, and thereby use that information to obtain extracellular and intracellular information, specifically the size of a local population and the length of flagella, respectively.

Obviously this work remains speculative, but it does have the potential to significantly expand knowledge of the topology and dynamics of intra- and extracellular information networks.

3.1.18. Multicellular Information Dynamics

The manner by which information dynamics govern development and function of *multicellular* living structures remains largely unexplored. It represents an exciting future challenge for IT. Clearly, the stable synchronization of large numbers of cells requires complex, but accurate, information flow. The need for increased information transmission and reception in multicellular organisms is evident in the genome of the cyanobacterium *Nostoc punctiforme*, which forms a simple filamentous multicellular structure and contains 7400 genes. Even the unicellular cyanobacterium *Synechocystis* contains 3200 genes [55].

Insight into at least some of the principles of intercellular information dynamics can be obtained in rudimentary bacterial communication during *quorum sensing* [56, 57]. Typically information is carried by small molecules (autoinducers) produced and excreted into the environment at a regular rate by each bacterium. The molecule freely diffuses through the bacterial wall and, typically, binds to a regulator protein in the cytoplasm. The regulator/autoinducer complex then binds to a specific DNA promoter segment upstream of the protein that catalyzes formation of the autoinducer. This produces a positive feedback loop [58]. Once a critical concentration of the autoinducer is exceeded, the signal is dramatically amplified by the autocatalytic pathway. In turn, this produces a population phase change in which all of the individuals act in unison. Here the content of the message is simply "I am here." The resulting environmental information is the summation of these messages. These are conveyed to each individual through a transmembrane molecular flux that is coupled to an intracellular positive feedback loop.

Similar but more complex information dynamics is observed in aggregates of the amoeba *Dictyostelium* and has been extensively modeled [59, 60]. This is a creative reaction to an imminent state of starvation. First, the individual amoebae must accurately detect information that starvation is imminent. Second, individuals in the population must be instructed to aggregate into a fruiting body through a chemotactic relay system. The system is based on detection and production of cAMP, which tells individuals to undergo a phase transition from proliferation to differentiation and to then migrate in a specific direction.

The information dynamics in these rudimentary multicellular networks seem to be based on pathways that integrate small diffusing molecular messages with the intracellular interactome. In other words, nodes from one individual link to nodes within others via a diffusing messenger. It will be interesting to see if these interactions form a *scale-free* network that echoes the organizational principles within cells.

Formation of a multicellular society from a single germ cell requires sufficient information content in the initial cell to form a complex multicellular organism. In the case of humans, the latter consists of approximately $10^{14.}$ cells. Understanding this synthesis is undoubtedly the greatest challenge for information theory. Indications are that organizational principles of developmental biology include multiple diffusing information carriers and specific receptors. These build on communication strategies observed in rudimentary multicell societies. However, it appears that more complex multicellular organisms have also developed additional mechanisms of information exchange between cells in contact. This exchange allows a flow of information across cell membranes via direct interactions of membrane proteins. These interactions also occur through intercellular channels that are formed by gap junctions, and probably involve a bidirectional flux of ions and small molecules. Since these links are dependent on cell contact, they may well exhibit Poisson rather than power law statistics. This suggests that complex multicellular organisms may use a combination of contact-mediated links between nodes of next-nearest neighbors, and longer distance, nonrandom links through diffusing information carriers and specific receptor molecules. This combination of random and scale-free networks may produce information dynamics substantially different from that of the cellular interactome.

3.1.19. Information and Disease

Ultimately, the goal of IT is to use insights into intra- and extracellular information dynamics in order to understand disease and suggest therapeutic strategies [61]. It seems clear that biological information networks are designed to be adaptive and robust and, therefore, resistant to failures due to environmental perturbations or direct attack. Disease must represent some, at least transient, failure of the system due to mutations or attacks for which it was unprepared.

Information degradation in disease appears to exhibit both simple linear and highly complex nonlinear dynamics. For example, sickle cell disease is a straight-forward result of biallelic mutations that alter a specific node in the interactome (the hemoglobin protein). Similarly, some infectious agents (HIV for example) selectively attack specific nodes of specific cell types, producing a complex but relatively straightforward series of events in the cells and organism. However, given the complexity of the informational networks it is not surprising that many diseases appear to result from a disruption of multiple nodes and interactions, such that consistent correlation with a single gene mutation or external stimulus will *not* be possible.

The multistep transition from normal tissue through a variety of intermediates to invasive cancer represents an interesting model for these dynamics. Carcinogenesis requires accumulating genetic mutations and chromosomal defects [62]. In fact, Loeb and others hypothesize an increased mutation rate as a *necessary* condition for cancer formation [63].

Clearly, a sufficient degradation of the genome will result in disruption of critical information pathways. It therefore appears that somatic evolution of the *malignant* phenotype requires a loss of cellular information. However, while random genetic mutations appear necessary for carcinogenesis, two observations indicate that the information dynamics are more complex than by this simple picture:

1. A constant mutation rate, in the absence of modifying constraints, will result in an "error catastrophe" [64–66] in which genomic information decays to the point that it insufficient to maintain life, even disordered, nonfunctioning carcinogenic life.
2. Despite this effect, the cancer phenotype exhibits apparent *nonrandomness*, in that differentiated cell functions are progressively (semideterministically) lost, while genes necessary for proliferation remain functional (and often upregu-lated) even in the most advanced cancers. It is not clear how this occurs.

Using IT, it has been demonstrated that information degradation from the mu-tator phenotype is highly constrained by Darwinian selection during somatic evo-lution of the malignant phenotype [46, 47]. These selection dynamics will pre-serve the genomic information content necessary for growth, while the information content of gene segments that *suppress* growth of such tumor suppressor genes will be rapidly degraded. As described above, genes that encode the differenti-ation function will have negative fitness values, and are subject to progressive degradation. These dynamics allow tumor cells to preserve information that is necessary for survival and proliferation, while progressively losing information that encodes *cellular function* in a multicellular society (and the energy demands for maintaining and transcribing that information). That is their dual strategy for winning out.

It results that the global information content of cancer cells asymptotically approaches the *minimum* necessary to maintain proliferation. Interestingly, the assumption of a minimal information state leads to a unique prediction for the *growth law* governing an in situ (in the patient) tumor mass. Using EPI, it is found that under "free field" growth conditions, i.e., unconstrained by host response, substrate limitation, or any curative program, the growth law has the power law form t^γ (Eq. (3.43)), with t the age of the tumor and the exponent $\gamma = 1.62$. This prediction is in remarkably good agreement with six clinical studies which show that small human breast cancers do empirically exhibit power law growth, and with an exponent of 1.73 ± 0.23 [46, 47]. See Sections 3.2.10–3.2.12.

Clearly, IT has potential for defining the informational basis of disease, and this application remains a fascinating future challenge.

3.1.20. Conclusions

Living systems, uniquely in nature, exist at the dynamical interface of information and thermodynamics. The information storage, processing, and communication effects of these systems distinguish biology from all other sciences. The past decade has seen remarkable technological developments in this direction, giving great insight into the informational bases of life. Meanwhile, information theory— the study of information storage, communication, and processing—has similarly made great strides in organizing these data into conceptual frameworks. These provide novel insights into the underlying structure of information networks and the mechanisms by which information is converted to work and function. These theoretical developments have been accomplished using both traditional Shannon methods and new mathematical techniques such as network analysis, measurement theory, EPI, and game theory. It is likely that, in coming years, these tools will be applied to many other important biological problems, including information flow among cellular and subcellular components of multicellular organisms and the disruption of biological information to aid in treating disease. The results will almost certainly be fascinating, and the prospects are exciting.

3.2. Applications of IT and EPI to Carcinogenesis

3.2.1. Summary

Cellular information dynamics during somatic evolution of the malignant phenotypes are complex and poorly understood. The accumulation of random genetic mutations and, therefore, the *loss* of genomic information, appears necessary for carcinogenesis. However, additional effects can be inferred as well: *Unconstrained* mutagenesis and information loss would ultimately produce a state of cellular degradation incompatible with life. The cancer would spontaneously die off. Since this often does not occur, what constraints on information loss exist? In fact the stability of some genomic segments, such as those controlling proliferation and metabolism, indicates the *presence* of mutational constraints.

By applying Information Theory in the form of EPI, we demonstrate next that the phenotypic characteristics and growth patterns of cancer populations are properties that can be largely predicted. These emerge from the nonlinear dynamics of accumulating, random genetic mutations, coupled with certain tissue selection factors. Interestingly, a maximum loss of transgenerational information is demonstrated in cancer genomic segments: These encode negative (or neutral) evolutionary properties. This is most evident in the progressive de-differentiation, or loss of function, that occurs during carcinogenesis. Its end can be an *information catastrophe*, causing the loss of differentiation or function that defines cancer growth. These cells exhibit a decoherent morphology of cell structure and function.

The ultimate end of such information catastrophe is cell death. Hence, in the absence of any other mitigating effects, these cancer cells would ultimately perish. However, in fact certain cell selection pressures preserve genomic information-retention properties; these confer selective *growth* advantages. This is even in the presence of a high background mutation rate. Thus, phenotypic traits characteristically retained by tumor populations can be identified as critical selection parameters that favor their proliferation.

The information model of carcinogenesis is tested by applying EPI analysis to predict tumor growth dynamics. The prediction is that cellular proliferation due to information degradation produces power law tumor growth, with an exponent of 1.62. Data from six published studies that use sequential mammograms to measure the volume of small, untreated human breast cancers demonstrates such power law tumor growth, with a mean exponent value of 1.73 ± 0.23. Thus, the EPI prediction of 1.62 is well within the experimental range of results. The EPI approach gives rise to other predictions as well, including exponential (rather than power law) growth for tumor cells that instead grow in vitro. A final prediction is a root-mean-square (rms) error of 30% in the estimated time of onset of the cancer. The latter predictions are likewise supported by experimental observation.

The nonlinear dynamics of stochastic information loss constrained by somatic evolution indicates that, contrary to the usual dogma, cancers are structures that emerge out of multiple, fundamentally *random*, genetic pathways. They are *not* associated with any predictable, fixed sequences of genomic alterations.

3.2.2. Introduction

Diverse cellular phenotypes arise out of two complementary processes: the translation of specific subsets of inherited information from the genome and the ability of cells to receive and process information in the environment [67]. The active information content of a cell is a time-dependent summation of translated intracellular and acquired extracellular information. This information state controls the morphology and function of that cell as well as its interaction with the external environment. Abnormalities in cellular information content will result in disease, as clearly exemplified by several thousand genetically linked disorders [68].

Less clear is the role of disordered cellular information in cancer. While carcinogenesis typically requires multiple genetic mutations [69–71], the precise mechanisms by which this information loss produces the malignant phenotype remains largely unknown. This is due primarily to the absence of a prototypical malignant genotype—Instead, cancer cells typically possess hundreds and even thousands of genomic errors. *Virtually every different tumor has its own pattern of genetic mutations* [71, 79]. Thus, unlike classical genetic diseases, no single genotype is present in cancer cells. Finally, there is not even a well-defined correspondence between the genetic mutations present in cancer populations and the cellular characteristics of the malignant phenotype.

Clearly, the large number of genetic mutations found in cancer cells indicates a critical role being played by genomic information. However, while cellular information *loss* through random accumulating genetic mutations appears necessary for carcinogenesis [73–75] it is *not sufficient*. As pointed out earlier in the chapter, the absence of information-catastrophe and non-randomness of mutations indicates other important constraints.

Thus, cancer, as a "disease of the genes" [76], is a complex disorder of the storage, processing, and propagation of cellular information in the genome. This general class of phenomenon can be addressed by the informations of Shannon (3.10) and Fisher (1.50). As we saw, they are both special cases of the more general K–L information (3.1). Their seminal contributions have spawned extensive research in the physical sciences [9, 18] and modest application in biology [2].

In this section we develop mathematical models of cellular and extracellular information dynamics in carcinogenesis. The aim is to understand the dynamical forces that guide the evolution of cellular information content toward genomic instability and invasive cancer growth (carcinogenesis). By applying Shannon information to carcinogenesis, we demonstrate the interactions of stochastic mutation events and environmental selection pressures [77, 78] that produce the cellular characteristics of the cancer.

Note that this approach is not meant to be analogous to the linear model of sequential genetic changes in colorectal tumorigenesis described by Fearon and Vogelstein [79]. That is, we do not address the role of specific defects in oncogenes and tumor suppressor genes, the sequence of these changes, the causes of tumor genomic instability, or the influence of nonneoplastic cells during carcinogenesis.

Following this, we use Fisher information to analyze tumor population growth dynamics based on perturbations in environmental information produced by genomic alterations during carcinogenesis. *Physical information* is the difference between the two Fisher information quantities (see Chapter 1 or below). Using the EPI approach we accurately predict tumor growth dynamics in vivo and in vitro, based solely upon changes in the environmental information content. Again, cancer is a disease of information.

3.2.3. Bound and Free Intracellular Information

The character of information in biological systems can be described as bound or free [60]. *Bound* information is encoded in the genome, serving as a reservoir to be passed from one generation to the next. *Free* information is contained in nonnucleotide organic polymers, such as proteins, lipids, fatty acids, and polysaccharides. These regulate cellular morphology and function. Free information is dependent in a complex way on the translation of specific subsets of the bound information and their subsequent interactions. Free information is also present in the extracellular space in multicellular organisms. In this case, cellular function is influenced by both the intra- and extracellular free information.

The *bound* genetic information which, by the EPI approach of Chapter 1, is rigorously a Fisher source information J, can be approximated by an appropriate Shannon information [60]. This latter approach, due to Kendal and others, measures the quantity of information of a string of symbols as in DNA [5, 6, 80]. If there are γ equally probable arrangements for the symbols of a coded message, then the information content is the displacement from randomness, defined by $k \ln \gamma$. With $k = 1$, the entropy (3.8) results once again. If instead k is taken to be the Boltzmann constant, then the thermodynamic entropy results in many cases (not further discussed).

By comparison, in the analysis of carcinogenesis given in Sections 3.2.7 to 3.2.11, both the bound and free information quantities will be described by the *Fisher* informations I and J in a rigorous EPI approach. We continue for now with the Shannon-based analysis of the bound information.

An increase in the Shannon entropy implies an increase in disorder and loss of information. However, there are several limitations to the application of Shannon information to biological systems. An example is as follows:

As we mentioned in Section 3.1.11, a codon is a triplet message—a sequence of three of the nucleotides A, T, C, and G that encodes an amino acid. The first limitation arises out of an unknown amount of degeneracy in codon structure. This leads to an unknown level of redundancy in the genome which, in turn, must reduce the 6-bit level of Shannon information previously found. Second, while Shannon information may measure the quantity of information, it does not measure its *quality*. That is, a small change in the information content of a gene (e.g., due to a single base pair mutation) may catastrophically alter the function of a gene product; or, by contrast, even multiple errors in a less critical segment of the gene may have *no* biological consequences.

Despite these limitations, Shannon information has been shown to be a useful and reasonably accurate measure of genetic information [60]. Furthermore, because Shannon information is both *additive* and *conserved*, it is well suited to estimating changes in information content over time. It is interesting and important that the Fisher information has these properties as well.

3.2.4. Information Dynamics Before and After Reproduction, With and Without Mutation

Let a cellular genotype consist of G genes, $g = 1, 2, \ldots, G$. Suppose any gene contains m codons, and let $r_j^i \equiv r_j^i(t)$ be the probability of occurrence of the jth codon configuration for the ith codon, at the time t. Let these configurations occur independently. Then its entropy H^i obeys Eq. (3.8),

$$H^i = -\sum_{j=1}^{N} r_j^i(t) \ln r_j^i(t). \tag{3.16}$$

Limit $N = 4^3 = 64$, which is the number of possible combinations of A, T, C, and G in each such codon triplet. Suppose that $r_j^i = r_j$; i.e., the probability of the jth codon configuration is independent of codon number i. Then by (3.16) the information $H^i = H$, a single value. Suppose the gene contains m codons in all, and these supply information independently. Then, the total information I_g stored in the gene obeys additivity, and is simply the sum

$$I_g = \sum_{i=1}^{m} H^i = mH. \tag{3.17}$$

(Actually, there is degeneracy in the third position of the triplet code, so that this slightly overestimates the genetic information content, as previously discussed.)

The bound information content I_c of the cell c is estimated to be the total information content over all the genes. Then, again by additivity of their contributions,

$$I_c = \sum_{g=1}^{G} I_g. \tag{3.18}$$

Suppose, next, that reproduction takes place at a time $(t + \Delta t)$. Then, the entropy $\bar{H}^i$ of the *trans*generational encoded information depends upon $\bar{r}_j^i(t + \Delta t)$, the probability of occurrence of the jth codon configuration of the ith codon, as in Eq. (3.16). Using also, as before, $\bar{H}^i = H$, the information in a codon obeys

$$\bar{H} = -\sum_{j=1}^{N} \bar{r}_j^i(t + \Delta t) \left[\ln \bar{r}_j^i(t + \Delta t) \right]. \tag{3.19}$$

In the absence of mutations, all values of $\bar{r}_j^i(t + \Delta t) = r_j^i(t)$. Then, by Eqs. (3.16) and (3.19), $\bar{H} = H$ and there is no change in the entropy. However, in the presence of mutations the two probabilities are not equal, so the two entropies are likewise not equal. Again assuming the codons to independently contribute information, the mutated level of information/gene obeys $\bar{I}_g = m\bar{H}$, and it causes an information *change* for gene g to occur, of size

$$\Delta I_g = m(H - \bar{H}). \tag{3.20}$$

In the evolutionary environment that develops during carcinogenesis [27, 28], the information content of each gene may contribute to the selective growth advantage u of the cell. This, in turn, determines its ability to proliferate. Let u_g define this *growth advantage* for the gene g, where u_g is zero, positive, or negative, indicating no growth advantage, definite advantage, or definite disadvantage, respectively. Presumably, cells with the highest values of u_g in the tissue will proliferate at the expense of those with lower values, in a kind of zero-sum game. Suppose that

u_g is some unknown function

$$u_g = f(I_g) \tag{3.21}$$

of the the information content I_g and that after a generation the growth advantage generated by gene g becomes

$$u_g(t + \Delta t) = u_g(t) + \Delta u_g = f(I_g + \Delta I_g). \tag{3.22}$$

The change in growth advantage Δu_g can either increase, decrease, or leave unchanged the cellular growth advantage. Assuming additivity, the total growth advantage of a cell c with N genes is

$$u_c = \sum_{g=1}^{N} u_g. \tag{3.23}$$

Additivity is undoubtedly an oversimplification, because the complex interactions of gene products will likely produce nonlinear dynamics.

Building on this, we may define the proliferative capacity P_c of each cell population c as some increasing function h of the genetic growth information u_c of each population as compared to the mean value $\bar{u}$ over all of the M cellular populations present,

$$P_c = h(u_c - \bar{u}), \quad \bar{u} \equiv 1/M \sum_{m=1}^{M} u_m. \tag{3.24}$$

3.2.5. Limits of Information Degradation in Carcinogenesis

We view the interaction of mutation rate, evolutionary competition, and transmitted information in carcinogenesis by applying the limit developed by Eigen and Schuster [66]. In our previous notation, this is

$$I_{c_{\max}} = \alpha^{-1} \ln(u_c - \bar{u}), \tag{3.25}$$

where $I_{c_{\max}}$ is the log of the maximum number of "signals," or the genetic information, that can be successfully passed on to a new generation, $u_c - \bar{u}$ is the competitive advantage of the population compared to the average of all the local populations, and α is the mutation rate. Rearranging and combining with (3.24) gives a proliferative capacity

$$P_c = h(u_c - \bar{u}) = h(e^{\alpha I_{c_{\max}}}). \tag{3.26}$$

We now have explicitly linked the proliferation of transformed populations to mutation rate and maximum information transfer. Environmental parameters are indirectly included in (3.26), since they modify the specific competitive advantage of one or all of the populations (i.e., change u_c and $\bar{u}$). Again, note that the time frame is that of one or more cellular generations, thus smoothing stochastic events that might transiently reverse the otherwise monotonic-appearing multistep cellular progression from normal to cancer.

3.2.5.1. Mutation Rates for Various Gene Proliferation Types

The functional relationship of information loss, mutation rate, and selective growth advantage can be applied to individual genetic segments via (3.21). This indicates that, in an environment where cellular proliferation is dependent on its fitness, and the genome is randomly degraded by mutations, the apparent mutation rate will be indicated by the gene contribution to cellular fitness, as follows:

First, genes that are *essential* to survival and proliferation (contributing maximally to cellular fitness) will have an apparent mutation rate of 0, since any individual with a loss of function mutation in one of these critical genes will not survive. (Note, however, that mutations that increase or upregulate the expression of these genes may be common.) This prediction is supported by carcinogenesis literature showing a gain of function mutations in growthpromoting oncogenes, such as K-RAS, and amplification of glucose transport genes [34–36].

Second, genes that actively *limit* cellular proliferation (i.e., tumor suppressor genes) contribute *negatively* to cellular fitness (again, defined by cellular proliferation). These components of the genome should have a large apparent mutation rate, since the ensuing loss of their function increases cellular proliferation.

Finally, genes with *zero growth* advantage (these, e.g., would be genes that encode differentiated cellular *functions*) contribute minimally to the overall fitness of the cell. Therefore, by (3.23) and (3.25), these should exhibit an intermediate apparent mutation rate since their loss does not enhance or diminish cellular proliferation.

The net effect of these dynamics is a time-dependent decline in $I_{c_{max}}$. This is apparent clinically in progressive morphological and functional drift *away* from the original differentiated phenotype that is observed during carcinogenesis [30] (see Figure 3.1). Interestingly, the unconstrained accumulation of mutations that produces de-differentiation may eventually produce sufficient genetic noise that a differentiation phase transition [29] results. This new cell state, in the absence of sufficient transgenerational information to retain a "memory" of the original differentiated state, will exhibit nonphysiologic combinations of phenotypic traits, such as the expression of endothelial cell traits in melanoma and breast cancer [31, 32]. These cells will, in turn, produce morphologically disorganized and poorly functioning tissue typical of invasive cancers [33] (Figure 3.1).

An interesting test of this model is observed when environmental selection forces are perturbed by the administration of chemotherapy. This new environmental selection parameter unfortunately increases the value of u for genes controlling the multidrug *resistance* phenotype. The predicted amplification and gain of function mutations in these genes during the subsequent evolution of the cancer population has been observed [37, 38, 81].

3.2.5.2. Angiogenesis

Angiogenesis is the growth of an augmented blood supply by the developing tumor. Angiogenesis probably represents an evolutionary phenomenon similar to the preceding. Early cancers rely on diffusion and the native blood supply of substrate delivery. As the tumor grows, angiogenesis becomes a critical survival parameter. It has been shown that, as predicted by this model, the angiogenic phenotype is found far more frequently in "older" tumors and develops far more quickly in the presence of a high mutation rate [39].

Specific environmental selection pressures can be inferred by common phenotypic traits found in transformed (cancer cell) populations. For example, most transformed cells can withstand severe conditions of nutrient deprivation. From this we can infer that an enhanced ability to obtain and utilize substrate (nutritive environment) efficiently is strongly *selected for* in the Darwinian environment of carcinogenesis [40].

Thus, the somatic evolution of the malignant phenotype during carcinogenesis can be summarized as follows. If I_g denotes the information content of a gene, then, for genes that diminish cellular proliferation (obeying $u_g < 0$)

$$\frac{dI_g}{dt} \leq 0 \quad \text{at all times } t. \tag{3.27}$$

In contrast, for genes that increase cellular proliferation ($u_g \geq 0$), the apparent mutation rate is dependent on the overall cellular fitness:

$$\frac{dI_g}{dt} < 0 \quad \text{if } I_g > I_{g_{\max}} \tag{3.28}$$

or

$$\frac{dI_g}{dt} = 0 \quad \text{if } I_g = I_{g_{\max}}, \tag{3.29}$$

where $I_{g_{\max}}$ represents some *minimal* information level that allows the organism to remain functional. In other words, for genes with high u_g the time-dependent function describing the information loss has an extremum at $I_{g_{\max}}$, because cells with information levels that are lower than this threshold will not survive. This allows the cells to maintain a minimal information level necessary for life (i.e., they continue to proliferate and generate energy) despite the marked drift in morphology and function that result from genomic instability.

3.2.6. The Evolving Microenvironment and Resulting Mutation Rate

We briefly address the critical role of an *unstable* microenvironment in carcinogenesis. Cellular mutations that produce perturbations in the microenvironment will, by Eq. (3.26), alter the competitive advantage of individual cell populations u_c and, consequently, the average advantage $\bar{u}$ over all the populations. For example, if the competitive macroenvironment of carcinogenesis results in $d\bar{u}/dt > 0$, clonal expansion ($dP_c/dt > 0$) will be restricted to populations obeying $du_c/dt > d\bar{u}/dt$. In fact, this accelerating rate of competitive phenotypes is consistent with observations of malignant progression, where increasingly aggressive cellular populations emerge over time during carcinogenesis [41]. Other perturbations, as well, may be stimulated by the presence of tumor cells, including a host immune response. If, for example, cytotoxic T cells detect a specific surface antigen expressed on some of the tumor cells, the u_c of that tumor subpopulation will decline. However, as noted below, the forces of mutation and selection may simply result in new, resistant populations as the total tumor population continues to evolve into progressively more diverse phenotypes [42].

In an evolving microenvironment, by Eq. (3.26) and the increasing nature of function h with its argument, populations with $du_c/dt < 0$ will not survive (an obvious exception being the situation in which $\bar{u}$ is decreasing at a greater rate). Conversely, those with $du_c/dt > 0$ will proliferate. In this way, the high mutation rate found in cancer is favored because it will produce a greater cellular diversity and increased probability of a proliferative phenotype. This is demonstrated in bacterial studies that have found a 5000-fold increase in the mutation rate when culture conditions became more *restrictive* and are allowed to evolve over time [82]. Clinical studies have demonstrated that mutation rate increases during tumor progression [44]. Loeb and others [63] have hypothesized that the acquisition of a mutator phenotype is a critical and necessary event to carcinogenesis. Our analysis supports this hypothesis.

3.2.7. Application of EPI to Tumor Growth

We now turn our attention to the dynamics of altered environmental information caused by the degradation of the genome in transformed cells. As noted earlier, normal tissue will contain an extracellular distribution of organic molecules $p(x, t)$ that remains stable with time under physiologic conditions. However, the presence of a tumor will perturb this distribution, depending upon the number of tumor cells present. Thus, changes in $p(x, t)$ will depend upon the number of tumor cells present and the deviation of their internal information content from that of the tissue of origin. This new (cancer) distribution function $p_c(x, t)$ will increasingly differ from $p(x, t)$ with time. This difference between the distribution functions actually conveys information regarding the onset and progression of carcinogenesis.

This model prediction of time-dependent environmental change as a consequence of cellular information degradation during carcinogenesis can be tested by applying EPI [18] to derive estimates of tumor growth kinetics. This calculation is carried through next.

In particular we seek the form taken by the marginal probability law $p(t)$ of $p_c(x, t)$. This marginal law also represents, by the law of large numbers [13], the relative number of cancer cells anywhere in the space x of points within an affected organ. The time-dependence of $p(t)$ then defines as well the growth with time of the cancer.

As we are seeking the form of PDF $p(t)$, we will use EPI for this purpose. Since the time t is the Fisher coordinate of p, this indicates that a time value is to be estimated by the EPI measurement procedure. The time value of import in cancer growth is the *age* of the cancer, as discussed below.

3.2.7.1. Fisher vs Shannon Informations

The EPI approach [18, 83–85] utilizes *Fisher information* rather than *Shannon information H* or *I* (as in the preceding). However, there is a tie-in between the two informations. Both relate to the cross entropy measure K_{KL}, according to choice of the particular reference function $r(t)$. As shown at Eq. (3.3), the Fisher information is proportional to K_{KL} for a shifted reference function $r(t) \equiv p(t + \Delta t)$, as

$$I = - \lim_{\Delta t \to 0} \frac{2}{\Delta t^2} K_{KL}(p(t)/p(t + \Delta t)) \equiv \lim_{\Delta t \to 0} \frac{2}{\Delta t^2} \int dt\, p(t) \log \frac{p(t)}{p(t + \Delta t)}.$$
(3.30a)

And, as indicated in Eq. (3.10), the Shannon information is proportional to K_{KL} for a reference function that is a product of marginal probabilities.

The two measures also differ qualitatively. Shannon information measures the degree to which a random variable is uniformly (maximally) random, that is, resembles its uniform reference function, by Eqs. (3.5) and (3.6). This in turn measures the ability of a system to *distinguish signals*, as in Eq. (3.20) where it is used to distinguish genotypes. By comparison, Fisher information measures the degree to which a required parameter may be *known*; see Section 1.2.1.1 (Chapter 1). As will be discussed, the age t of a growing tumor is a particularly important parameter for purposes of describing cancer growth.

3.2.7.2. Brief Review of EPI

EPI is described in Chapter 1 (Sections 1.4 to 1.6). However, a quick review at this point should be useful. The working hypothesis of EPI is that the data in any measurements result from a flow of information—specifically, Fisher information—that proceeds from an *information source* in object space to an *information sink* in data space. The information source is the effect that underlies the data. Its information level is denoted as J. The data information is called I, and the information flow is symbolized by the transition $J \to I$. The flow physically consists of material particles that carry the information. Each phenomenon generally utilizes different messenger particles. (For example, in an optical microscope the particles are photons.) The source information J represents the Fisher information that is intrinsic to the measured phenomenon. Assuming that we are dealing with an isolated, passive system, the received information can at best equal the source level. That is, $I \leq J$. The level J corresponds as well to the bound—as opposed to free—information that was referred to earlier.

In our particular, application J represents the total extracellular Fisher information that is produced by a cancer cell. The age of a cancer mass is a particularly important determinant of its state of development. Knowledge of the age is also important for purposes of deciding upon a particular regimen for suppressing the cancer (be it chemotherapy, surgery, etc.) Hence the informations I, J are about the time, in particular the time θ at which the carcinogenesis process began in a given organ. The information J may be carried by several biological intermediates.

Of these, protons appear to be the dominant messengers, since increased glucose uptake and excessive secretion of acid into the extracellular spaces is observed in the vast majority of clinical cancers [86, 87]. Furthermore, experimental demonstration of the proton as a biological messenger between macromolecules has recently been published [88]. Other carriers of information may exist as well, such as increased interstitial pressure or decreased oxygen or glucose concentrations.

The totality of such information that reaches a normal cell due to all neighboring cancer cells is what we call I. Thus I is the total information that is provided to the functioning cell by its neighboring cancer cells about the onset time θ of carcinogenesis. Empirically, the received information tends to suppress the receiver, i.e., the normal cell. We now apply EPI to predict expected tumor growth dynamics.

3.2.8. Fisher Variable, Measurements

Let θ be the age of the individual at which carcinogenesis began in a given organ, and t be the time *since* it began (i.e., the age of the tumor). At the time $(\theta + t) \equiv \tau$ some quantity of information (e.g., lactate concentration) is observed to be present in the environment that the clinician recognizes to signal the presence of cancer, the event C. He records the time τ, but does not know either of its components θ or t. However, if the clinician observes as well the size of the tumor, then he or she can use this to estimate its age t, by some known (approximate) relation between size and age. For example, if the tumor is small t ought to be small. With τ and t then known, the estimate $\theta \equiv \tau - t$ may be formed. Thus, the observed time τ and the observed tumor size provide information about θ. EPI derives likelihood laws (Chapter 1). The likelihood law that describes this scenario is $p(\tau|\theta)$, the probability of observing the existence of a tumor at time τ in the presence of the fixed value of θ. Hence, EPI will be used to find this law. This likelihood law is equivalently $p(t)$, since it will not change shape with a shift in the absolute position θ (Eq. (1.48), Chapter 1).

To review, time τ is *the time at which a cancer is observed in a given organ*, time t is its unknown (random) component since carcinogenesis began, $p(t)$ is the probability density for the latter, and EPI can be used to find this density. The elapsed time t is the Fisher variable of the problem.

Note that actually $p(t) = p(t|C)$, the time conditional upon the presence of cancer. The reverse probability $P(C|t)$ is the PDF for cancer at the time t. By the law of large numbers, $P(C|t)$ is also proportional to the relative mass of the cancer at that time. This is the total mass of cancer cells within the organ divided by the total organ mass, assuming cancer cells and functioning cells have about the same mass density. Knowledge of this relative mass as a function of time is often of prime interest. In fact it can be computed from a known $p(t|C)$ by the use of Bayes' rule of statistics,

$$P(C|t) = \frac{p(t|C)}{p_0(t)} P_0(C). \tag{3.30b}$$

Here $P_0(C) = $ const. is the occurrence rate of cancer of the given organ within the general population, and $p_0(t)$ is uniform over the total observation time T. That is, observations by the clinician are made at a constant rate over time. Hence, by (3.30b) the desired $P(C|t) \propto p(t|C)$, the latter to be the EPI output law.

3.2.9. Implementation of EPI

3.2.9.1. Recourse to Probability Amplitude

The EPI approach will establish the answer $p(t)$ by first determining its real probability amplitude function $q(t)$, where, by Eq. (1.52),

$$p(t) = [q(t)]^2. \tag{3.31}$$

3.2.9.2. Self-Consistency Approach, General Considerations

Here we seek an EPI *self-consistent* solution for $q(t)$ (Section 1.5.3). This is to satisfy the basic condition that the information loss $(I - J)$ in transit from the phenomenon to the data be an extreme value

$$I - J = \text{extrem.} \tag{3.32}$$

through choice of the function $q(t)$. The solution $q(t)$ to the problem obeys the Euler–Lagrange Eq. (1.6),

$$\frac{d}{dt}\left(\frac{\partial(i - j)}{\partial \dot{q}}\right) = \frac{\partial(i - j)}{\partial q}. \tag{3.33}$$

The new functions $i(t)$ and $j[q, t]$ are defined as information *densities*,

$$I \equiv 4 \int dt\, i(t) \quad \text{and} \quad J \equiv 4 \int dt j[q, t], \tag{3.34}$$

where

$$i(t) \equiv \dot{q}^2. \tag{3.35}$$

In general, the indicated differentiations in (3.33) result in a differential equation in the unknown function $q(t)$. This can be linear or nonlinear, depending upon case. A linear example from physics is the Schroedinger wave equation—a linear, second-order differential equation [18, 85]. A nonlinear example from genetics is the law of genetic change [84]. Biological growth processes, including cancer growth, are generally nonlinear [84].

By the *self-consistency* approach, condition (3.33) is to be supplemented by a *zero-condition*

$$i - \kappa j = 0, \quad \kappa \geq 0 \tag{3.36}$$

as a constraint. This merely states that $I = \kappa J$, by Eq. (1.73), or $I \leq J$ as mentioned before. The simultaneous solution $q(t)$ to the combined problem (3.33), (3.36) is the self-consistent EPI solution.

Quantity κ in (3.36) is a positive constant that is also to be found.

Equations (3.34) and (3.35) together define the Fisher information level I in received data for a temporally shift-invariant system. From (3.36), the constant $\kappa \equiv i/j$ represents the *efficiency* with which information j is transformed into data information i. We next find the self-consistent solution to the problem (3.33), (3.36).

3.2.10. EPI Solution: A Power Law

The amplitude function $q(t)$, constant κ, and functional $j[q, t]$ are the effective unknowns of the problem. They are solved for by carrying through the self-consistency approach. The solution is found in Appendix A as

$$q(t) = \left(\frac{1+\kappa}{A}\right)\left(\frac{At}{\kappa}\right)^{\kappa/(1+\kappa)}, \qquad (3.37)$$

where $A = $ const. This is a power law solution. The power $\kappa/(1+\kappa)$ is as yet unknown. We find it next.

3.2.11. Determining the Power by Minimizing the Information

Without loss of generality, the solution (3.37) may be placed in the form

$$q(t) = Et^{\alpha}, \quad \alpha = \kappa/(1+\kappa), \quad E = \text{const.} \qquad (3.38)$$

The power is now α. It is important to find this power, since it governs the rapidity of growth of the cancer.

Since form (3.38) grows unlimitedly with t, for the corresponding $q^2(t) \equiv p(t)$ to be normalized, the cancer growth must be presumed to exist over a finite time interval $t = (0, T)$. This is of course the survey time over which cancers are observed. The normalization requirement then gives a requirement

$$\int dt\, p(t) = E^2 \int_0^T dt\, t^{2\alpha} \equiv 1, \quad \text{or} \quad E^2 = (2\alpha + 1)T^{-(2\alpha+1)} \qquad (3.39)$$

after the integration.

The Fisher information conveyed by the cancer over the entire survey period $(0, T)$ is by (3.34), (3.35)

$$I = 4\int_0^T dt\, \dot{q}^2, \quad p \equiv q^2. \qquad (3.40)$$

Through the *game corollary* (Section 1.6.6, of Chapter 1), EPI requires that a free field solution for $q(t)$ convey *minimum Fisher information* through choice of any free parameter. Here by (3.38) the parameter is the power $2\alpha \equiv \gamma$ in $p(t)$. It is interesting that this prediction of a minimum is also *independently* part of the information model for cancer growth previously mentioned (Section 3.1.19). That is, cancer growth follows the game corollary in this life-and-death conflict (one hesitates to call it a "game") between the observer and nature.

The *free field* solution implies the absence of a field due to some source, imposed externally, that modifies the evolution of $q(t)$. Examples of external fields might include clinical treatment such as some regimen of chemotherapy, tissue constraints such as inadequate blood supply producing tumor necrosis, host immune response, and physical barriers such as adjacent bone or cartilage. In fact, here we are studying a *freely* growing in situ tumor. This then defines a free field scenario. Thus, *we require the Fisher information for the solution q(t) to be a minimum*. The minimum will be through choice of the free parameter of the process, which is here α. (Note that the observation interval T is already fixed, and E is fixed in terms of α and T by the second Eq. (3.39).)

Substituting (3.38) into (3.40) and using the second Eq. (3.39), gives, after integration,

$$I = \frac{4}{T^2}\left(\frac{2\alpha+1}{2\alpha-1}\right)\alpha^2. \tag{3.41}$$

This is to be minimized through choice of α. Differentiating it $\partial/\partial\alpha$ and setting $\partial I/\partial\alpha = 0$ results in the algebraic equation

$$4\alpha^2 - 2\alpha - 1 = 0, \quad \text{or} \quad \alpha = \frac{1}{4}(1 \pm \sqrt{5}) \tag{3.42}$$

as roots. However, the negative square root choice when placed in (3.41) gives a negative value for I, which is inconsistent since our I must be positive (see (3.40)). Thus, the answer is a power law solution

$$p(t) = E^2 t^{2\alpha} \equiv F t^\gamma, \quad \gamma = \frac{1}{2}(1 + \sqrt{5}) \approx 1.62, \quad F = \text{const.} \tag{3.43}$$

3.2.12. *Experimental Verification*

This predicted value (3.43) of the power γ can be compared to published data on growth rate determined by sequential mammographic measurements of tumor size in human breast cancers [89–94]. *The values of γ were 1.72, 1.69, 1.47, 1.75, 2.17, and 1.61 (mean 1.73 with standard deviation 0.23).* The tumors in these studies seem well suited to test our prediction because they were small (less than 2 cm in diameter), untreated cancers growing within the breast. That is, the tumors were not subject to any apparent clinical or tissue constraints so that the conditions for a free field solution are probably met.

Also, the power law prediction (3.43) was experimentally confirmed, as follows: The most convenient clinical data for the purposes of confirming (3.43) would directly observe values of $p(t)$, i.e., cancer mass versus time. On the other hand, the clinical data [89–94] do not consist of values of $p(t)$, but rather values of $p(v)$— the occurrence of tumor volumes (or masses) irrespective of time. However, such data can in fact be used. It is shown in [95] that if $p(t)$ obeys a power law (3.43), then $p(v)$ likewise obeys a power law, with a *different* but related power,

$$p(v) = B v^{\frac{1-\gamma}{\gamma}}, \quad B = \text{const.} \tag{3.44}$$

Given the value of γ in (3.43), this becomes

$$p(v) = Bv^{-\frac{1}{2}(3-\sqrt{5})}. \tag{3.45}$$

The exponent is approximately value -0.382. Hence, on a log–log basis, (3.45) predicts a straight-line relation with a slope of -0.382. The data in [89–94] are values of $p(v)$ versus tumor size (apparent diameter). We cubed the diameters to give equivalent volumes v. Figure 3.5 shows, on a log–log basis, the theoretical solution (3.45) with, for comparison, all points of the clinical studies in references [89–94] that are quantitatively tabulated. (One point in each of the three studies [89–91] is ignored because the indicated tumor sizes for these points is the indefinite value 2+.) The agreement between the theoretical curve and the experimental data is visually good. Moreover, as we saw above, the theoretical curve is well within one standard deviation of the experimentally determined values.

Of course, we must also emphasize the many factors that may produce clinical variation from the predicted growth dynamics. These include: (1) Altered tumor cell volume. We assumed that the tumor cell volume is exactly equal to the volume of the normal cells, but some heterogeneity in tumor cell size is frequent. (2) Power law growth is predicted only for tumors unconstrained by external factors (i.e., a "free field") and thus represents a relative maximum value. That is, the predicted dynamics are only applicable for *unimpeded* in situ replication of the tumor cells. If, instead, tumor cell expansion is constrained by external factors, such as inadequate vascular supply, host response, spatial considerations, or clinical therapy, slower growth will be observed. (3) Tumors may contain significant volumes of noncellular regions such as necrosis and blood vessels or large numbers of nonneoplastic cells such as endothelium, fibroblasts, macrophages, and lymphocytes.

It is encouraging that, even despite these possible complications, the predicted tumor growth law agrees well with the experimental data.

3.2.13. Implications of Solution

3.2.13.1. Efficiency κ

By the second equation (3.38), $\kappa = \alpha/(1 - \alpha)$. Then by the positive root (3.42)

$$\kappa = 4.27. \tag{3.46}$$

This is the total efficiency due to the unknown number S of cancer cells that are communicating with the functioning cell. Since the efficiency for any single cell can be at most unity, this implies that there are *at least* $S = 4$ neighboring cancer cells that are interacting with the functioning cell. That S is rather large means that *a relatively large amount of lactic acid bathes the healthy cell.* This forms a toxic environment that may eventually kill the normal cell, leading to compromise of functioning tissue and death of the host. Hence, the message conveyed by the information messengers of the cancer channel is "die!" This acid-based mechanism of tumor invasion has been previously proposed and modeled [86].

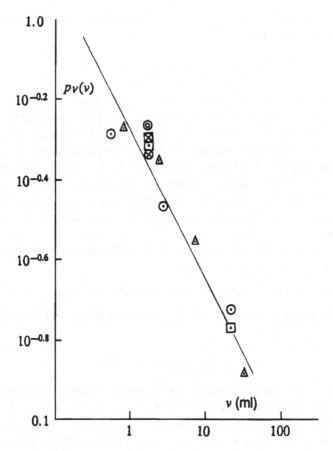

FIGURE 3.5. Tumor growth dynamics based on the information-degradation model of carcinogenesis as predicted by EPI (solid straight line) compared with experimental data from six studies: Points $\triangle$ = from Tabar et al. [89]; $\square$ = from Fagerberg et al. [90]; $\odot$ = from Thomas et al. [91]; $\boxtimes$ = from De Koning et al. [92]; $\circledcirc$ = from Peer et al. [93]; and $\otimes$ = from Burhenne et al.; [94]. The agreement is very good.

3.2.13.2. Fibonacci Constant

We note that the power γ of the growth process (3.43) is precisely the Fibonacci "golden mean." This number is often shown to characterize healthy growth, e.g., the growth of a colony of rabbits or the arrangement of seeds on a pinecone. Here it characterizes decidedly unhealthy growth, taking on a more ominous role.

It is interesting to compare the cancer growth (3.43) with ideal, Fibonacci growth. The latter describes the relative increase in an *ideally breeding* (no deaths, constant fitness) population $p_F(t)$ from one generation to the next,

$$\frac{p_F(t + \Delta t)}{p_F(t)} \equiv \gamma. \tag{3.47}$$

By comparison, (3.43) shows that the relative increase in cancer mass from one generation to the next obeys

$$\frac{p(t + \Delta t)}{p(t)} = \left(\frac{t + \Delta t}{t}\right)^{\gamma} \approx 1 + \gamma \frac{\Delta t}{t}, \qquad (3.48)$$

to first order in Taylor series. The increment Δt is the time between generations, $\Delta t \ll t$. Thus, by (3.48) the growth increment $\gamma(\Delta t/t)$ characterizing the cancer *decreases* with the time t, while by (3.47) an ideally growing Fibonacci population maintains a *constant* relative increase γ with time. The prediction is therefore that a clinical cancer grows at a less than maximum rate. This agrees with empirical data which show that (a) in vitro cancer cells maintained in three-dimensional culture under ideal conditions grow exponentially [96], i.e., follow full Fibonacci growth, whereas (b) in situ cancer grows according to power law growth, i.e., *much slower* than Fibonacci growth.

Because of the ideal growth effect in (3.47), a Fibonacci population grows in time according to a pure exponential law

$$p_F(t) = B \exp(t/\tau), \quad \tau = (\gamma + 1)\Delta t, \quad B = \text{const.} \qquad (3.49)$$

Use of this probability law in (3.40) gives an information level for ideal growth of value

$$I_F = \frac{1}{(\Delta t)^2(\gamma + 1)^2}, \quad \gamma \approx 1.62. \qquad (3.50)$$

By contrast the cancer population obeyed the information (3.41) with $\alpha = (1/4)(1 + \sqrt{5})$. Comparing (3.41) with (3.50) shows that, since $\Delta t \ll T$, necessarily $I \ll I_F$. That is, *in situ cancer transmits much less Fisher information about the time than does ideally growing tissue.* The similarity of this prediction to the extensive data on telomere dysfunction in malignant populations is quite evident [97–99]. Since the telomeres of *malignant* populations do not typically shorten normally with time, they convey incorrect information regarding their "age" [99] and are less likely to demonstrate the cellular effects of senescence. Thus, the results of the analytic solutions are consistent with the critical roles of telomerase and telomere dysfunction in promoting the malignant phenotype [97–99].

3.2.13.3. Uncertainty in Onset Time

A Fisher information level I has a direct bearing on estimated error. This is via the Cramer–Rao equality (1.37), according to which the minimum mean-squared error $e_{\min}^2$ in the estimate obeys $e_{\min}^2 = 1/I$. By this formula, the information level (3.41), (3.42) gives rise to a prediction that the error $e_{\min}^2$ in the onset time θ for breast cancer is 30% of the total time over which such cancers are observed. This result is derived as Eq. (3.A24) in Appendix A. This error is quite high and indicates that cancer has an innate ability to mask its onset time. The clinical consequence of this observation is that the size of the cancer at the time of first observation may *not* necessarily correlate well with the age of the cancer. This may place a theoretical limit on the *value* of screening, assuming that the development state of a cancer is dependent upon its age rather than size. This also has ramifications to the choice of treatment for the cancer. This choice depends, in part, upon accurate knowledge of its state of development. Hence the 30% error figure suggests that any *estimate* of the state of development should use observed size as its basis, rather than apparent age.

3.2.14. Alternative Growth Model Using Monte Carlo Techniques

Aside from the use of data provided by clinical observations, is there another way to check these largely theoretical predictions? The answer is that there is, through the use of a Monte Carlo calculation which numerically simulates the random cancer growth. This is described next. Since the approach is not information-based, we only show its results. See [100] for full details.

The complex dynamical interaction of intracellular information and environmental selection for cellular proliferation can be modeled by a Monte Carlo (M.C.) calculation. This approach is independent of the EPI approach taken above, and serves both to check it and to establish possible new effects as indicated by the M.C. numerical outputs. The M.C. model tracks the growth history of all existing cells, both breeding and nonbreeding, as they contribute to the total tumor mass over successive generations $n = 1, 2, \ldots$. The model is statistical, and follows a six-step algorithm. The rules of the algorithm are contrived to define a "simplest" model, in accordance with the EPI finding above that cancer exhibits minimal information or complexity. Thus, a correspondence

M.C. model complexity $\Longleftrightarrow$ EPI based complexity

is presumed in forming the Monte rules, where both complexities are *minimal.*

The rules are modeled to comply with the following biological realities:

Cancer cell proliferation is controlled by the fitness of the cancer cells in comparison with that of the functioning (differentiated) populations.

Successful cancer populations channel all substrate into proliferation at the expense of differentiated functions.

The proviso is, of course, that the gene segments critical for cell survival and mitosis remain intact. Thus, a cellular population "breeds" as long as adequate substrates exist and as long as the cell maintains sufficient internal information to process these materials and convert them into new individuals. This leads to a required input parameter as follows.

The steady accumulation of genomic mutations causes some cells to inevitably experience loss of critical gene segments for allowing them to proliferate. Consequently they stop "breeding," with high probability (fixed as an input parameter b).

The results of the M.C. calculation, shown in Figure 3.6, predict that "ideal" tumors in vivo will exhibit power law growth with an exponent of about 1.6. This is in remarkable agreement with the entirely separate analysis of tumor growth based

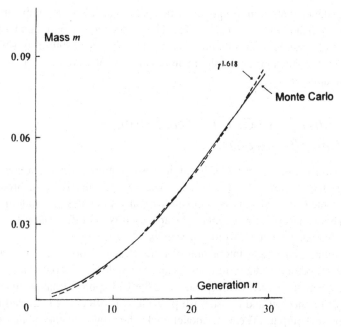

FIGURE 3.6. Tumor mass as a function of generation number for first 30 generations using EPI and M.C. simulations. The solid curve is the average of 200 randomly generated histories of tumor growth. For comparison, the ideal power law (3.43) resulting from EPI is shown as the dashed curve. The areas under the two curves are normalized to unity. The ability for the Monte Carlo approach to simulate the theoretical result is visibly good, and is substantiated by an rms difference of 0.0007 between the two curves.

on information degradation using EPI, as outlined above, and in good agreement with clinical measurements of the growth of a small breast cancer. Interestingly, the M.C. simulations also show that nearly 40% of tumor populations fail to grow beyond the first generation because a mutation results in failure of clonal expansion when the total population is small, and lacks sufficient phenotypic diversity to compensate. The clinical implications of this result for growth dynamics in benign tumors, and possible strategies for tumor prevention, will be the subject of future investigations.

3.2.15. Conclusions

We have developed a broad, conceptual framework for the intra- and extracellular dynamics of the malignant phenotype. The results generally agree with data, where such data exist. Where data do not exist, they make testable predictions (see below).

We conclude that living cells maintain a stable ordered state by using energy to establish a steep, stable transmembrane gradient of entropy and information. During carcinogenesis, accumulating mutations degrade genomic information, resulting in increasing cellular and tissue disorder characteristic of invasive cancer. However, additional growth control parameters can be inferred because *unconstrained mutagenesis will monotonically decrease intracellular information until it is insufficient to maintain life.* Here we use various quantitative methods to examine the information dynamics during somatic evolution of the malignant phenotype.

In general we find that, during carcinogenesis, Darwinian competition among different phenotypically distinct populations constrains the information loss from accumulating mutations. The interactions between genomic information degradation and clonal selection generate the forces encouraging tumor development and growth.

As with other investigations of scientific phenomena, such as in physics (and especially in this book), we utilize mathematical models from information theory. We find that phenotypic properties and growth patterns of cancer cells are emergent phenomena, growing out of and controlled by nonlinear interactions. These are outgrowths of random genomic mutations and time-dependent somatic evolutionary forces.

We demonstrate that, even in the presence of a mutator phenotype, the *observed* mutation rate in each gene is dependent on its contribution to cellular fitness. Genes that encode information promoting cell proliferation (i.e., oncogenes) are constrained to maintain a high transgenerational information content and, so, remain stable or exhibit only gain of function mutations even in the presence of a mutator phenotype. Genes that inhibit proliferation (tumor suppressors) are subject to maximal degradation and information loss. Information that is neutral in determining cellular proliferation (differentiation genes) exhibits an intermediate mutation rate. The net effect is that accumulating mutations due to CIN or MIN, even if randomly distributed, will degrade the genome unevenly. (CIN = chromosomal instability, MIN = microsatellite instability)

The information degradation in segments of the genome that *do not confer growth advantage* is predicted to alter differentiated phenotypes. Specifically, selective degradation of genes that encode for differentiated function, and do not contribute to cellular proliferation, will invariably produce the de-differentiation characteristic of carcinogenesis. Furthermore, if the mutation rate increases above some threshold value, an *error catastrophe* results (as classically described by Eigen and Schuster [66]). This produces a phase transition from a differentiated state into an unpredictable state characterized by nonphysiologic combinations of phenotypic traits and chaotic dysfunctional tissue. These are characteristics that are often observed in highly aggressive clinical cancers.

Concurrent with these effects, subsets of the genome can exist that, in fact, *do confer selective growth advantage.* These are shown to remain relatively stable because selection forces generated by somatic evolution *constrain* their apparent mutation rates. This allows a cancer cell to maintain an information minimum (Section 3.2.11) that is, nevertheless, compatible with life. That is, there is compatibility despite underlying genomic instability.

Testable predictions from this model include the following:

1. Accumulated mutations during carcinogenesis will always be fewer in genes controlling critical cellular functions than in those controlling differentiation function and morphology; this divergence will increase with time.

2. Traits retained by most malignant phenotypes will be parameters that confer the greatest selective growth advantage; these are, therefore, optimal targets for therapeutic intervention.

Modeling cancer growth by use of the EPI approach leads to predictions which, for the most part, have already been verified clinically. These are (adding to the list)

3. In situ cancer growth follows a simple power law (3.43). The power of the power law is defined by postulating that it suffers maximal information degradation.

4. Remarkably, the power is precisely the Fibonacci golden mean, a constant heretofore used to describe healthy growth in nature. The dynamics are also postulated to arise out of an unstable genome. Results are quantitatively and qualitatively in good agreement with clinical and experimental observations.

Furthermore, this approach demonstrates mathematically the critical role played by degradation of cellular information in producing uncertainty in the age of the cancer. This is consistent with extensive literature regarding the biological importance of telomere dysfunction and prolonged cell survival in tumor growth. In particular, the following prediction is made:

5. The minimum mean-squared error in the estimated onset time θ for breast cancer is 30% of the typical time during which tumors are clinically observed. This large degree of uncertainty could place a fundamental limit on the efficacy of cancer screening, if the onset of metastases correlates with the age of the cancer rather than with its size.

Of course, as with any mathematical model of a complex system, such as carcinogenesis, the analysis is necessarily limited by a number of *simplifying assumptions*. These are (a) certain assumed values for ion concentrations (Section 3.1.17); (b) that a cancer mass devolves toward a state of minimum information (Section 3.2.11); (c) that codons independently contribute information; and (d) that the volume of tumor cells about equals that of normal cells, whereas, in practice, tumors may contain significant volumes of noncellular regions, and tumor cells are often larger than normal cells. Furthermore, in addition to necroses, tumors include blood vessels and (often) large numbers of nonneoplastic cells such as endothelium, fibroblasts, macrophages, and lymphocytes (Section 3.2.12). Finally, (e) the power law dynamics are only predicted for *unimpeded* in vivo replication of the tumor cells. If, instead, tumor cell expansion is constrained by external factors, such as inadequate vascular supply, host response, spatial considerations, or clinical therapy, *slower growth* will be observed.

We conclude that the genetic pathways in carcinogenesis are complex and unpredictable, as with most nonlinear dynamical systems. Hence, *fixed*, linear genetic sequences leading to sporadic human cancer [79] will likely be *uncommon*. (Note that congenital neoplastic syndromes such as retinoblastoma, neurofibromatosis, and familial polyposis syndrome may be exceptions.) This is consistent with studies demonstrating heterogenous genetic pathways in the pathogenesis of breast and renal cancers [72, 101]. Finally, our model adds a cautionary note to a wide range of strategies in cancer therapy. The same mutational and selection forces that produced the initial malignant population remain available to foil these therapies through *rapid evolution* of resistant phenotypes.

3.3. Appendix A: Derivation of EPI Power Law Solution

3.3.1. General Solution

The use of Eqs. (3.34) and (3.35) in (3.33) and (3.36) gives an Euler–Lagrange solution

$$\frac{d}{dt}\left(\frac{\partial(\dot{q}^2 - j[q, t])}{\partial \dot{q}}\right) = \frac{\partial(\dot{q}^2 - j[q, t])}{\partial q} \tag{3.A1}$$

and a zero solution

$$\dot{q}^2 - \kappa j[q, t] = 0. \tag{3.A2}$$

Explicitly carrying through the differentiations in Eq. (3.A1) gives

$$2\frac{d^2 q}{dt^2} = -\frac{\partial j}{\partial q}. \tag{3.A3}$$

Equations (3.A2) and (3.A3) must be simultaneously solved for j and q. Note that each equation involves both j and q. The aim is to come up with only *one equation* in *one* of the unknown functions q or j.

Differentiating d/dt (3.A2) gives

$$2\dot{q}\frac{d^2 q}{dt^2} = \kappa\left[\frac{\partial j}{\partial t} + \frac{\partial j}{\partial q}\dot{q}\right]. \tag{3.A4}$$

Equations (3.A3) and (3.A4) must be solved simultaneously.

Using identity (3.A3) on the left-hand side of (3.A4) gives a requirement

$$-\dot{q}\frac{\partial j}{\partial q} = \kappa\left[\frac{\partial j}{\partial t} + \frac{\partial j}{\partial q}\dot{q}\right]. \tag{3.A5}$$

Taking the square root of (3.A2) gives

$$\dot{q} = \sqrt{\kappa}\sqrt{j[q, t]}, \tag{3.A6}$$

where the $+$ sign was chosen because we are seeking a solution that grows with the time. Substituting (3.A6) into (3.A5) gives directly

$$-\sqrt{\kappa}\sqrt{j}\frac{\partial j}{\partial q} = \kappa\left[\frac{\partial j}{\partial t} + \frac{\partial j}{\partial q}\sqrt{\kappa}\sqrt{j}\right]. \tag{3.A7}$$

After a rearrangement of terms and division by $\sqrt{\kappa}$ this becomes

$$\sqrt{j}\frac{\partial j}{\partial q}(1+\kappa) + \sqrt{\kappa}\frac{\partial j}{\partial t} = 0. \tag{3.A8}$$

This accomplishes our immediate aim of obtaining one equation in the single unknown function j.

Let us seek a separable solution

$$j[q,t] \equiv j_q(q)j_t(t), \tag{3.A9}$$

where functions $j_q(q)$ and $j_t(t)$ are to be found. Substituting (3.A9) into (3.A8) gives a requirement

$$\sqrt{j_q}\sqrt{j_t}\,j_q'\,j_t(1+\kappa) + \sqrt{\kappa}j_qj_t' = 0, \tag{3.A10}$$

where the derivatives indicated by primes are with respect to the indicated subscripts (and arguments) q or t. Multiplying the equation by $j_q^{-1}j_t^{-3/2}$ gives

$$j_q^{-1/2}j_q'\,(1+\kappa) + \sqrt{\kappa}j_t^{-3/2}j_t' = 0. \tag{3.A11}$$

The first term on the left-hand side is only a function of q and the second one is only a function of t. The only way they can add up to zero (the right-hand side) for all q and t is for the first to equal a constant and the second to equal the negative of that constant. Thus,

$$j_q^{-1/2}j_q'\,(1+\kappa) = A, \quad \sqrt{\kappa}j_t^{-3/2}j_t' = -A. \tag{3.A12}$$

The first Eq. (3.A12) can be placed in the integrable form

$$j_q^{-1/2}dj_q = A(1+\kappa)^{-1}dq. \tag{3.A13}$$

Integrating and then squaring both sides gives

$$j_q(q) = \frac{1}{4}\left(\frac{Aq}{1+\kappa} + B\right)^2, \quad A, B = \text{const.} \tag{3.A14}$$

The second Eq. (3.A12) can be placed in the integrable form

$$j_t^{-3/2}dj_t = -A\kappa^{-1/2}dt. \tag{3.A15}$$

Integrating this, then solving algebraically for j_t, gives

$$j_t(t) = 4\left(\frac{At}{\sqrt{\kappa}} + C\right)^{-2}, \quad C = \text{const.} \tag{3.A16}$$

Using solutions (3.A14) and (3.A16) in Eq. (3.A9) gives

$$j[q,t] = \left(\frac{\frac{Aq}{1+\kappa} + B}{\frac{At}{\sqrt{\kappa}} + C}\right)^2. \tag{3.A17}$$

Having found j we can now use it to find q. Using (3.A17) in (3.A6) gives directly

$$\dot{q} = \frac{\frac{Aq}{1+\kappa} + B}{\frac{At}{\kappa} + D}, \quad D = C/\sqrt{\kappa}. \tag{3.A18}$$

This is equivalent to the differential form

$$\frac{dq}{\frac{Aq}{1+\kappa} + B} = \frac{dt}{\frac{At}{\kappa} + D}. \tag{3.A19}$$

This may be readily integrated, giving logarithms of functions on each side. After taking antilogarithms, the solution is

$$q(t) = \left(\frac{1+\kappa}{A}\right)\left(\frac{At}{\kappa} + D\right)^{\kappa/(1+\kappa)} - (1+\kappa)\frac{B}{A}. \tag{3.A20}$$

This is a general power law solution in the time t. Two of the constants A, B, D may be fixed from boundary-value conditions, as follows:

3.3.2. Boundary-Value Conditions

At $t = 0$ the cell is presumed not to be cancerous. Cancer has its onset at $t = \epsilon$, $\epsilon > 0$, a small increment beyond 0. This has two boundary-value effects:

(a) At $t = 0$, the information j about cancer in the given tissue is zero, $j(0) = 0$.
(b) At $t = 0$ the cancer has zero mass, $q(0) = 0$.

Using these conditions in (3.A17) gives the requirement $B = 0$. Using the latter and condition (b) in (3.A20) gives either $D = 0$ or $\kappa = -1$. But the latter is ruled out a priori by EPI, since $\kappa \geq 0$ represents an efficiency measure of information transmission. Hence, $D = 0$. With this choice of constants, (3.A20) becomes

$$q(t) = \left(\frac{1+\kappa}{A}\right)\left(\frac{At}{\kappa}\right)^{\kappa/(1+\kappa)}. \tag{3.A21}$$

This is the power law we sought. Its square, the probability law $p(t)$, is therefore also a power law.

3.3.3. Error in the Estimated Duration of the Cancer

The time t is the time *since* the unknown onset time θ of the cancer condition. However, the time of observation of the tumor is not t, but rather τ, where $\tau = \theta + t$. The question is, how well can θ be estimated based upon knowledge of τ? This amounts to determining the position of the origin of the growth curve $p(t)$ based upon one observed event at the randomly displaced time value $\theta + t$.

The Cramer–Rao inequality (1.37) states that, any estimate of θ that is correct on average (unbiased) suffers an rms error e that can be no smaller than

$$e_{min} = 1/\sqrt{I}. \tag{3.A22}$$

In our case I has a value given by (3.41) and (3.42) as

$$I = \frac{11 + 5\sqrt{5}}{2T^2}. \tag{3.A23}$$

Using this in (3.A22) gives a minimum rms error of size

$$e_{min} \approx 0.3T. \tag{3.A24}$$

As a relative error, this is quite large, amounting to 30% of the total time interval T over which cancer events are detected. Thus, cancer has an innate ability to mask its onset time. This contributes to its prevalence, and hence might be considered an evolutionary tactic.

4
Information and Thermal Physics

ANGELO PLASTINO AND ANGEL RICARDO PLASTINO

Thermal physics constitutes a fundamental part of our understanding of nature. In this chapter we will be concerned with the connection

$$\text{thermal physics} \Leftrightarrow \text{information physics.}$$

Interest in information physics (IP) has been rekindled in the last 15 years, mainly on account of

(1) the quantum communication explosion [1],
(2) the wide use of information measures (IMs) other than Shannon's [2, 3], and
(3) the Wheeler *"it from bit"* dictum [4] that has evolved into a "program" that intends to derive the whole of physics from information theory (IT) [5]. More specifically, Wheeler has predicted that in the future we will be able to understand and express all of physics in the language of information, and stated that "all things physical are information-theoretic in origin and this is a *participatory universe*" [4].

Several additional factors have also contributed, although to a lesser degree, to this renewed interest and increased IP activity. One may cite among them

- a sort of climbing Mount Everest "attitude" (just because it is there. . .);
- a search for the "universe's essence." In pre-Socratic fashion this was the idea of Thales (water), Anaximander (air), etc. [6]. Information would constitute such an "essence." Note that information is not energy, nor is it an entropy (see below), although it has properties related to these;
- the fascinating possibility of regarding the universe as a computer running a cosmic Turing algorithm [7].
- the tantalizing concept advanced by Stephen Hawkins of a wave function for the universe.

The central IP concept is that of IM [8–11]. We must at the very beginning categorically state that an IM is *not necessarily* an entropy! Unnecessary controversies arise from this absolutely wrong assimilation. At most, in some *special* cases, an association IM $\Leftrightarrow$ entropy can be made.

Now, *the pivotal idea* of IT is that of giving quantitative form to the everyday concept of *ignorance* [8–11]. If, in a given scenario, N distinct outcomes ($i = 1, \ldots, N$) are possible, then *three* situations may ensue [8–11]:

- Zero ignorance: Predict with certainty the actual outcome.
- Maximum ignorance: There is no prior knowledge of preferential values of the N outcomes. They are equally likely.
- Partial ignorance: We are given the probability distribution $\{P_i\}$; $i = 1, \ldots, N$ of the outcomes.

In this chapter we will attempt to integrate thermodynamics within a "Fisher program." In particular, we will discuss a reformulation of thermal physics in terms of Fisher information. This is partitioned into two parts:

(1) On the one hand, we will show that the structure of classical mechanics, together with Gibbs' postulates for statistical mechanics, is able to describe thermal equilibrium at a given temperature T for any classical Hamiltonian via a universal form for Fisher's measure.

(2) On the other hand, we will see that one can insert the whole of thermal physics, both equilibrium and off-equilibrium ones, within the scope of the "Wheeler program" (wp), using Fisher's information measure (FIM). A Schroedinger-like differential equation provides us with a "Fisher-based *thermodynamics*." Remarkably, this mathematical Schroendinger wave equation (SWE) is also found to provide a basis for econophysics (Chapter 2) and population growth in ecological systems (Chapter 5).

We will also show that a theory that minimizes Fisher's measure (MFM), subject to appropriate constraints, is analogous to use of Jaynes' MaxEnt.

Finally, the Fisher–Schrodinger approach that is given has also been successfully applied to the description of the electrical conductivity in gases, viscosity in fluids, and propagation of sound waves in an ideal gas [12–14]. These applications conclusively verify that a Fisher-tool is at hand for dealing with problems of both equilibrium and off-equilibrium thermodynamics.

4.1. Connections Between Thermal and Information Physics

4.1.1. Summary

We wish now to describe the formal structure of a connection between thermal and information-physics. In order to gain an adequate perspective, we schematically review Descartes' ideas regarding the construction of a scientific theory so as to proceed according to his methodology. Our basic tool is a variational approach in which the leading role is played by the concept of information. This, in turn, forces us to remind the reader of basic ideas concerning the theory of probabilities.

4.1.2. Introduction

An important idea of IP is that of extremizing some information measure in order to obtain via such procedure an underlying probability distribution that will govern the pertinent physics. This idea can be said to originate in the work of Bernoulli and Laplace (the fathers of probability theory) [15], on the rationale that *the concept of probability refers to a state of knowledge.* An information measure quantifies the information (or ignorance) content of a probability distribution [15]. If our state of knowledge is appropriately represented by a set of, say, M expectation values, then the "best," least unbiased probability distribution is the one that

- reflects just what we know, without "inventing" unavailable pieces of knowledge [8–10], and, additionally,
- maximizes ignorance: the truth, all the truth, *nothing but* the truth.

Such is, for instance, the Jaynes' MaxEnt rationale [8–10]. *Here* we will review a parallel "Fisher rationale" for thermal physics that has been concocted in the last years by Frieden, Plastino, Plastino, and Soffer (see, for instance, [16] and references therein). Since ignorance diminishes as Fisher information increases [17], the following correspondence ensues:

$$\text{MaxEnt} \Leftrightarrow \text{Minimum Fisher information.}$$

This is in the sense that both maximizing entropy and minimizing Fisher information signify the same qualitative scenario, that of maximum disorder or ignorance. In order to discuss the Fisher end of this correspondence, we first briefly review some elementary epistemological concepts that are needed in order to give Fisher's information its proper place in the scheme of things.

4.1.3. A Scientific Theory's Structure

All scientific disciplines must undergo a preliminary *descriptive* stage (long series of careful and detailed observations, searching for patterns and regularities) followed by several (tentative) mathematical descriptions of the ensuing detected patterns. Eventually, the following Galileo–Descartes program ensues [18–22] (as an example, see below the Gibbs' axioms for statistical mechanics [23]):

- *I: advance a set of axioms (not unique!)* and derive from them
- *II: theorems and laws* that are relations among quantities of interest, from which one should be able to make
- *III: predictions*, that is, new relationships, never detected or imagined before (e.g., the planet Neptune or Hertzian waves), that are to be subjected to
- *IV: experimental verification.* If verification fails,
- *V: go to I.*

A kind of "wheel" arises that is supposed to revolve forever, as long as science exists. The WP's goal is to derive all of physics from a set of axioms of purely information-theoretic origin [5].

The WP referred to above that intends to derive the whole of physics from information theory, has received a tremendous impetus from the extreme physical information (EPI) principle of Frieden and Soffer [17, 24], discussed at length in Chapter 1 and elsewhere in this book. The EPI principle has incorporated into the WP such disparate subjects as general relativity and electromagnetism, among many others [17]. Here we wish to address the issue of incorporating thermal physics into the WP, in the wake of the work reported in [12–14, 25]. This goal has been achieved using a special, approximate use of EPI: MFI subject to appropriate constraints, also discussed in Chapters 1 and 2.[1]

This minimum Fisher information (MFI) principle has well fit the Wheeler program by rederiving thermal physics on entirely a Fisher basis, i.e., without the need for the use of Boltzmann entropy (except for purposes of comparison). In order to insert the Fisher measure into the Wheeler program, we will review below some basic concepts regarding Fisher's information measure (FIM). Before doing that, however, let us mention some details of the current (intense) activity revolving around the Fisher ⇔ physics connection.

4.1.4. Fisher-Related Activity in Contemporary Physics

The last years have witnessed a great deal of activity revolving around physical applications of FIM (see for instance, as a rather small and by no means exhaustive sample [17, 24, 26–33]). Frieden and Soffer [24] have shown that Fisher's information measure provides one with a powerful variational principle, that of EPI, which yields most of the canonical Lagrangians of theoretical physics [17, 24]. Additionally, the Fisher information I has been shown to provide an interesting characterization of the "arrow of time," alternative to the one associated with Boltzmann's entropy [17, 34, 35].

Our focus in this work is that of reviewing a recent approach to thermal physics, based upon Fisher's measure (a kind of "Fisher-MaxEnt") [12–14, 25], which exhibits definite advantages over conventional textbook treatments. More specifically, we will discuss some general FIM-features in relation with both macroscopic thermodynamics and microscopic statistical mechanics. This chapter consists of two parts, with different goals:

(1) On the one hand, we wish to show how the structure of classical mechanics becomes translated into kind of "Fisher-based counterpart" that is able to describe thermal equilibrium at a given temperature T for any classical Hamiltonian. This will be done within the traditional axiomatics of Gibbs' [23].

(2) Fully insert thermal physics within the scope of the Wheeler program using Fisher's information measure. We will see that a Schrodinger-like differential equation results, yielding a "Fisher-based *thermodynamics*." Here we will work within the framework of a different axiomatics, of a purely information theoretic nature.

---

[1] Note that in Chapter 1 MFI has been incorporated into EPI by regarding it as a solution of type (C), the least accurate approximation to a rigorous (based upon a unitary transformation to get J) EPI approach, where the search for J is abandoned (which is tantamount to leaving the strictly microscopic level) in favor of macroscopic empirical data to model J. In other words, what people regard as constraints on the Fisher information minimization are regarded as a source information J in EPI, but one gotten purely from empirical, macroscopic data. Since macroscopic data typify thermal physics, this is a sound procedure.

4.1.5. Brief Primer on Fisher's Information Measure

R.A. Fisher advanced, already in the twenties [36], a quite interesting information measure I (for details and references see, for instance, Chapter 1 and [17, 26]). Consider a $(\theta, \mathbf{z})$ "scenario" in which we deal with a system specified by a physical parameter θ, while $\mathbf{z}$ is a stochastic variable ($\mathbf{z} \in \mathfrak{R}^M$) and $f_\theta(\mathbf{z})$ is the probability density for $\mathbf{z}$ (that depends also on θ). One makes a measurement of $\mathbf{z}$ and has to best infer θ from this measurement, calling the resulting estimate $\tilde{\theta} = \tilde{\theta}(\mathbf{z})$. The basic question is how well θ can be determined. Estimation theory [36] states that the best possible estimator $\tilde{\theta}(\mathbf{z})$, after a very large number of $\mathbf{z}$-samples is examined, suffers a mean-square error e^2 from θ that obeys a relationship involving Fisher's I, namely, $I e^2 = 1$. This Fisher information measure I is of the form

$$I(\theta) = \int d\mathbf{z} \, f_\theta(\mathbf{z}) \left\{ \frac{\partial \ln f_\theta(\mathbf{z})}{\partial \theta} \right\}^2 . \tag{4.1}$$

This "best" estimator is the so-called *efficient* estimator. Any other estimator exhibits a larger mean-square error. The only caveat to the above result is that all estimators be unbiased, i.e., satisfy $\langle \tilde{\theta}(\mathbf{z}) \rangle = \theta$. Thus, the error e^2 has a lower bound: No matter what parameter of the system one chooses to measure, the error has to be larger than or equal to the inverse of the Fisher information associated with the concomitant experiment. This result,

$$I e^2 \geq 1, \tag{4.2}$$

is referred to as the Cramer–Rao inequality [17, 36].

A particular I-case is of great importance: that of translation families [17, 27], i.e., distribution functions (DF) whose *form* does not change under θ displacements. These DF are shift-invariant (à la Mach, no absolute origin for θ). In such a scenario, in the one-dimensional instance, Fisher's information measure I adopts the somewhat simpler appearance I_τ [17]:

$$I_\tau = \int dz \, f(z) \left[\frac{d}{dz} (\ln f(z)) \right]^2 . \tag{4.3}$$

4.2. Hamiltonian Systems

4.2.1. Summary

We restrict ourselves here to classical considerations and construct thermal physics à la Fisher from Hamiltonian dynamics plus Gibbs' axioms for statistical mechanics. Special systems that are considered are those with quadratic Hamiltonians, the free particle, N-harmonic oscillators, and the paramagnetic system, where the Langevin function naturally arises as a manifestation of the information.

4.2.2. Classical Statistical Mechanics à la Fisher

In this first part of the present chapter we will show how classical Hamiltonian considerations give rise to a "Fisher statistical mechanics." This is done, perhaps surprisingly, entirely within the strictures of Gibbs' axiomatics, whose postulates read [23] as follows:

(1) *Ensemble postulate*: the system at equilibrium is represented by an appropriately designed ensemble,
(2) equal a priori probabilities for cells in phase space,
(3) the phase space probability distribution depends *only* on the system's Hamiltonian, and
(4) this dependence is of *exponential* form.

Now, Fisher's measure is additive [17] under independence. If one is dealing with phase space, the above referred to stochastic variable z is an $(M = 2n)$-dimensional vector, with $2n$ degrees of freedom, n "coordinates" r, and n "momenta" p. As a consequence, if one wishes to estimate phase-space "location," one confronts an $I(z)$-measure and [37]

$$I(\mathbf{z}) = I(\mathbf{r}) + I(\mathbf{p}), \quad \mathbf{z} = (\mathbf{r}, \mathbf{p}). \tag{4.4}$$

Here we clearly encounter an additive situation.

The interesting relation (4.4) holds because Fisher information obeys additivity under independence, and the coordinates r and p are indeed independent in classical mechanics. By comparison, in quantum mechanics it would not hold, since of course, there these two quantities become noncommuting operators.

In order to demonstrate how classical Hamiltonian considerations can lead to a "Fisher statistical mechanics" in three dimensions (setting $n = 3N$), we need to focus attention upon a classical system governed by a time-independent Hamiltonian $\mathcal{H}$ with $6N$ degrees of freedom:

- $3N$ "generalized coordinates" q_i and
- $3N$ "generalized momenta" p_i.

In what follows, we closely follow the work reported in [38]. It will prove convenient to redefine z as a "state-vector"

$$\mathbf{z} \equiv (q_1, q_2, \ldots, q_{3N}, p_1, p_2, \ldots, p_{3N}), \tag{4.5}$$

which specifies a point in the so-called "phase-space Γ." (See the related concept of an *economic* phase space in Chapter 2.) For each degree of freedom Hamilton's celebrated canonical equations hold

$$\dot{q}_i = \frac{\partial \mathcal{H}(\mathbf{z})}{\partial p_i} \tag{4.6}$$

$$\dot{p}_i = -\frac{\partial \mathcal{H}(\mathbf{z})}{\partial q_i}, \tag{4.7}$$

with $i = 1, \ldots, 3N$.

We assume that the system of interest is in thermal equilibrium at the temperature T, and that it is appropriately described by Gibbs' exponential canonical probability distribution, according to his postulates 3 and 4 [39, 40].

All our results in this first part below depend upon this critical assumption. Thus, at point $\mathbf{z}$, this canonical ensemble probability density $\rho(\mathbf{z})$ reads

$$\rho(\mathbf{z}) = \frac{e^{-\beta \mathcal{H}(\mathbf{z})}}{Z} \tag{4.8}$$

with $\beta = 1/kT$ and k Boltzmann's constant. If h denotes an elementary cell in phase space, we write, with some abuse of notation [39, 40],

$$d\Gamma = d\mathbf{z} \equiv \frac{1}{(3N)! h^{3N}} \prod_{i=1}^{3N} dq_i \, dp_i, \tag{4.9}$$

for the pertinent integration measure, while the partition function is of the form [39, 40]

$$Z = \int d\mathbf{z} \, e^{-\beta \mathcal{H}(\mathbf{z})}. \tag{4.10}$$

Given the canonical equations of motion (4.6), (4.7), and the probability density form (4.8), it is rather straightforward to ascertain that

$$-kT \frac{\partial \ln \rho(\mathbf{z})}{\partial p_i} = \dot{q}_i \tag{4.11}$$

$$-kT \frac{\partial \ln \rho(\mathbf{z})}{\partial q_i} = -\dot{p}_i. \tag{4.12}$$

Equations (4.11) and (4.12) are just the above referred-to canonical equations of Hamilton in a different guise. We emphasize this point because the expressions in the left-hand-sides of (4.11 and 4.12) enter the definition of Fisher's measure, if one expresses I_τ as a functional of the probability density ρ. Indeed, this FIM has the form (4.3), so that, in view of (4.4), it acquires the form

$$I_\tau = \sum_{i=1}^{3N} \int d\mathbf{z} \, \rho(\mathbf{z}) \left\{ a \left[\frac{\partial \ln \rho(\mathbf{z})}{\partial p_i} \right]^2 + b \left[\frac{\partial \ln \rho(\mathbf{z})}{\partial q_i} \right]^2 \right\}. \tag{4.13}$$

Dimensional balance [37] requires that one introduce above the coefficients a and b. These may depend on the specific Hamiltonian of the system one is interested in (see examples below). Note now that, from Equations (4.6), (4.7) and (4.11), (4.12),

$$\frac{\partial \ln \rho(\mathbf{z})}{\partial p_i} = -\beta \frac{\partial \mathcal{H}(\mathbf{z})}{\partial p_i} \tag{4.14}$$

$$\frac{\partial \ln \rho(\mathbf{z})}{\partial q_i} = -\beta \frac{\partial \mathcal{H}(\mathbf{z})}{\partial q_i}, \tag{4.15}$$

so that we can rewrite Fisher's measure in the fashion

$$I_\tau = \beta^2 \sum_{i=1}^{3N} \left\{ a \left\langle \left(\frac{\partial \mathcal{H}(\mathbf{z})}{\partial p_i} \right)^2 \right\rangle + b \left\langle \left(\frac{\partial \mathcal{H}(\mathbf{z})}{\partial q_i} \right)^2 \right\rangle \right\}, \qquad (4.16)$$

where the pointy brackets $\langle \mathcal{G} \rangle$ stand for the classical expectation value

$$\int d\Gamma \, \rho(\mathbf{z}) \, \mathcal{G}(\mathbf{z}). \qquad (4.17)$$

It is clear that derivatives of I_τ with respect to the generalized coordinates and momenta vanish identically, since we have integrated over them. Accordingly, *the Poisson brackets of I_τ with the Hamiltonian vanish and the information I_τ becomes a constant of the motion.* This means that any joint measurement of the p's and q's carries a constant momentum-coordinate information over time, which is also consistent with the constancy of the size of h, the elementary cell size in phase space. Thus, this is a result of *uniformity of information and that of error over all of phase space.* That is, there is no preferred region of phase space insofar as accuracy goes, a kind of a "repeat" of the Mach principle for the p and q coordinates except, now, with respect to information rather than with respect to origins.

In view of Equations (4.11) and (4.12) Fisher's measure (4.13) can be further simplified, and one finds

$$I_\tau = \beta^2 \sum_{i=1}^{3N} \left\{ a \left\langle \dot{q}_i^2 \right\rangle + b \left\langle \dot{p}_i^2 \right\rangle \right\}. \qquad (4.18)$$

We see that I_τ automatically incorporates the symplectic structure of classical mechanics. Equations (4.16) and (4.18) *are important results*, since they give the FIM-form for *any* Hamiltonian system, that is, a universal "quadratic" form in the average values of "squared velocities" and "squared forces." Note that the purely quantum mechanical I_τ is proportional to the kinetic energy [17]. Its classical counterpart is seen here to incorporate the "squared forces" additional term.

Compare now (4.18) with its Shannon's logarithmic information measure counterpart

$$S = -k \int d\Gamma \rho \ln \rho, \qquad (4.19)$$

for the canonical probability distribution (4.8). One easily obtains [40]

$$S = k(\ln Z + \langle \mathcal{H} \rangle) \Rightarrow TS = kT \ln Z + \langle \mathcal{H} \rangle \Rightarrow F = \langle \mathcal{H} \rangle - TS, \qquad (4.20)$$

which is merely a statement concerning the *macroscopic*, thermodynamic definition of Helmholtz' free energy

$$F = -kT \ln Z. \qquad (4.21)$$

Notice that all reference to our system's specificities concerning its dynamics on the microlevel clearly present in (4.18), has vanished in this orthodox, textbook version of things. The reason underlying such different behavior of the two information measures S and I_τ can be attributed to the *global* character of the former (leading after extremization to a *fixed* exponential functional form) versus the *local* nature of the latter, which after extremizing leads to a differential equation (see Section 1.2.2.12 or [24] for details). This is but one example of the Fisher-based analysis over a conventional Boltzmann one.

4.2.3. Canonical Example

Assume we wish to describe a classical system of N identical particles of mass m whose Hamiltonian is separable with respect to the coordinates $\mathbf{r}_i$ and momenta $\mathbf{p}_i$ of the individual particles of the system

$$\mathcal{H} = \mathcal{T} + \mathcal{V}. \tag{4.22}$$

A special important instance is that of

$$\mathcal{H}(\mathbf{r}, \mathbf{p}) = \sum_{i=1}^{N} \frac{\mathbf{p}_i^2}{2m} + \sum_{i=1}^{N} V(\mathbf{r}_i), \tag{4.23}$$

where $(\mathbf{r}_i, \mathbf{p}_i)$ are the coordinates and momenta of the ith particle and $V(\mathbf{r}_i)$ is a central single-particle potential. The vector state here reads

$$\mathbf{z} = (\mathbf{r}, \mathbf{p}) = (\mathbf{r}_1, \mathbf{r}_2, \ldots, \mathbf{r}_N, \mathbf{p}_1, \mathbf{p}_2, \ldots, \mathbf{p}_N), \tag{4.24}$$

specifying a point in the phase-space Γ. The first term on the right-hand side of (4.16) is in fact the same for all systems for which $\mathcal{T}$ is the conventional ("momentum squared") kinetic energy. Notice that cyclic canonical coordinates, i.e., those from which the Hamilonian does not explicitly depend upon, do not contribute to I_τ.

4.2.4. General Quadratic Hamiltonians

In many important cases (a set of multiple harmonic oscillators, for instance) [39] one faces Hamiltonians of the type

$$\mathcal{H}_L = \sum_{i=1}^{N} \left(A_i P_i^2 + B_i Q_i^2 \right); \quad \langle \mathcal{H}_L \rangle = \sum_{i=1}^{N} \left(A_i \langle P_i^2 \rangle + B_i \langle Q_i^2 \rangle \right), \tag{4.25}$$

with (P_i, Q_i) generalized momenta and coordinates, respectively. These lead to $2DN$ differential equations ($2D$ is the dimension of the "single i−cell" in phase space), the so-called linear Hamiltonian problem [41]. According to the preceding considerations on dimensional balance, we have now, for the dimensionless quantity I_τ

$$I_\tau = \beta^2 \sum_{i=1}^{N} \left[\tilde{A}_i \left\langle \left(\frac{\partial \mathcal{H}_L}{\partial P_i} \right)^2 \right\rangle + \tilde{B}_i \left\langle \left(\frac{\partial \mathcal{H}_L}{\partial Q_i} \right)^2 \right\rangle \right]$$

$$= 4\beta^2 \sum_{i=1}^{N} \left[\tilde{A}_i \langle P_i^2 \rangle + \tilde{B}_i \langle Q_i^2 \rangle \right] = 4 \frac{\beta^2}{\beta_0} \langle \mathcal{H}_L \rangle.$$

In order to obtain the last equality, we need to consider dimensions. It is clear that

$$\dim(A_i P_i^2) = \dim(B_i Q_i^2) = \dim(kT) = \dim(1/\beta)$$
$$= \text{energy dimensions } (k \equiv \text{Boltzmann's constant}).$$

Without loss of generality we can take it that both A_i and B_i are dimensionless quantities for all i. Their "tilde-counterparts" must then have dimensions of inverse energy (IE). Define $\beta_0 = 1/k$ as the concomitant IE-unit. In order to achieve a proper dimensional balance, we must have $\tilde{A}_i = (1/4\beta_0) A_i$ and $\tilde{B}_i = (1/4\beta_0) B_i$ for all i, so that, summing up,

$$I_\tau = 4\frac{\beta^2}{\beta_0} \langle \mathcal{H}_L \rangle. \tag{4.26}$$

Now, we know that the equipartition theorem holds for $\langle \mathcal{H}_L \rangle$ [39]. Thus, the mean value of the energy $\langle \mathcal{H}_L \rangle = DN/2\beta$, which when used in (4.26) gives for the Fisher measure

$$I_\tau = 2ND\frac{\beta}{\beta_0}. \tag{4.27}$$

As one would expect, *the Fisher information steadily diminishes as the temperature grows*. The divergence at zero temperature is of no importance in the present context, since classical mechanics is not valid at low enough temperatures.

4.2.5. Free Particle

We consider now N identical, noninteracting particles of mass m. The classical expression for the Hamiltonian is

$$\mathcal{H} = \sum_{j=1}^{3N} \frac{p_j^2}{2m}, \tag{4.28}$$

where we have once more assumed that our problem is three-dimensional. We make here the obvious choice $a = m/\beta_0$ in order to obtain a dimensionless Fisher measure (of course, $\beta_0 = 1/kT_0$, with T_0 a reference temperature). Since

$$\langle \dot{p}_i^2 \rangle = 0; \quad \text{and} \quad \langle \dot{q}_i^2 \rangle = 2\langle \mathcal{H} \rangle/(Nm),$$

we have by (4.18) and the preceding,

$$I_\tau = 2\frac{\beta^2}{\beta_0} \langle \mathcal{H} \rangle. \tag{4.29}$$

Then, using the well-known classical theorem of equipartition of energy [39] $\langle \mathcal{H} \rangle = 3N/2\beta$, one obtains the Fisher measure

$$I_\tau = 3N\frac{\beta}{\beta_0}. \tag{4.30}$$

4.2.6. N Harmonic Oscillators

We discuss now, specifically, a system of N identical harmonic oscillators whose Hamiltonian is

$$\mathcal{H} = \sum_{j=1}^{3N} \left[\frac{p_j^2}{2m} + \frac{m\omega q_j^2}{2} \right], \tag{4.31}$$

(assumed to be three-dimensional). It is easy to calculate the classical mean values in (4.18), which have the form

$$\langle \dot{q}_i^2 \rangle = 3m/\beta \tag{4.32}$$

$$\langle \dot{p}_i^2 \rangle = 3m\omega/\beta. \tag{4.33}$$

In this case the proper choice is $a = m/\beta_0$ and $b = 1/m\omega\beta_0$, so that the information (4.18) takes the appearance

$$I_\tau = 6N \frac{\beta}{\beta_0}, \tag{4.34}$$

in agreement with (4.27). In these last two examples we appreciate again the fact that a Fisher equipartition emerges, β/β_0 per degree of freedom.

4.2.7. A Nonlinear Problem: Paramagnetic System

Consider N magnetic dipoles, each of magnetic moment μ, in the presence of an external magnetic field $\mathbf{H}$ of intensity H. These N distinguishable (localized), identical, mutually non-interacting, and freely orientable dipoles give rise to a dipole-potential energy [39]

$$\mathcal{H} = -\sum_{j=1}^{N} \mu_j \cdot \mathbf{H} = -\mu H \sum_{j=1}^{N} \cos \theta_j, \tag{4.35}$$

where θ_j gives the dipole orientation with respect to the field direction. Since there is no interaction between our N particles, and both I_τ and the entropy S are additive quantities, it is permissible to focus attention on just one generic particle (whose canonical conjugate variables we call (θ_i, p_{θ_i})) and, at the end of the calculation, multiply by N. Hamilton's canonical equations yield then the non-linear equation

$$\dot{p}_{\theta_i} = \mu H \sin \theta_i. \tag{4.36}$$

Notice that p_{θ_i} does not appear in $\mathcal{H}$, i.e., $\partial \mathcal{H}/\partial p_{\theta_i} = \dot{\theta}_i = 0$, which entails, of course, $\langle \dot{\theta}_i \rangle = 0$. Thus, in choosing the "dimensional-balance" constants entering the definition (4.18) of I_τ we need only to care about b, with

$$b = \frac{1}{(\mu H \beta_0)}. \tag{4.37}$$

Accordingly, the Fisher information (4.18) is

$$I_\tau = \frac{\mu H \beta^2 N}{\beta_0} \langle \sin^2 \theta \rangle. \qquad (4.38)$$

One can now easily evaluate the above mean value using $(\sin \theta \, d\theta \, d\varphi)$ for the elemental solid angle, so that

$$\langle \sin^2 \theta \rangle = \frac{1}{Z} \int_0^{2\pi} \int_0^{\pi} e^{\beta \mu H \cos \theta} \sin^3 \theta \, d\theta \, d\varphi. \qquad (4.39)$$

Also, here the partition function per particle has the form [39]

$$Z = 4\pi \frac{\sinh(x)}{x} \qquad (4.40)$$

with $x = \mu H / kT$. Computing the integral (4.39) in explicit fashion, we obtain

$$I_\tau = 2N \frac{\beta}{\beta_0} L(x), \qquad (4.41)$$

an interesting result since $L(x)$ is the well-known *Langevin function* [39]

$$L(x) = \coth x - \frac{1}{x}. \qquad (4.42)$$

Since $L(x)$ vanishes at the origin, so does I_τ for infinite temperature, a result one should expect on more general grounds [37].

Notice that the very Langevin function that dominates paramagnetism directly characterizes the information as well. One might then be tempted to assert, following Wheeler's dictum "it from bit", that $L(x)$ is an outgrowth of the information. One thus may be permitted to wonder, then, whether paramagnetism is an outgrowth of information per se.

Once again, equipartition of Fisher information ensues. Since the differential equations are not linear in the conjugate (canonical) variables, the information per degree of freedom does not equal that of (4.27).

4.2.8. Conclusions

Working entirely within Gibbs' classical tenets [23], we have found that.

(1) The mathematical form of the information measure of Fisher's I for a Gibbs' canonical probability distribution incorporates important features of the intrinsic structure of classical mechanics.
(2) It has a universal form in terms of (generalized) "forces" and "velocities."
(3) If the system of differential equations associated to Hamilton's canonical equations for $2ND$ degrees of freedom is linear, the amount of Fisher information per degree of freedom is $I_\tau = 2ND(\beta/\beta_0)$, proportional to the *inverse* of the temperature, to the dimensionality of the concomitant phase space $2D$, and to the number of particles N. We have equipartition of information!

(4) This in turn clearly reaffirms the intuitive notion of temperature as an information-loss factor.

(5) Equipartition of I has been seen to hold also for a paramagnetic system.

(6) In this last instance a *nonlinear* system of differential equations is involved. Some physicists have in the past stated categorically that Fisher information works just because it gives rise to second-order, linear differential equations. (That is, this arises because Fisher's I is quadratic in the gradient of the amplitude function, and therefore in the Euler–Lagrangian equations, one simply gets terms that are linear in the second derivative w.r.t. time, space, etc., of these terms.) On this basis it would seem not to accommodate non-linear ones. However, to the contrary, the present work shows that it even applies in these non-linear cases.

4.3. The Place of A Fisher Thermal Physics

4.3.1. Summary

We embark now into a different project, leaving Gibbs' axiomatics and trying to forge a Fisher one for thermal physics. We will to this end replace Jaynes' MaxEnt by a principle of MFI [16]. A "Schrodinger link" between information theory and both equilibrium and off-equilibrium thermodynamics will be constructed.

4.3.2. Preliminaries

We will deal now with a more general perspective, using new, purely information-theoretic axiomatics. In this way we show how thermal physics, in particular, can be incorporated into the Wheeler program via Fisher's, see also Chapters 1 and 2 in this regard.

Essentially, we replace Jaynes' MaxEnt by a principle of MFI [16]; this is also a type (C) EPI solution (see Chapter 1). We have in this second part a twofold purpose:

- To remind the reader, for didactic reasons, of the main theoretical structures that have been built over the years around the science of thermodynamics; and
- to show that there is a Schrodinger link between information theory and both equilibrium and off-equilibrium thermodynamics, where the link arises out of "Fisher minimization" or MFI.

We begin our considerations with a short review of the orthodox, standard thermal theory.

4.3.3. The Standard Macroscopic Theory

4.3.3.1. Macroscopic Thermodynamics

Thermodynamics can be thought of as a formal logical structure whose *axioms* are empirical facts, which gives it a privileged, rather unique status among all scientific disciplines. The postulates enumerated below can be shown to be equivalent to the celebrated three laws of thermodynamics [42]:

- *Postulate 1:* A unique value of the internal energy E is associated with each system's state.
- *Postulate 2:* Equilibrium states are uniquely determined by (i) E and (ii) an appropriate set of extensive parameters A_i, $i = 1, \ldots, n$.
- *Postulate 3:* For every system there exists a state function S, a function of E and the A_i, that always grows if internal constraints are removed. S is a monotonous (growing) function of E.
- *Postulate 4:* Both S and $[(\partial E/\partial S)]_{A_1,\ldots,A_n}$ vanish for the state of minimum energy and are ≥ 0 for all other states.

4.3.3.2. Legendre Structure

It is a basic thermodynamic fact that if one happens to know either

$$S = S(E, A_1, \ldots, A_n) \qquad (4.43)$$

or

$$E = E(S, A_1, \ldots, A_n), \qquad (4.44)$$

one is in possession of a *complete* thermodynamic description of a system [42]. It is often experimentally more convenient to work with *intensive* variables than with the *extensive* ones appearing in either (4.43) or (4.44). (Note that intensive variables are functions of position, like temperature or volume, while extensive variables depend on the size of the system, and obey additivity.) In order to accommodate such empirical circumstance within the theoretical structure of thermodynamics, a series of mathematical transformations becomes necessary. After first defining $S \equiv A_0$, we will associate an intensive variable P_j to the extensive A_i ($i = 0, 1, \ldots, n$) in the fashion

$$P_j = [(\partial E/\partial A_j)]_{S,A_1,\ldots,A_{j-1},A_{j+1},\ldots,A_n}. \qquad (4.45)$$

Using this notation, the all-important concept of temperature T arises (theoretically) from the definition

$$T \equiv P_0 = [(\partial E/\partial S)]_{A_1,\ldots,A_n}. \qquad (4.46)$$

Any one of the possible Legendre transforms that replaces any s extensive variables by their associated intensive ones,

$$L_{r_1,\ldots,r_s} = E - \sum_j P_j A_j, \quad j = r_1, \ldots, r_s, \qquad (4.47)$$

is seen mathematically to contain the same information as S [42]. The Legendre transform $L_{r_1,\ldots,r_s}$ is a function of $n - s$ *extensive* and s *intensive* variables and replaces, information-wise, either (4.43) or (4.44), which allows one to speak of the *Legendre invariant structure of thermodynamics*.

4.3.4. Statistical Mechanics

This discipline has as a goal constructing *microscopic realizations* of thermodynamics. These are often referred to as "mechanical analogues" of thermodynamics,
and we will see below how to construct one of them on the basis of Fisher information. As we have remarked above, the first complete formal such microscopic
theory was built up by Gibbs (1902) [23,40], based on the incredibly imaginative
notion of the *ensemble*. One focuses attention not on the physical system itself, but
on a very large collection of copies of it, which exists only in our imagination [18].
According to Gibbs, this notion makes sense only at equilibrium [23].

In the late 1960s, Jaynes used information theory to reformulate statistical mechanics [8] on the basis of his celebrated maximum entropy principle
(MaxEnt) [9,10]. The new formulation has but one axiom, i.e., MaxEnt. Moreover,
no appeal is made either to the notion of ensemble or to that of equilibrium. The
latter concept gets defined in purely information terms as that very special situation in which the observer knows only the expectation value of a constant of the
motion.

4.3.5. Axioms of Information Theory

IT is a mathematical discipline founded by Shannon [43], which associates a
degree of knowledge (or ignorance) to any normalized probability distribution
$p(i)$, $(i = 1, \ldots, N)$, determined by a functional of the $\{p(i)\}$, called the information measure I.

Khinchin derived information theory from an appropriate axiomatics in 1952
[44], thereby formalizing Shannon's theory:

(1) I is a function *only* of the $p(i)$;
(2) I is an absolute maximum for the uniform probability distribution;
(3) I is not modified if an $(N + 1)$th event of probability zero is added;
(4) This last axiom establishes a composition rule for independent events. This
    postulate is not as simple and self-evident as the preceding ones, and thus
    deserves special comment. Consider two subsystems
    - $[\Sigma^1, \{p^1(i)\}]$ and
    - $[\Sigma^2, \{p^2(j)\}]$, of a total composite system
    - $[\Sigma, \{p(i, j)\}]$,
    such that

$$p(i, j) = p^1(i) \, p^2(j). \qquad (4.48)$$

Consider further the conditional PDF $Q(j|i)$ of the event j in system 2 for
fixed i in system 1. To this PDF, one associates the information measure $I[Q]$.
Clearly,

$$p(i, j) = p^1(i) \, Q(j|i), \qquad (4.49)$$

and then Khinchin postulates *4th Khinchin's axiom:*

$$I(p) = I(p^1) + \sum_i p^1(i) I(Q(j|i)). \tag{4.50}$$

Out of the four axioms, one finds that Shannon's measure

$$I_S\{p(i)\} = -\sum_{i=1}^{N} p(i) \ln [p(i)] \tag{4.51}$$

is unique, in the sense that it is the only one fulfilling them.

4.4. Modern Approaches to Statistical Mechanics

4.4.1. Summary

Here we enter the realm encompassing developments in statistical mechanics of the last 45 years, beginning with the work of Jaynes. The Legendre structure of Jaynes' formulation, and non-Shannon information measures, are also discussed.

4.4.2. Jaynes' Reformulation

Statistical mechanics and thereby thermodynamics can be formulated on the basis of information theory if a statistical operator $\hat\rho$ is known. In quantum mechanics, this operator carries all the information an observer may possess about the system at hand. In classical mechanics you have, instead of the statistical operator, a probability density distribution ρ over phase space. The question is how to determine $\hat\rho$ (or ρ). According to Jaynes [9, 10], these quantities are obtained by recourse to the maximum entropy principle (usually abbreviated as MaxEnt) [9,10]. We give here, for brevity's sake, only the quantum version:

MaxEnt: Assume that your prior knowledge about the system is given by the M expectation values $\langle A_1\rangle, \ldots, \langle A_M\rangle$ and you are dealing with an information measure $I_S[\hat\rho]$.

Axiom: Then, the statistical operator $\hat\rho$ is uniquely fixed by extremizing $I(\hat\rho)$, subject to the constraints given by the M conditions $\langle A_j\rangle = Tr[\hat\rho\,\hat A_j]$.

Here we are, as customary, introducing M Lagrange multipliers λ_i, intimately related to the $\langle A_i\rangle$. Only this axiom is needed, whereas the original Gibbs' formulation of statistical mechanics requires four postulates [23]. Additionally, since no reference to the ensemble notion (that only makes Gibbsean-sense at equilibrium) is made, Jaynes' approach also encompasses non-equilibrium situations, while the Gibbs' formulation is unable to cope with these instances. In the process of reformulating à la Jaynes statistical mechanics, one discovers that $I_S \equiv S$ [9, 10], the equilibrium Gibbs' entropy, if our prior knowledge (the $\langle A_i\rangle$) refers to *extensive* quantities. However, the Jaynes' entropy retains its validity in off-equilibrium instances also. The MaxEnt-Shannon measure $I_S = I_{MaxEnt}$, once determined, *yields complete thermodynamic information with respect to the system of interest* [9, 10].

4.4.3. Legendre Structure in Jaynes' Formulation

For simplicity, we will employ only a classical notation. The MaxEnt PDF reads [9, 10]

$$f(x) = \exp\left\{-\left[\Omega + \sum_{i=1}^{M} \lambda_i A_i(x)\right]\right\} \tag{4.52}$$

$$\ln[f(x)] = -\left[\Omega + \sum_{i=1}^{M} \lambda_i A_i(x)\right], \tag{4.53}$$

where, for purposes of normalization,

$$\Omega(\lambda_1, \ldots, \lambda_M) = \ln\left[\int dx \left(\exp\left[\sum_{i=1}^{M} \lambda_i A_i(x)\right]\right)\right]. \tag{4.54}$$

Also, since

$$[\partial\Omega(\lambda_1, \ldots, \lambda_M)/\partial\lambda_j] = -\langle A_j \rangle, \quad j = 1, \ldots, M \tag{4.55}$$

holds, one derives immediately the important relationship [cf. 45]

$$I_S = -\langle f \ln f \rangle = \Omega + \sum_{i=1}^{M} \lambda_i \langle A_i \rangle. \tag{4.56}$$

One easily derives from the latter the so-called Euler theorem [9, 10],

$$\left[\frac{\partial I_S}{\partial\lambda_i}\right] = \sum_{k} \lambda_k \left[\frac{\partial\langle A_k \rangle}{\partial\lambda_i}\right]. \tag{4.57}$$

It also follows that

$$dI_S = \sum_{i=1}^{M} \lambda_i \, d\langle A_i \rangle, \tag{4.58}$$

leading to

$$[\partial I_S/\partial\langle A_i \rangle] = \lambda_i \quad \text{with } I_S = I_S(\langle A_1 \rangle \cdots \langle A_M \rangle). \tag{4.59}$$

The Legendre transform of I_S is [9, 10]

$$\Omega = \Omega(\lambda_1, \ldots, \lambda_M) = I_S - \sum_{i=1}^{M} \lambda_i \langle A_i \rangle \left[\frac{\partial I_S}{\partial}\langle A_j \rangle\right]. \tag{4.60}$$

This verifies another quite important relation, already derived above, namely,

$$[\partial\Omega/\partial\lambda_j] = -\langle A_j \rangle, \quad j = 1, \ldots, M. \tag{4.61}$$

The above equations neatly express the Legendre structure of thermodynamics.

Unfortunately, the paradigm based on the I_S measure seems to be inadequate to deal with many interesting physical scenarios (self-gravitating systems are typical ones [16], see also [17]). Therefore alternative information measures to Shannnon's are called for.

4.4.4. Non-Shannon Information Measures

The theory of dynamical systems and information theory were connected by Kolmogorov and Sinai [46, 47].

Subsequent investigations have categorically proved the usefulness for discussing dynamical systems of information measures different from that of Shannon's [48], which depend upon a real parameter q.

One of them was the cause of great activity during the 1990s, and has not abated yet: *Tsallis measure* [2, 3, 49–55]

$$I_T\{p(i), q\} \equiv \frac{\left(\sum_{i=1}^{N} p(i)\left[1 - p(i)^{q_1}\right]\right)}{q_1}. \tag{4.62}$$

Here $p(i)$ is the probability law of a system, q is a parameter, and $q_1 \equiv q - 1$, and it is to be stressed that $I_T\{p(i), 1\} = I_S$, that of Shannon, in the sense of a $\lim q \to 1$. Information I_T has proved to be an appropriate measure for many physical systems [49–55].

An important property of I_T is that it is *nonextensive*. Let two independent systems A and B be governed by, respectively, the probability laws $\{p_A(i)\}$ and $\{p_B(j)\}$. Then, using the abbreviation $I_T^{(q)}(p_A, p_B) \equiv I_T^{(q)}(A, B)$, we have

$$I_T^{(q)}(A, B) = I_T^{(q)}(A) + I_T^{(q)}(B) + (1 - q)I_T^{(q)}(A)I_T^{(q)}(B). \tag{4.63}$$

As $q \to 1$, additivity is approached in the preceding equation. Thus, parameter q measures the degree of non-additivity (also called "nonextensivity") of the two systems. The equation is also equivalently a *composition* relation that replaces Khinchin's 4th postulate [52].

4.4.5. Legendre Structure Preserved by a Change of Measure $I_S \to I_T$

A.R. Plastino and A. Plastino were able to show [56] that the Legendre structure of the preceding paragraphs is preserved if, in Jaynes' information theory approach to statistical mechanics, one replaces I_S by I_T. The ensuing Tsallis thermostatistics is nowadays being hailed as the possible basis of a theoretical framework appropriate to deal with nonextensive settings. The recent application of I_T to an increasing number of physical problems is beginning to provide a picture of the kind of scenarios where the new formalism is useful, in addition to that of self-gravitating systems. We may cite, for example, the two-dimensional pure electron plasma, or the behaviors of (i) dissipative low-dimensional chaotic systems, (ii) self organized critical systems, and many others [49]. Tsallis entropy has also been advanced as the basis of a thermostatistical foundation for all *power law* phenomena [49] (cf. Chapter 8, for the corresponding role played by Fisher information in power law phenomena).

4.4.6. Still More General Measures

The preceding result can be still generalized further, as shown in [56]. Indeed, it holds for *any* reasonable global information measure [57]. Global nature implies that the measure depends only on the probability distribution, but not on its derivatives. Local measures, as Fisher's, are a different story. For a given probability distribution $\rho(x)$, the most general global information measure I_G is an arbitrary function $f(\rho)$ such that

$$I_G = \int dx f(\rho)$$

possesses a definite concavity property. As always in our work, the prior information is assumed to take the form

$$\langle A_i \rangle = \int dx \rho A_i(x), \quad i = 1, \ldots, M.$$

The MaxEnt variational problem now reads

$$\delta_\rho \left[\int dx (f(\rho) - \alpha\rho - \sum_i \lambda_i A_i \rho) \right] = 0,$$

and leads to a density ρ^* that maximizes our information measure, subject to the pertinent constraints, which must obey the relation

$$f'(\rho^*) = \alpha + \sum_i \lambda_i A_i, \tag{4.64}$$

where prime denotes derivatives, as usual. We need now the concept of "inverse function" $g \equiv y^{-1}$ of a given function $y(x)$, such that $(gy)(x) = (yg)(x) \equiv (1)x = x$; $\forall x$, where the "strange" multiplicative function "1" makes its appearance $(1z = z; \ \forall z)$. Thus, $y = f(x) \Rightarrow x = g(y)$. For instance, let $y = \ln(x)$. Then, $g = \exp(x)$. Of course, $\ln[\exp(x)] = 1x = x$ and $x = \exp(y)$. In the present situation we need the inverse function g of the first derivative f' of f,

$$g \equiv [f'(\rho^*)]^{-1} \Rightarrow (f'g)(\rho^*) = (gf')(\rho^*) = 1(\rho^*) = \rho^*, \quad \rho^* = g(f'). \tag{4.65}$$

For example, if one chooses the entropy measure $f(x) \equiv x \ln x$, then $g(x) \equiv f'(x) = 1 + \ln x$ so that $g(f') = \exp(f'(\rho^*) - 1)$. As a check, then the operation of applying $g \equiv f'^{-1}$ to ρ^* reads as follows: $[f'(\rho^*)]^{-1}([f'(\rho^*)]) = \exp[f'(\rho^*) - 1] = \exp[1 + \ln\rho^* - 1] = \rho^*$, QED.

Returning to the general problem, we now have (cf. Eq. (4.64)) as a solution for the PDF,

$$\rho^* = g(f') = g\left(\alpha + \sum_i \lambda_i A_i\right),$$

where function g is known from before as the inverse function $[f'(\rho^*)]^{-1}$. Also, as a consequence, our general measure adopts the appearance

$$I_G = \int dx f(\rho^*) = \int dx f\left\{g\left(\alpha + \sum_i \lambda_i A_i\right)\right\},$$

with f and g known. Also, the associated Legendre transform is here

$$L(\{\lambda_i\}) = I_G - \sum_i \lambda_i \langle A_i \rangle,$$

and one easily ascertains that [21]

(1) $\partial I_G / \partial \langle A_i \rangle = \lambda_i$,
(2) $\partial L / \partial \lambda_i = -\langle A_i \rangle$, the Euler theorem, and
(3) $\partial I_G / \partial \lambda_i = \sum_k \lambda_k [\partial \langle A_k \rangle / \partial \lambda_i]$.

Again, we reobtain our all-important Legendre structure. Remember that f remains unspecified! What we have shown is only that, whatever this form may be (within reasonable limits), we are sure to recover the Legendre structure. If one wishes for a definite ρ^*, f has to be made explicit before. The celebrated Boltzmann form for ρ^* is the correct one for systems characterized by either (1) short-range interactions, (2) very short memories, and/or (3) physical configuration spaces in which any region is (in principle) accessible [2, 3]. There are many instances, though, in which one of these conditions fails to be fulfilled. There is ample evidence that, in such cases, the Boltzmann distribution is not correct [2, 3]. Historically, the first example in which an alternative, more correct form was encountered (for a physical situation) is by Plastino and Plastino [51].

4.5. Fisher Thermodynamics

4.5.1. Summary

FIM belongs to a fundamentally *different* family of information measures than those just discussed. The latter depend only on the probability distribution (according to Khinchin's tenets), and are therefore *global* measures. FIM, instead, (1) depends on *derivatives* of the probability distribution as well, and (2) is consequently a *local* measure (Section 1.2.2.12 or [17]). We now show that FIM minimization is equivalent to MaxEnt as a foundation for thermodynamics. An epistemological viewpoint is taken.

4.5.2. FIM Concavity and Second Law

From hereon, for simplicity we denote the stochastic variable entering Fisher's measure as a generic variable x. This will later become a *Fisher* variable, in particular. As stated already many times, in dealing with such measure, we are interested only in the special case of *translation families*: monoparametric families of distributions of the form $g(x - \mu)$, which are known up to a shift parameter μ. All members of the family possess identical shape, and the Fisher measure I_τ has the form Eq. (1.50) Chapter 1,

$$I_\tau[g(x)] = \int dx \left[\frac{1}{g(x)} \right] \left\{ \left[\frac{\partial g}{\partial x} \right] \right\}^2. \tag{4.66}$$

This measure is intimately related to many of the fundamental equations of theoretical physics. On the basis of such a measure, Frieden, Plastino, Plastino, and Soffer have constructed an à la Fisher thermodynamics [12–14, 25]. First of all, they showed in [25] that I_τ exhibits an all-important *concavity* property, that guarantees the stability of matter. If we let

$$P = p_1 + p_2, \tag{4.67}$$

the concavity property can be expressed as

$$aI_\tau(p_1) + bI_\tau(p_2) \geq I_\tau(ap_1 + bp_2) = I_\tau(P), \tag{4.68}$$

which is indeed obeyed by the FIM [17], [25]. The second law of thermodynamics is also Fisher-reproduced because, as predicted by Frieden [58] and explicitly demonstrated by Casas, Plastino, and Plastino, I_τ yields an arrow of time, indeed a much stronger one than as provided by Shannon's measure [45], namely, $(dI/dt) \leq 0$. This is so because the Fisher arrow works for a wide class of dynamical systems, while Shannon's does so only for Hamiltonian ones, a very special and restricted, although quite important, class of systems [45].

4.5.3. Minimizing FIM Leads to a Schrodinger-Like Equation

Following [25], let the relevant PDF be of the form

$$f(x) \equiv |\psi|^2, \tag{4.69}$$

that is, we express it in terms of an amplitude ψ, with

$$\int dx f(x) = 1 \tag{4.70}$$

and x denoting the relevant variable at hand (a velocity, for instance). As an example of the epistemological viewpoint, we assume that we are in possession of a prior information in the form of the expectation values of M extensive quantities

$$\langle A_1 \rangle, \ldots, \langle A_M \rangle. \tag{4.71}$$

These have been "learned" by their observation as mean values

$$\langle A_i \rangle = \int dx\, A_i(x) f(x). \tag{4.72}$$

In this notation, the associated Fisher measure (let us call it simply I in this subsection) for translational families has the form (4.66)

$$I[f] = \int dx f(x) \left(\frac{[\partial f/\partial x]}{[f(x)]} \right)^2. \tag{4.73}$$

We wish to minimize I subject to the constraints posed by our prior information (4.71) (M Lagrange multipliers λ_i) plus normalization $\langle 1 \rangle = 1$ (Lagrange multiplier α). Following the procedure of Section 1.1.2.4 (Chapter 1), the problem becomes

$$\delta_f \left\{ I[f] - \left[\sum_{i=1}^{M} \lambda_i \langle A_i(x) \rangle \right] - \alpha \langle 1 \rangle \right\} = 0. \tag{4.74}$$

This entails dealing with the Lagrangian

$$\mathcal{L} = \left[\frac{f'^2}{f} \right] - \left[\sum_{i=1}^{M} \lambda_i A_i f \right] - \alpha f \tag{4.75}$$

and with its associated Euler–Lagrange equations

$$\{\partial/\partial x\}[\partial \mathcal{L}/\partial f'] = [\partial \mathcal{L}/\partial f]. \tag{4.76}$$

We are then straightforwardly led to [25]

$$\left\{ \left[\frac{\partial \ln (f)}{\partial x} \right] \right\}^2 + 2 \left(\left[\frac{\partial^2 \ln (f)}{\partial x^2} \right] \right) + \alpha + \sum_{i=1}^{M} \lambda_i A_i = 0. \tag{4.77}$$

It is easy to see now that

$$\left[\frac{\partial I}{\partial \lambda_i} \right] = \left[\frac{\partial}{\partial \lambda_i} \right] \int dx \left(\frac{f'^2}{f} \right) = \sum_{k}^{M} \lambda_k \left[\frac{\partial \langle A_k \rangle}{\partial \lambda_i} \right], \tag{4.78}$$

which is the Euler theorem that we have already encountered above. Additionally, we have, as an immediate result of our variational problem,

$$\left[\frac{\partial I}{\partial \langle A_j \rangle} \right] = \sum_{i}^{M} \left[\frac{\partial I}{\partial \lambda_i} \right] \left[\frac{\partial \lambda_i}{\partial \langle A_j \rangle} \right], \tag{4.79}$$

so that

$$\left[\frac{\partial I}{\partial \langle A_j \rangle} \right] = \sum_{i}^{M} \sum_{k}^{M} \lambda_k \left[\frac{\partial \langle A_k \rangle}{\partial \lambda_i} \right] \left[\frac{\partial \lambda_i}{\partial \langle A_j \rangle} \right] \left[\frac{\partial I}{\partial \langle A_j \rangle} \right], \tag{4.80}$$

and, finally,

$$\left[\frac{\partial I}{\partial \langle A_j \rangle} \right] = \sum_{k}^{M} \lambda_k \sum_{i}^{M} \left[\frac{\partial \langle A_k \rangle}{\partial \lambda_i} \right] \left[\frac{\partial \lambda_i}{\partial \langle A_j \rangle} \right] = \lambda_j. \tag{4.81}$$

As for f, the Euler–Lagrange treatment leads to the equation

$$\left[\frac{\partial \ln (f)}{\partial x}\right]^2 + 2\left(\left[\frac{\partial^2 \ln (f)}{\partial x^2}\right]\right) + \alpha + \sum_{i=1}^{M} \lambda_i A_i = 0, \qquad (4.82)$$

which, in terms of the amplitude ψ takes the form [25]

$$-\left(\frac{1}{2}\right)\psi''(x) - \left(\frac{1}{8}\right)\left[\sum_{i=1}^{M} \lambda_i \langle A_i(x)\rangle\right]\psi(x) = \left(\frac{\alpha}{8}\right)\psi(x). \qquad (4.83)$$

This is a Schrodinger-like equation, with an "information potential"

$$U(x) = -\left[\frac{1}{8}\right]\left[\sum_{i=1}^{M} \lambda_i \langle A_i(x)\rangle\right], \qquad (4.84)$$

and an "energy" eigenvalue equal to $\alpha/8$.

4.5.4. FIM Legendre Transform Structure

Does FIM possess a Legendre structure? In order to answer this important question we need first of all the Fisher-Legendre transform $L(\lambda_1, \ldots, \lambda_M)$,

$$L(\lambda_1, \ldots, \lambda_M) = I(\langle A_1\rangle, \ldots, \langle A_M\rangle) - \left[\sum_{i=1}^{M} \lambda_i \langle A_i\rangle\right]. \qquad (4.85)$$

This has the property

$$\left(\frac{\partial L}{\partial \lambda_k}\right) = \sum_i \left\{\left[\frac{\partial I}{\partial \langle A_i\rangle}\right]\left[\frac{\partial \langle A_i\rangle}{\lambda_k}\right] - \lambda_i \left[\frac{\partial \langle A_i\rangle}{\partial \lambda_k}\right]\right\} - \langle A_k\rangle, \qquad (4.86)$$

i.e.,

$$\left(\frac{\partial L}{\partial \lambda_k}\right) = -\langle A_k\rangle. \qquad (4.87)$$

From the above equation, together with (4.81), we see that with I, L, and the $\langle A_i\rangle$, we can indeed construct a Legendre structure, which is thermodynamics' essential feature. This enables us to speak of a Fisher-thermodynamics arising from a single, information-theoretic axiom: I-minimization with appropriate constraints. The following comparison is of didactic value.

4.5.5. Shannon's S vs Fisher's I

We now *compare* corresponding structures of the conventional entropy (Shannon measure) and the Fisher measure. Using the notation $I_S \equiv S$ [40, 42] and FIM $\equiv I_\tau$, these depend upon prior knowledge as

$$S = S(\langle A_1\rangle, \ldots, \langle A_M\rangle) \;||||||\; I_\tau = I_\tau(\langle A_1\rangle, \ldots, \langle A_M\rangle). \qquad (4.88)$$

Also, regarding Legendre transform structure, with Ω the Legendre transform of S,

$$S = \Omega + \sum_{i=1}^{M} \lambda_i \langle A_i \rangle \;\;\|\|\|\|\|\; I_\tau = L + \sum_{i=1}^{M} \lambda_i \langle A_i \rangle. \tag{4.89}$$

As for differential changes

$$dS = \sum_{i=1}^{M} \lambda_i \, d\langle A_i \rangle \;\Rightarrow\; \left[\frac{\partial S}{\partial \langle A_i \rangle} \right] = \lambda_i, \tag{4.90}$$

while

$$dI_\tau = \sum_{i=1}^{M} \lambda_i \, d\langle A_i \rangle \;\Rightarrow\; \left[\frac{\partial I_\tau}{\partial \langle A_i \rangle} \right] = \lambda_i. \tag{4.91}$$

Further,

$$\Omega = \Omega(\lambda_1, \dots, \lambda_M) = S - \sum_{i=1}^{M} \lambda_i \langle A_i \rangle, \tag{4.92}$$

and

$$L = L(\lambda_1, \dots, \lambda_M) = I_\tau - \sum_{i=1}^{M} \lambda_i \langle A_i \rangle, \tag{4.93}$$

Finally, for all $j = 1, \dots, M$, we have

$$[\partial S / \partial \langle A_j \rangle] = \lambda_j \quad \text{and} \quad [\partial \Omega / \partial \lambda_j] = -\langle A_j \rangle, \tag{4.94}$$

$$[\partial I_\tau / \partial \langle A_j \rangle] = \lambda_j \quad \text{and} \quad [\partial L / \partial \lambda_j] = -\langle A_j \rangle. \tag{4.95}$$

The complete parallelism of the $S-$ and $I_\tau-$ structures is apparent.

4.5.6. Discussion

First of all, note that nothing prevents us, in the Fisher case, to consider that the expectation values $\langle A_i \rangle_t$ have been gathered *at the time t*, as made explicit by the attached subindex. This seemingly innocent fact will later open the door for discussing time evolution.

It is clear that the main result of the preceding subsections is that of establishing, via Fisher's I_τ, a "connection" between thermodynamics and a Schrodinger-like equation. It results that both thermodynamics and quantum mechanics can be expressed by recourse to a formal SWE, out of a common informational basis. The physical meaning of this SWE is quite flexible. In the Fisher instance its "potential function" $U(x)$ originates in prior data $\langle A_k \rangle_t$, to be chosen according to the application under consideration. The $\langle A_k \rangle_t$ are introduced into the theory as *empirical* inputs gathered at the time t. This time-"liberty" can be construed as regarding both non-equilibrium and equilibrium thermodynamics as obtained from the constrained Fisher extremizing process, whose output is a Schrodinger-like wave equation (see the following subsection for the pertinent details concerning this temporal aspect). Any SWE has multiple solutions. Equilibrium thermodynamics would correspond to the ground-state (gs) solution [25,58]. We discuss below the possibility that non-equilibrium thermodynamics would correspond to admixtures of this gs with excited state solutions [12–14]. One should emphasize, summing up, that

- No specific potential has been assumed, as is appropriate for a generalized view of thermodynamics.
- The specific $A_k(x)$ one is to use depends upon the nature of the physical application one is envisioning; that could be of either a classical or a quantum character.
- Our Schrodinger-like equation poses a boundary value problem, generally with multiple solutions, in contrast with the *unique* solution attached to the Gibbs–Shannon entropy. This ultimately traces from the *local* nature of the FIM as compared with the global nature of Shannon entropy, as mentioned previously.
- The solution leading to the lowest I-value is the equilibrium one [25]. As will be discussed below, admixtures of excited solutions yield non-equilibrium states [12–14].

Our strategy below for discussing time evolution is to show that the SWE-treatment is equivalent to the well-known method of moments (MOM) advanced by Grad for tackling the Boltzmann transport equation (the concomitant literature is vast. See, for instance [59–63]).

More explicitly, our Fisher measure I_τ is now built up, more generally, with a PDF of space *and time*,

$$p = |\psi|^2, \quad \psi \equiv \psi(x|t). \tag{4.96}$$

We further the possibility of accommodating a temporal variable in our formalism by writing

$$I(t) \equiv I_\tau(t) = \int dx \left[\frac{(\partial p/\partial x)^2}{p} \right], \quad \text{for } p = p(x|t). \tag{4.97}$$

Thus, p is a positive-definite, normalized PDF $p(x|t)$, evaluated at (conditional upon) the time t. Also, our a priori known mean values are to be now regarded as having been measured *at the time t*,

$$\langle A_k \rangle_t = \int dx\, A_k(x)\, p(x|t), \quad k = 1, \ldots, M. \tag{4.98}$$

As we have seen above, the identification (4.96) then leads one to the equation

$$-\left(\frac{1}{2}\right) \psi'' - \left(\frac{1}{8}\right) \sum_k^M \lambda_k(t) A_k\, \psi = \frac{\alpha\psi}{8}, \tag{4.99}$$

where $\alpha/8$ plays the role of an energy eigenvalue and

$$\sum_k \lambda_k A_k(x) \tag{4.100}$$

forms an effective, time-dependent potential function

$$U = \left(\frac{1}{8}\right) \sum_{k}^{M} \lambda_k(t) A_k, \quad U = U(x, t). \tag{4.101}$$

The solution leading to the lowest I-value is the equilibrium one [25,58]. Now we ask: Can we choose other "excited" solutions and from them describe a meaningful time evolution?

4.6. The Grad Approach in Ten Steps

4.6.1. Summary

The approach of Grad [59–63] is a hallmark approach to computing non-equilibrium gas dynamics. It is described succinctly in the following.

4.6.2. Steps of the Approach

We appeal now, for an answer, to the Grad technique for dealing with Boltzmann's celebrated transport equation for gases so as to build up non-equilibrium solutions for it, the equilibrium solution being the Maxwell–Boltzmann probability distribution. One is to focus attention on the non-equilibrium state of a gas and proceed to derive a solution, the MOM-Grad one [59–63], in the 10 steps enumerated below:

- Consider a *fixed* time t, large compared to the time of initial randomization.
- Additionally, t should be small compared to the macroscopic relaxation time τ^* for attaining an overall situation ruled by the Maxwell–Boltzmann law f_0 on velocities.
- Local equilibrium: At each point of the vessel containing the gas, one assumes that a state of affairs arises that is close to the *local* equilibrium state described by $f_0 =$ Maxwell–Boltzmann's law on velocities. The effect of collisions is not neglected by this assumption. On the contrary, it is regarded as very important, since one considers that collisions quickly restore local equilibrium. However, this effect is not explicitly used, since no appeal to scattering cross-sections is made.
- The local equilibrium hypothesis allows for expanding the non-equilibrium distribution $f(x|t)$ we are looking for as

$$f(x|t) = f_0 + \epsilon \xi(x, t), \tag{4.102}$$

where $\epsilon \ll 1$. One is to find the function ξ.
- The unknown function $\xi(x, t)$ is itself expanded as a series of (orthogonal) Hermite–Gaussian (H–G) polynomials $H_i(x)$ with coefficients $a_i(t)$ at the fixed time t, so that one is entitled to write

$$f(x, t) = \sum_i a_i(t) H_i(x), \tag{4.103}$$

- H–G polynomials are orthogonal with respect to a Gaussian kernel (the *equilibrium distribution*). They constitute indeed the only set of functions, orthogonal and complete with respect to a Gaussian kernel function.
- Because of the orthogonality, the unknown coefficients $a_i(t)$ relate linearly to appropriate (unknown) moments of f over velocity space (x-space).
- Substituting the expansion (4.103) for f into the transport equation and integrating over all velocities yields now a set of first-order differential equations in the moments (which are generally a function of the fixed time value t).
- These are now solvable subject to *known initial conditions*, like our expectation values. The moments now become known (*including any time dependence*).
- As a consequence, the coefficients $a_i(t)$ of

$$f(x, t) = \sum_i a_i(t) H_i(x) \qquad (4.104)$$

are also known, which gives f at t. According to Grad, the solution of the ensuing system of equations would be equivalent to the exact solution of Boltzmann's equation, *if only enough a priori information were available*.

We show below that these 10 steps are equivalently accomplished by dealing with our SWE.

4.7. Connecting Excited Solutions of the Fisher-SWE to Nonequilibrium Thermodynamics

4.7.1. Summary

It is shown that non-equilibrium states of a classical system of particles are described by higher, "excited-state" solutions of the Fisher-SWE equation. This is complete analogy with the mechanics of a quantum system, whose excited states are likewise described by the higher-state solutions of its SWE. Applications to the phenomenon of viscosity are made and compared with corresponding solutions by Grad's approach.

4.7.2. Establishing the Connection

The Fisher-based SWE solutions $\psi_n(x, t)$ for $n > 0$, i.e., the excited solutions, can always be cast as a superposition of H–G polynomials, because these provide us with a Hilbert-space basis. For convenience, it is better to slightly change notation here with respect to the one used in the preceding paragraphs. H–G polynomials, that we will call now $\mathcal{H}_i(x)$, are the product of a fixed Gaussian function ϕ_0 of x times a Hermite polynomial $H_i(x)$. Thus, $\mathcal{H}_i(x) = \phi(x) H_i(x)$ and

$$\psi_n(x, t) = \sum_i b_{in}(t) \mathcal{H}_i(x) = \phi_0 \sum_i b_{in}(t) H_i(x), \qquad (4.105)$$

where, fortunately enough, ϕ_0 turns out to coincide with the ground state wave function of the harmonic oscillator.

The number of coefficients $b_{ni}(t)$ depends upon how far from equilibrium we are. At equilibrium there is only one such coefficient (that for the lowestorder H–G). The Fisher coefficients $b_{in}(t)$ are computed at the fixed time t at which our input data $\langle A_k \rangle_t$ are collected. Our functions $\psi_n(x, t)$ are to be connected to the Grad distribution $f(x, t)$ of the preceding subsection via the squaring operation $|\psi_n(x, t)|^2$. Notice that the square of an expansion in H–G polynomials is likewise a superposition of such polynomials, with coefficients $c_{in}(t)$:

$$\psi_n^2(x, t) = \phi_0 \sum_i c_{in}(t) \, H_i(x), \quad n = 1, 2, \ldots. \qquad (4.106)$$

We argue now that, for fixed n, the Grad coefficients $a_i(t)$ of (4.103) and our $c_{in}(t)$ are equal. Why?

1) The Grad coefficients are certainly *computed* at a *fixed* time t. Their momenta are evaluated at that time. Likewise ours (the $\langle A_k \rangle$) can be regarded as velocity momenta at that time as well.
2) The difference between the Grad coefficients and the SWE ones is one of physical origin, as follows:
   - While Grad *solves for* the velocity moments at the fixed time t and the ensuing M_{Grad} moments are computed using the Grad's a_i,
   - we, instead, collect *as experimental inputs* these velocity moments (at the fixed time t).
3) Thus, if the M_{Grad} moments *coincide with our experimental inputs*, necessarily the $a_i(t)$ and the $c_{in}(t)$ have to coincide as well.
4) The Grad moments at the time t are physically correct by construction, since one *solves for them via use of the Boltzmann transport equation*. The premise of our Fisher approach is that its input constraints are correct: They come *from experiment*.
5) Thus, the Fisher approach necessarily gives exactly the same solutions *at the fixed (but arbitrary) time t as does the Grad approach* and, for fixed n, our $c_{in}(t)$ have to coincide with Grad's $a_i(t)$ and our $p(x|t) = |\psi|^2$ with Grad's $f(x|t)$.

In [12–14] explicit examples are given that validate the line of reasoning expounded above. We discuss one of them in some detail below, in order to provide an illustration of the somewhat abstruse formalism above expounded.

4.7.3. Application: Viscosity

4.7.3.1. Boltzmann Equation in the Relaxation Approximation

In this viscosity example we closely follow the work of [13] and focus attention upon the celebrated transport equation of Boltzmann's [40] by considering a three-dimensional gas in which the effect of molecular collisions is always to restore a *local equilibrium* situation described by a very special PDF that depends on the positions and velocities $\mathbf{r}$, and $\mathbf{v}$, respectively, namely, the Maxwell–Boltzmann PDF $f_0(\mathbf{r}, \mathbf{v})$ [40]. In other words, we assume that

(1) if the molecular distribution is disturbed from the local equilibrium so that the actual PDF f is different from f_0, then
(2) the effect of collisions is simply to restore f to the local equilibrium value f_0 exponentially, with a relaxation time τ of the order of the average time between molecular collisions.

In symbols, for fixed $\mathbf{r}$, $\mathbf{v}$, f changes as a result of collisions according to

$$f(t) = f_0 + [f - f_0] \exp(-t/\tau), \tag{4.107}$$

with t the time and τ an appropriate relaxation constant. In these conditions, the ensuing Boltzmann equation for f becomes [40]

$$\frac{\partial f}{\partial t} + \sum_{i=1}^{3} \left[v_i \frac{\partial f}{\partial x_i} + \dot{v}_i \frac{\partial f}{\partial v_i} \right] = -\frac{f - f_0}{\tau}, \tag{4.108}$$

a *linear* differential equation for f. We pay attention now to a situation slightly removed from equilibrium: $f = f_0 + f_1$ with $f_1 \ll f_0$, so that (4.108) becomes

$$\frac{\partial f}{\partial t} + \sum_{i=1}^{3} \left[v_i \frac{\partial f}{\partial x_i} + \dot{v}_i \frac{\partial f}{\partial v_i} \right] = -\frac{f_1}{\tau}. \tag{4.109}$$

The left-hand side of (4.109) is small, since the right-hand side is, by definition, small. As a consequence, we can evaluate it by neglecting terms in f_1 and write

$$\frac{\partial f_0}{\partial t} + \sum_{i=1}^{3} \left[v_i \frac{\partial f_0}{\partial x_i} + \dot{v}_i \frac{\partial f_0}{\partial v_i} \right] = -\frac{f_1}{\tau}. \tag{4.110}$$

Since f_0 is the Maxwell–Boltzmann PDF, independent of time ($\partial f_0/\partial t = 0$), we finally get the so-called Boltzmann equation in the relaxation approximation [40]

$$\sum_{i=1}^{3} \left[v_i \frac{\partial f_0}{\partial x_i} + \dot{v}_i \frac{\partial f_0}{\partial v_i} \right] = -\frac{f_1}{\tau}. \tag{4.111}$$

4.7.3.2. Generalities on Viscosity

Imagine, in a gas, a plane with its normal pointing along the z-direction. The fluid below this plane exerts a mean force per unit area (stress) $\mathbf{P}_z$ on the fluid above the plane. Conversely, the gas above the plane exerts a stress $-\mathbf{P}_z$ on the fluid below the plane. The z-component of $\mathbf{P}_z$ measures the mean pressure $\langle p \rangle$ in the fluid, i.e., $P_{zz} = \langle p \rangle$. When the fluid is in equilibrium (at rest or moving with uniform velocity throughout), then $P_{zz} = 0$ [40]. Consider a non-equilibrium situation in which the gas does not move with uniform velocity throughout. In particular, imagine that the fluid has a constant (in time) mean velocity u_x in the x-direction such that $u_x = u_x(z)$. For specific examples see, for instance, [40]. Now any layer of fluid below a plane $z = constant$ will exert a tangential stress P_{zx} on the fluid above it. If $\partial u_z/\partial z$ is small, one has [40]

$$P_{zx} = -\eta \frac{\partial u_z}{\partial z}, \tag{4.112}$$

where η is called the viscosity coefficient. This phenomenon was first investigated by Maxwell, who showed that, for a dilute gas of particles of mass m moving with mean velocity $\langle v \rangle$,

$$\eta \propto n\langle v \rangle ml, \tag{4.113}$$

where n is the number of molecules per unit volume and l is the mean free path [40]. Now consider any quantity $\chi(\mathbf{r}, t)$ whose mean value is

$$\langle \chi(\mathbf{r}, t) \rangle = \frac{1}{n(\mathbf{r}, t)} \int d^3 v f(\mathbf{r}, \mathbf{v}, t) \chi(\mathbf{r}, t), \tag{4.114}$$

with $n(\mathbf{r}, t)$ the mean number of particles, irrespective of velocity, which at time t are located between $\mathbf{r}$ and $\mathbf{r} + d\,\mathbf{r}$. If $\chi(\mathbf{r}, t) \equiv \mathbf{v}(\mathbf{r}, t)$, the above relation yields the mean velocity $\mathbf{u}(\mathbf{r}, t)$ of a molecule located near $\mathbf{r}$ at time t. Velocity $\mathbf{u}(\mathbf{r}, t)$ is the mean velocity of a flow of gas at a given point, i.e., the (macroscopic) hydrodynamical velocity. The particular velocity $\mathbf{U}$ of a molecule relative to the gas is defined as [40]

$$\mathbf{U} = \mathbf{v} - \mathbf{u}, \tag{4.115}$$

so that

$$\langle \mathbf{U} \rangle = 0. \tag{4.116}$$

If one is interested in transport properties, the fluxes of various quantities become the focus of attention. Consider the net amount of the quantity χ that is transported

(1) per unit time and
(2) per unit area *of an element of area oriented along* $\hat{\mathbf{n}}$

by molecules with velocity $\mathbf{U}$ due to their random movement back and forth across this element of area. The χ-associated flux $\mathcal{F}_n$ generated in this way is

$$\mathcal{F}_n(\mathbf{r}, t) = \int d^3 v\, f(\mathbf{r}, \mathbf{v}, t) \, [\hat{\mathbf{n}} \cdot \mathbf{U}] \, \chi(\mathbf{r}, t) = n \, \langle [\hat{\mathbf{n}} \cdot \mathbf{U}] \chi \rangle. \tag{4.117}$$

Something apparently wrong with what appears to be a subscript n that is too far below the line. Also in line preceding.

For the present discussion we have $\chi = mv_x$ and $\hat{\mathbf{n}} \cdot \mathbf{U} = n\,U_z$. The ensuing flux gives then, precisely, P_{zx} [40]. Since u_x does not depend upon the velocity and $\langle u_x \rangle = 0$,

$$P_{zx} = nm\langle U_z v_x \rangle = nm\langle U_z[u_x + U_x] \rangle = nm\langle U_z U_x \rangle. \tag{4.118}$$

A simple phenomenological line of reasoning that utilizes the so-called path integral approximation yields then [40]

$$P_{zx} = -\frac{n\tau}{\beta}\frac{\partial u_x}{\partial z}, \quad \eta = \frac{n\tau}{\beta}, \tag{4.119}$$

where τ is the average time between molecular collisions (relaxation time) and $\beta = 1/kT$.

Assume now, as discussed above, that the effect of collisions is just to produce a local equilibrium distribution *relative* to the gas moving with a mean velocity u_x at the location of each collision. The relevant equilibrium Maxwell–Boltzmann PDF is

$$f_0(\mathbf{r}, \mathbf{v}, t) = g(U_x, U_y, U_z) = g(U)$$
$$U_x = v_x - u_x(z), \quad U_y = v_y, \quad U_z = v_z$$
$$g(U) = n\left[\frac{m\beta}{2\pi}\right]^{3/2} \exp\left[-\beta mU^2/2\right]. \tag{4.120}$$

This PDF satisfies Eq. (4.108). When a mean velocity gradient exists, so that u_x is such that its derivative with respect to z does not vanish, (4.120) no longer complies with (4.108). Since the situation is time-independent, the ensuing (new) PDF can not depend upon time, but will depend on z (the direction of the velocity gradient). There are no external forces, so that $\dot{\mathbf{v}}$ vanishes. As a consequence, (4.108) reduces to

$$v_z\frac{\partial f}{\partial z} = -\tau^{-1}(f - f_0). \tag{4.121}$$

One assumes that $\partial v_x/\partial z$ is small enough that $\partial f/\partial z$ is also small, so that

$$f = f_0 + f_1, \quad f_1 \ll f_0. \tag{4.122}$$

As a result, we find that

$$f_1 = -\tau v_z\frac{\partial f_0}{\partial z}. \tag{4.123}$$

It is clear from (4.120) that

$$\frac{\partial f_0}{\partial z} = \frac{\partial g}{\partial U_x}\frac{\partial U_x}{\partial z} = -\frac{\partial g}{\partial U_x}\frac{\partial u_x}{\partial z}, \tag{4.124}$$

while

$$\frac{\partial g}{\partial U_x} = -m\beta g U_x, \tag{4.125}$$

a relation that we will use below. Here, it will become clear that we need simply write

$$\frac{\partial f_0}{\partial z} = -\frac{\partial u_x}{\partial z}\frac{\partial g}{\partial U_x}, \tag{4.126}$$

so that

$$f_1 = \tau \, v_z \frac{\partial u_x}{\partial z} \frac{\partial g}{\partial U_x} = -m\beta \, \tau \, v_z \, U_x \frac{\partial u_x}{\partial z} \, f_0 \tag{4.127}$$

and, finally,

$$f = f_0 + \tau v_z \frac{\partial u_x}{\partial z} \frac{\partial g}{\partial U_x} = f_0 \left\{ 1 - m\beta \, \tau \, v_z \, U_x \frac{\partial u_x}{\partial z} \right\}. \tag{4.128}$$

Now, the zx-component of the stress is

$$P_{zx} = m \int d^3 v f \, U_x U_z. \tag{4.129}$$

As f_0 depends only on the absolute value of $\mathbf{U}$ the above integral vanishes if one replaces f by f_0 in the preceding integral by symmetry reasons. Thus,

$$P_{zx} = m \frac{\partial u_x}{\partial z} \int d^3 v v_z \tau \frac{\partial g}{\partial U_x} U_x U_z. \tag{4.130}$$

According to (4.114) $v_z = U_z$, so that, assuming that the relaxation time does not depend upon velocity [40]:

$$P_{zx} = m \, \tau \frac{\partial u_x}{\partial z} \int d^3 v \frac{\partial g}{\partial U_x} U_x \, U_z^2$$
$$= m \, \tau \frac{\partial u_x}{\partial z} \int \int dU_y \, dU_z U_z^2 \int dU_x \frac{\partial g}{\partial U_x} U_x. \tag{4.131}$$

Consider

$$I_{xx} = \int_{-\infty}^{\infty} dU_x \frac{\partial g}{\partial U_x} U_x. \tag{4.132}$$

Since $\partial g / \partial U_x = m\beta \, g$, we have

$$I_{xx} = -m\beta \int dU \, g \, U_x^2. \tag{4.133}$$

As a consequence, using the equipartition theorem [40]

$$P_{zx} = -m^2 \, \beta \, \tau \frac{\partial u_x}{\partial z} \int d^3 U f_0 \, U_z^2 U_x^2 = -m^2 \, \beta \, \tau \frac{\partial u_x}{\partial z} n \, \langle U_z^2 \rangle \, \langle U_x^2 \rangle \tag{4.134}$$

$$= -m^2 \, \beta \tau \frac{\partial u_x}{\partial z} n \left(\frac{kT}{m} \right)^2 = -\frac{\partial u_x}{\partial z} \frac{n \, \tau}{\beta}. \tag{4.135}$$

Summing up, our phenomenological results write:

1) for the coefficient of viscosity η

$$\eta = n \, \tau / \beta, \tag{4.136}$$

in agreement with (4.119), and, via (4.122) and (4.127), for the phenomenological distribution function f,

2) $$f = f_0 \left[1 - U_z U_x \tau m \beta \frac{\partial u_x}{\partial z} \right]. \tag{4.137}$$

We will try to reproduce them below with the two microscopic techniques under discussion here, i.e., Grad's and SWE-Fisher's.

4.7.4. Comparison with the Grad Treatment

Let us see now how the Grad approach is able to describe the above phenomenology via a treatment of the Boltzmann equation in the relaxation approximation. Recall that the first two Hermite polynomials are

$$H_0 = 1; \quad H_1 = \frac{1}{\sqrt{2}} \, 2x, \tag{4.137}$$

and, with

$$\phi(x, \omega) = \left[\frac{\omega}{\pi} \right]^{1/4} \exp\left[-x^2/2 \right], \tag{4.138}$$

the first two members of the Gauss–Hermite basis (of $\mathcal{L}^2$) are

$$\psi_0 = H_0 \phi; \quad \psi_1 = H_1 \phi. \tag{4.139}$$

Since we have for the z-component of our PDF (cf. Eq. (4.120))

$$n[m\beta/(2\pi)]^{1/2} \exp\left[-\beta m v_z^2/2 \right] = f_{0,z} = n \, \psi_0^2, \tag{4.140}$$

our variables x, ω in (4.137) and (4.138) translate as follows:

$$\omega = m\beta/2; \quad 2x = \sqrt{2\beta \, m} v_z, \tag{4.141}$$

which allows us to recast (4.137) as

$$H_0 = 1; \quad H_1 = \sqrt{\beta m} v_z. \tag{4.142}$$

However, we deal actually with a three-dimensional problem. The pertinent Gauss–Hermite basis is the set of functions

$$\left\{ \psi_0(v_x) \, \psi_0(v_y) \, \psi_0(v_z) \left[1 + \sum_{l,m,n} H_l(U_x) \, H_m(U_y) \, H_n(U_z) \right] \right\}, \tag{4.143}$$

where l, m, n run over all non-negative integers. Our data here is

$$P_{zx} = m \int d^3 v f U_z U_x, \tag{4.144}$$

and, according to (4.119),

$$P_{zx} = -\left(\frac{\tau n}{\beta} \right) \frac{\partial u_x}{\partial z} = \eta \frac{\partial u_x}{\partial z}. \tag{4.145}$$

In the present instance, in view of (4.144), the Grad recipe to find f should be

$$f(\mathbf{U}) = f_0(\mathbf{U})[1 + a\, H_1(U_x)\, H_1(U_z)] = f_0[1 + a\,\beta m\, U_x\, U_z], \qquad (4.146)$$

with the coefficient a to be determined from the relevant velocity-moment (4.144) and the prior knowledge expressed by (4.119). We thus evaluate (4.144) using the ansatz (4.146)

$$P_{zx} = m \int d^3 U \{ f_0[1 + a\beta m\, U_x\, U_z] \} U_z\, U_x. \qquad (4.147)$$

The integral $\int d^3 U \{ f_0\, U_x\, U_z \}$ vanishes by symmetry. Thus,

$$P_{zx} = a\, m^2\, \beta \int d^3 U \; \{ f_0\, U_x^2\, U_z^2 \}$$

$$= a\, m^2 \beta\, n \langle U_x^2 \rangle \langle U_z^2 \rangle = \frac{a\, m^2\, n\, \beta}{[\beta m]^2} = n\, \frac{a}{\beta}, \qquad (4.148)$$

where the equipartition theorem has been employed. Since (4.148) and (4.119) have to be equal,

$$a = -\tau\, \frac{\partial u_x}{\partial z}, \qquad (4.149)$$

and

$$f = f_0 \left[1 - U_x\, U_z \left(\tau\, \frac{\partial u_x}{\partial z}\, m\beta \right) \right], \qquad (4.150)$$

which is indeed identical to (4.7.3.2). Our Grad goal has been successfully achieved.

4.7.5. The Fisher Treatment of Viscosity

We first determine the ground state of our Fisher-SWE and, afterward, admixtures of it with excited states.

4.7.5.1. Ground State

We consider now that our prior knowledge is that provided by the equipartition result

$$\langle v_z^2 \rangle = \frac{1}{\beta m}, \qquad (4.151)$$

so that we have, for the z-component of the probability amplitude, the SWE

$$\psi_z'' / 2 + \lambda_1(t) \left(v_z^2 / 8 \right) \psi_z = -(\alpha/8)\psi_z. \qquad (4.152)$$

This is an SWE identical to that for the harmonic oscillator. We set

$$\lambda_1(t)/8 = \omega^2/2 \quad \text{and} \quad (\alpha/8) = -E, \qquad (4.153)$$

so that the problem becomes time-independent. Equation (4.151) entails, as discussed above, $\omega = \beta m/2$. Since the (Gaussian) ground state of our SWE reads

$$\psi_{0,z} = \left[\frac{\omega}{\pi}\right]^{1/4} \exp\left[-\omega v_z^2/2\right], \tag{4.154}$$

we obviously obtain then,

$$\psi_{0,z}^2 = f_{0,z}, \tag{4.155}$$

the z-component of the equilibrium PDF of the preceding subsections.

4.7.5.2. Admixture of Excited States

We assume now that we have the additional piece of knowledge (4.119) for P_{zx}. Our SWE obeys now $\psi = \psi_x \psi_y \psi_z$ (and, also, $\psi_0 = \psi_{0,x} \psi_{0,y} \psi_{0,z}$)

$$\psi''/2 + \omega^2 \left(v_z^2/2\right)\psi + aU_x U_z \psi = E\psi, \tag{4.156}$$

that can be treated perturbatively in view of our knowledge of the problem. $a \ll 1$ is here the perturbation coupling constant.

It is well known [64] that, if one perturbs the ground state of the one-dimensional harmonic oscillator wave function with a linear term, only the first excited state enters the perturbative series because of the selection rules [64]

$$\langle \psi_0 H_n(x)|x|\psi_0 H_m(x)\rangle = c_1 \delta(n, m+1) + c_2 \delta(n, m-1), \tag{4.157}$$

with c_1, c_2 appropriate constants. This entails that, for $n = 0$ (ground state), only $m = 1$ (first excited state) contributes. As a consequence, we can write (up to first order in perturbation theory) [64],

$$\begin{aligned}\psi = \psi_0 + \psi_1 &= [1 + bH_1(U_x) H_1(U_z)]\psi_0 \\ &= (1 + b\beta m U_x U_z)\psi_0, \end{aligned} \tag{4.158}$$

and, up to first-order terms as well,

$$\begin{aligned}\psi^2 &= [1 + 2b H_1(U_x) H_1(U_z)]\psi_0 \\ &= [1 + 2b(\beta m) U_x U_z]\psi_0. \end{aligned} \tag{4.159}$$

We evaluate now $\langle \psi|U_x U_z|\psi\rangle$. For symmetry reasons it is obvious that

$$\langle \psi_0|U_x U_z|\psi_0\rangle = 0. \tag{4.160}$$

Thus (cf. (4.158)),

$$\langle \psi|U_x U_z|\psi\rangle = 2b\,\beta\,m\,\langle \psi_0|U_x^2 U_z^2|\psi_0\rangle = 2b\,\beta\,m\,\langle U_x^2\rangle\langle U_z^2\rangle. \tag{4.161}$$

Using now the equipartition result $\langle U_x^2 \rangle \langle U_z^2 \rangle = n/(m\beta)^2$, we arrive (up to first order) at

$$P_{zx} = m \langle \psi | U_x U_z | \psi \rangle = 2bm^2\beta \langle \psi_0 | U_x^2 U_z^2 | \psi_0 \rangle \qquad (4.162)$$

$$= \frac{2nm^2b\beta}{m^2\beta^2} = \frac{2bn}{\beta} = \frac{an}{\beta}; \quad a = 2b, \qquad (4.163)$$

which coincides with the Grad result obtained in the preceding subsection. Thus,

$$\psi^2 = (1 + a(\beta m) U_x U_z)\psi_0^2, \qquad (4.164)$$

which is, again, *identical* to the ansatz (4.146). We have recovered the Grad result, which we know is the correct one.

5
Parallel Information Phenomena
of Biology and Astrophysics

B. Roy Frieden and Bernard H. Soffer

The realms of biology and astrophysics are usually regarded as distinct, to be studied within individual frameworks. However, current searches for life in the universe, and the expectation of positive results, are guiding us toward a unification of biology and astrophysics called *astrobiology*. In this chapter the unifying aspect of Fisher information is shown to form two bridges of astrobiology: (i) In Section 5.1 quarter-power laws are found to both describe attributes of *biology*, such as metabolism rate, and attributes of the *cosmos*, in particular its universal constants. (ii) In Section 5.2 we find that the Lotka–Volterra growth equations of biology follow from quantum mechanics. Both these bridges follow, ultimately, from the extreme physical information EPI principle and, hence, are examples of the "cooperative" universe discussed in Chapter 1. That is, the universe cooperates with our goal of understanding it, through participatory observation. The participatory aspect of the effect (i) is the observation of biological and cosmological attributes obeying quarter-power laws. In the Lotka–Volterra quantum effect (ii) the participation is the observation of a general particle member that undergoes scattering by a complex potential. This potential causes the growth or depletion of the particle population levels to obey Lotka–Volterra equations. Effectively, the *interaction potentials* of a standard Hartree view of the scattering process become corresponding *fitness coefficients* of the L–V growth equations. The two ostensibly unrelated effects of scattering and biological growth are thereby intimately related; out of a common flow of Fisher information to the observer.

5.1. Corresponding Quarter-Power Laws of Cosmology and Biology

5.1.1. Summary

On a case-by-case basis, it is found that the unitless cosmological constants both (i) obey quarter-power laws, i.e., may be approximated by the mass of the universe to a power that is an integer multiple of $1/4$, and (ii) have physical counterparts in corresponding biological attributes that likewise obey quarter-power laws; here the mass is that of the living creature under observation. This striking correspondence may be accounted for on the basis that the universe and biological ecologies have the common property of being exceedingly complex systems, hence sharing statistical properties such as power-law behavior. This power-law behavior is derivable in both cases by using EPI. Section 8.3 of Chapter 8 gives the derivation for specifically biological systems but, with appropriate change of language, derives the effect for cosmic systems, i.e., universes, as well. Both types of system obey the same invariance laws (Section 8.3.2) governing complexity that are used as inputs to EPI. The quarter-power law effect is also consistent with modern theories of an inflationary "multiverse" of universes that evolve according to a kind of natural selection, in the sense of Darwinian evolution of traits. The traits are precisely the unitless constants of each embryonic universe.

5.1.2. Introduction

Cosmology is ordinarily viewed as a purely physical effect that must ultimately be understood purely on the basis of known laws of physics. On the other hand, the manifestation of cosmology—the universe or, perhaps, the multiverse (Appendix)—is a highly complex system, and complex systems obey laws of organization that, for the most part, *cannot* be directly derived from the basic laws of physics. Well-known examples are chemical systems or weather systems. Others are the biological systems discussed in Chapters 3 and 8.

If cosmological systems are, then, not amenable to complete analysis using the basic laws of physics, what *can* be used for this purpose? The premise of this book is that *information*—a concept of complex systems—provides a basis for all dynamical laws of science. Therefore, it ought to be capable of deriving the basic laws of cosmology as well. This chapter will describe how *information, in the form of the EPI principle, can, in particular, account for the cosmological constants.*

Merely for brevity, and without any prejudice for plant life, a living creature is often termed an "animal" in the following.

A living creature, such as a tree, or a system of living creatures, such as a forest of trees, is a *biological system*. We investigate in Section 8.3 a curious, but exceedingly important, property of biological systems. This is that many of their observed attributes, such as metabolic rate or population density, obey *quarter-power laws*. That is, if an animal of the system (see above terminology) is measured for such an attribute, the value is approximately the creature's mass raised to an integer multiple n of $1/4$, $n = 0, \pm 1, \pm 2, \ldots$ (see first two columns of Table 5.2). As examples, for metabolic rate $n = 3$, while for the population density of trees in a forest, $n = -3$. It is established in Section 8.3 that the quarter-power effect holds for all attributes that obey the EPI principle. That is, such attributes exhibit in their measurements maximal Fisher information about their true values. Why should this be? Why should the values of "wet" biological attributes be defined by a "dry" concept such as *information flow*?

Possibly it is this. We discussed in Chapter 1 a connection of EPI to the anthropic principle, according to which our existence as carbon-based, intelligent creatures requires a universe that has certain physical constants and obeys certain laws. One prerequisite of our existence as *intelligent* animals is surely that we have the capability to *effectively* process information about the world about us, i.e., that we accomplish the EPI condition $I \approx J$ (see Sections 1.41–1.45). Or stated negatively, if instead our acquired information level obeyed $I \ll J$, most observations would not mirror reality, so that the nature of reality could not be known and intelligence rendered useless. With no selective advantage to intelligence, we would not have evolved into intelligent animals. Man is the intelligence-gathering animal par excellence only because intelligence gave us an evolutionary niche, and (arguably) still does. This is not to say that all animals tend toward intelligence. Many evidently do not, emphasizing instead running speed or swimming ability, etc. The condition $I \approx J$ defines an evolutionary niche for intelligence, but does not demand that all creatures fill it.

Furthermore, this ability should be aided by "Baldwin coevolution," [1],[1] whereby behavioral learning, say from parents, can bias the course of natural selection in the direction of learned traits. Such learning directly represents an increase in acquired information level I, and promotes positive feedback for the genetic change so as *accelerate* its utility and, hence, its propagation to further generations.

As we see in the *derivation of the quarter-power laws* in Section 8.3, *power* laws in particular are a *natural consequence* of such evolution. That is, *the same effect $I \approx J$ that defined an evolutionary niche for man and other intelligent life forms also gives rise to the quarter-power laws.* (In particular, the calculation of the ideal attribute values a in the Section 8.3 derivation follows from the *requirement* of optimum information-processing ability.)

Now we come to perhaps the broadest, and strangest, ramification of the effect. Section 8.3 shows that quarter-power laws hold, as well, *outside* the field of biology; that, in fact, such laws are a common (although not ubiquitous) property of *general* complex systems. Astonishingly, we will find empirically in this chapter that they hold for the largest such systems, i.e., *our universe*. Even moreso (!), they may hold over many universes, the so-called *multiverse* (Appendix) of modern theories of cosmology.

By the previous paragraphs, this could mean that a population of universes exists that evolves according to laws resembling Darwinian evolution. Is there evidence for this evolutionary process and, if so, to what extent can it be true? We shall show

---

[1] Ref. [1] gives discussion of "coevolution" or "genetic assimilation" proposed by M. Baldwin. As an example, newly arrived finches to the Galapagos learn to selectively feed on nectar in certain ways, and their offspring learn this feeding behavior by imitation, even before their beaks adapt to an optimal configuration for this purpose. Then any steps in the beak shapes of future generations that tend toward the optimal configuration are quickly used to advantage by the finches. Without the learning, these beak changes might be ignored and, hence, not benefit the young finches and not be passed on to future generations. In this manner, evolution is actually a synthesis of nature and nurture, i.e., genes and learning.

that a case can be made for the process on the basis of (a) empirical evidence, i.e. the known sizes of the universal physical constants; and (b) some modern theories of cosmology (Appendix).

5.1.3. Cosmological Attributes

We spoke of *biological attributes* in the preceding. These are of course confined to describing living systems. We now proceed to consider a much broader—in fact the broadest—possible class of attributes, those describing *the universe* or cosmos. These are the basic physical constants defining the fundamental scale sizes of the various phenomena of physics and astronomy. Examples are the charge e on the electron, gravitational constant G, speed of light c, etc. Table 5.1 shows all physical constants [2] that will be used in this chapter. This includes the basic constants and derived constants that are indicated functions of these. Most of the constants in the table of course have units.

5.1.3.1. Unitless Physical Constants

It turns out to be decisive to form *unitless* ratios [2] from the constants in Table 5.1. A subset of these is denoted as y_n and called "magic numbers" of cosmology. Well-known examples of magic numbers y_n are the ratio m_n/m_e of the nucleon (proton or neutron) mass to electron mass, and the fine structure constant $\alpha = 2\pi e^2/(\hbar c) \approx 1/137$ shown in Table 5.1. A third example, which turns out to be key to our approach, is the mass M of the universe in units of the nucleon mass.

TABLE 5.1. Basic and derived physical constants.

	Value
Basic constants	
Speed of light c	3.00×10^{10} cm/s
Charge on electron e	4.80×10^{-10} esu
Planck's constant $2\pi\,\hbar$	1.05×10^{-27} cm^2 g/s
Mass of electron m_e	9.11×10^{-28} g
Mass of nucleon (neutron or proton) m_n	1.66×10^{-24} g
Boltzmann constant k	1.38×10^{-16} erg/K
Gravitational constant G	6.67×10^{-8} cm^3(g/s^2)
Hubble length L	$9.25 \times 10^{28}/h$ cm ($h \sim 0.5$)
Microwave background density ρ_{bk}	4.60×10^{-34} g/cm^3
Mass of typical star ($\sim$2 solar masses) M_{star}	4.00×10^{33} g
Derived constants	
Fine structure constant $\alpha \equiv e^2\hbar c$	$0.73 \times 10^{-2} \approx 1/137$
Radius r $\sim GM_{star}/c^2$ of BH of mass M_{star}	2.96×10^5 cm
Nucleon (neutron or proton) radius $a_n \equiv \hbar/m_n c$	2.00×10^{-14} cm
Planck length $l_p \equiv (G\hbar/c^3)$	1.61×10^{-33} cm
Planck mass $m_p \equiv (\hbar c/G)^{1/2}$	2.18×10^{-5} g
Mass of universe $M \sim \rho L^3$	10^{58} g
Nucleon to electron mass ratio m_n/m_e	1836
Gravitational length GM/c^2 of universe	10^{30}

From Table 5.1 this has the approximate value

$$m \equiv \frac{M}{m_n} \approx 1.7 \times 10^{82} \sim 10^{80} \equiv N^2. \tag{5.1}$$

(Note that the symbol $\sim$ means *roughly of order*.) Quantity m is then a unitless mass m of the universe. Its colossal value 10^{80} defines what is called the "Eddington number." Hence the Eddington number is the mass of the universe in units of the nucleon mass.

For later use, from (5.1) the square root of the Eddington number is

$$N = 10^{40} = m^{1/2}. \tag{5.2}$$

The unitless constant (5.1) and others cited below are not merely of academic or historic importance. Many are built into *astronomical models* of the evolution of the universe. In fact, NASA's MAP satellite was designed to observe the cosmic microwave background, the first light released after the big bang, specifically *to determine these constants* to a high degree of accuracy. This will allow selection of the model that best fits the observed universe [3].

5.1.3.2. Dirac Hypothesis for the Cosmological Constants

The *Dirac large-number hypothesis* [4] defines the *particular* unitless constants that are called magic numbers y_n. By definition these have values that follow a simple mathematical power law relation in which the coefficients are of order unity,

$$y_n \equiv N^{n/2}, \quad n = 0, \pm 1, \pm 2, \ldots. \tag{5.3}$$

An order n generally defines a *multiple*, or cluster, of magic numbers. That (5.3) holds (approximately) with coefficients of order unity has been called the "cluster hypothesis" [2]. This cluster property is, in fact, what has led some authors to call these unitless constants magic numbers [2]. However, since the word "magic" has a negative connotation in science, we simply call them "cosmological constants" from this point on.

By Eqs. (5.2) and (5.3), *the cosmological constants can alternatively be described as obeying*

$$y_n = m^{n/4}, \quad n = 0, \pm 1, \pm 2, \ldots. \tag{5.4}$$

Since we had $m \sim 10^{80}$, numerically the sequence (5.4) has the values 1, $10^{\pm 20}, 10^{\pm 40}, \ldots$. Hence, the Dirac large-number hypothesis implies that there are relatively few in-between cosmological constants, e.g., with values $10^{10}, 10^{-30}, \ldots$. The number of unitless constants obeying (5.4) is impressively large [2].

It is curious that Eq. (5.4) represents the constants as, in particular, quarter-powers of the *mass* of the universe. This cosmological relation has apparently not been noticed before. Of central interest to this chapter is that, as found in Chapter 8, this precise dependence *is also shared with many attributes of biological systems*. In these biological applications, m in (5.4) is instead the mass of the particular creature whose attribute is being observed (see Section 5.1.2).

5.1.4. Objectives

The main objectives of this study are to show that:

(i) The cosmological constants y_n empirically obey the same power law (5.4) as do many biological attributes. Corresponding attributes are of the *same physical type*. Thus the correspondences are physically significant. Hence, they are are not merely Pythagorean (purely numerical). The correspondences range over an appreciable interval of values n (Section 5.1.6).

(ii) This similarity originates in the common nature of the two domains as *complex systems*, in particular systems that have undergone extensive dynamic evolution (Section 5.1.7).

(iii) Certain as yet undiscovered cosmological and biological attributes can be predicted (Section 5.1.7).

A further, albeit more speculative, objective is that

(iv) The quarter-power-law behavior shared by cosmology and biology supports certain inflationary models of the multiverse.

5.1.5. Biological Attributes

Curiously, the form of the cosmological Eq. (5.4)—a power law with powers that are multiples of $1/4$—holds as well for many (but not all) *attributes of biology*. These biological effects are by now well known (see [5–15]). As with the astronomical attributes, a fixed order n of a biological attribute actually contains a multiple, or *cluster, of attributes*. Examples of these are listed in the second column of Table 5.2 (for example, for order $n = -1$ there are five listed attributes). The power laws (5.4) in application to biology apply to both *internal and external* properties of living systems. That is, they apply to both intrinsic attributes of any one organism, such as its metabolic rate or the size of its brain, and extrinsic attributes governing properties of *entire populations* of organisms, such as the average grazing area for herbivores. Examples of these extrinsic laws are Bonner's law [5], Damuth's law [6], and Fenchel's law [7].

Remarkably, Eq. (5.4) continues this trend of biological systems to cosmological systems. Thus, the power law (5.4) applies from phenomena on the biological cell level, to phenomena on the level of individual organisms, to phenomena of entire populations of organisms, and finally, to astronomical phenomena on the cosmological level. (There is evidence that it holds on the level of elementary particles as well, as briefly noted below, but we do not pursue this issue in the chapter.) The principal distinction between biological and cosmological uses of Eq. (5.4) is that in biological uses, m is the mass of any observed organism, whereas in cosmological uses m is, by (5.1) and (5.2), a single mass, i.e., the mass of the universe. We return to this point below.

The biological attributes listed in column two of Table 5.2 are many $n/4$ examples that are summarized from [8–17]. The *cosmological* attributes that correspond to these are found below, and are shown in column three of Table 5.2.

TABLE 5.2. Some quarter-power laws that are common to biological and cosmological systems.

Power $n/4$ in law $y = m^{n/4}$	Biological attribute y (m = mass, organism)	Cosmological attribute y (m = Eddington mass)
−3/4 *Densities on area basis*	Average population density of trees in forest	Mass/surface area of spherical universe; Planck mass/surface area relative to universe mass/Planck area
−1/4 *Incremental rates of change*	Heart rate; RNA density; rate of increase of population; number of births/year of placental mammals, birds	Incremental star mass/universe; incremental BH length/Hubble = r/L (BH = black hole)
0 *Background scales*	Hemoglobin density in blood; temperatures of mammals, birds, certain fish	Fine structure constant α; nucleon mass scale: nucleon/electron; Gravitational length scale: gravitational/Hubble; cosmological temp., background; blackbody temp./photon $kT/(h(\nu))$
1/4 *Space- or timescales*	Life span; time between heartbeats; unicellular DNA length; average running speed	Expansion time: Hubble/BH = $(L/c)(r/c)$; mass scale: universe/star;
2/4 *Density-dependent effects*	Heat flow from mammal, bird	Relative length: universe/nucleon; Relative density: nucleon/universe; Relative force: electron/gravitational
3/4 *Maximum power "use"*	Metabolic rate; food consumption; brain size; cross-sectional area of aorta, tree trunk; power use of tree in forest	Mass−energy of star/nucleon; mass−energy of universe/Planck; length of universe/Planck length; $(Mc^2/a_n)/(M_{star}c^2/L)$; $(Mc^2/L)/(m_n c^2/r)$
4/4 *Total mass range*	Weight of heart, bird plumage, animal fur; energy expended/bird litter; home range area/mammal	Mass of universe/nucleon; mass density Planck/Nucleon
5/4 *Energy/progeny*	Energy expended/mammal litter	Number of nucleons/fractional star energy; Number of nucleons/Hubble sphere;
6/4 *Unknown property*	Predicted biological attribute(s)	Planck/microwave bakground density; Number of nucleons/fractional surface area BH

A cosmological attribute is placed in a given row if (a) *it obeys the same power* $n/4$ as corresponding biological attributes in that row, and (b) it shares a physical property *in common* with the corresponding biological attribute. This is a newly discovered effect, next developed. Historically, an attribute number $n/4$ has an explicit, known association with a given biological effect, since each number $n/4$ was found to specify a definite effect (say, metabolism rate). However, this is not the case for the cosmological numbers (5.4). In forming a "cluster" of order n, there is historically no known physical meaning for its member constants [2]; the only known meaning is the mathematical one, i.e., that they have about the same *numerical* value y_n. However, if the cosmological representation (5.4) is to have more than a numerical meaning, a level $n/4$ in Eq. (5.4) should have as well a *physical* meaning that is common to *all members* of the cluster. Finally, if representation (5.4) is to have physical meaning *on both levels, i.e., biological and cosmological*, then a level $n/4$ should define a physical attribute that is common to all listed attributes—*both biological and cosmological*— in the given row. It should be added that there are also attributes for a given power $n/4$ that do not obey the physical attribute. Our aim is to emphasize that there is a significant number that obey the physical attribute.

These common attribute meanings have been found, as worked out in Section 5.1.6. They constitute a chief finding of this chapter. The common attribute meanings are listed in italics in the first column (for example, for $n/4 = -3/4$, it is *Densities on an area basis*). However, there are occasional gaps, whereby a given attribute number $n/4$ defines a biological effect but not a cosmological effect (see case $n/4 = 3/4$ in Section 5.1.6), and vice versa. These are taken to constitute predictions of as yet undiscovered effects (aim (iii) of Section 5.1.4).

Aim (ii) of Section 5.1.4 is to establish that phenomena of biology and cosmology are statistically similar with respect to obeying the common power-law effect (5.4). To the extent that this effort succeeds, Eq. (5.4) will become a common organizer/specifier of system attributes covering the gamut of system sizes from the biological to the cosmological and, perhaps, including as well effects on the extreme microscale.

With this in mind, we reiterate that the mass m has a dual interpretation, depending upon whether the application is biological or cosmological. In the biological power laws, m represents the mass of a biological organism whose attribute values y are observed. In the cosmological power laws, it represents the mass of the universe as defined in Eq. (5.1). For an overall cosmological system consisting of many universes, i.e., the "multiverse," m will represent, in particular, the mass of a randomly sampled universe from the multiverse. Here the carryover from biology is complete, with the randomly sampled mass of an individual creature corresponding to the randomly sampled mass of an individual universe. The question of who/what is doing the "observing" in the cosmological case, is key.

Examples of the correspondences are developed next, in the order given in Table 5.2. The list is not meant to be exhaustive. The indicated biological attributes are taken to be givens, as in [8–17]. The aim is to find corresponding cosmological attributes.

5.1.6. *Corresponding Attributes of Biology and Cosmology*

Power $n/4 = -3/4$

The smallest power $n/4$ that describes biological attributes via Eq. (5.4) appears to be value $-3/4$. (We ignore trivially taking the reciprocal of large positive values of $n/4$ below.) However, there is no a priori reason to rule out smaller values as *eventual* possibilities.

Here the biological attribute in column two is a population density, i.e., a *number/area* (of trees). Aim (i) of Section 5.1.4 states that the corresponding cosmological attribute must be of the same physical type. It must therefore *likewise* be a number/area. To render it unitless as well, the attribute should be expressed relative to the number/area for some reference particle (e.g., the Planck particle).

We take the cosmological "area" to be the surface area of the universe, assuming the universe to be closed. We also assume that, for certain observations, a sizeable fraction of the mass of the universe effectively lies on its Schwartzschild surface. This would be the case if, for example, a universe that evolved out of a rotating black hole (BH) [18–23] were observed at an *exterior point* (as, e.g., Hawking radiation from near its surface). Up to 29% of the mass of such a universe is rotational energy [2], and therefore is associated with its event surface. This is located at one-half the Scharzschild radius [2] as the limit as rotation increases indefinitely. This event surface is also where the Hawking radiation effectively originates.

Broadly speaking, the observable universe spans the Hubble sphere, so that the Schwarzschild *radius* of the universe is approximated by L, the Hubble length. (Such BH models of the universe are further discussed below and in Appendix) Its surface area goes as the square of the radius and, therefore, as L^2. This will be used next.

To arrive at a unitless density, we construct the density y as the (mass/area) on the surface, relative to the (mass/area) of a reference particle. This is taken to be the "Planck particle," a hypothetical particle whose mass is, from Table 5.1, the Planck mass $m_p \equiv (\hbar c/G)^{1/2}$, and whose extension is the Planck length $l_p \equiv (G\hbar/c^3)^{1/2}$. A "Planck area" is, correspondingly, l_p^2. Thus, we construct as the attribute density

$$y \equiv \frac{(\text{mass/area})_{\text{univ}}}{(\text{mass/area})_{\text{Planck}}} \tag{5.5}$$

Using L^2 as the effective surface area of the universe (see above), Eq. (5.5) can be rearranged as

$$y = \left(\frac{M}{m_p}\right)\left(\frac{\text{Planck area}}{\text{area of universe}}\right) = \left(\frac{M}{m_p}\right)\left(\frac{l_p}{L}\right)^2 \tag{5.6a}$$

$$= m^{3/4}(m^{-3/4})^2 = m^{-3/4}. \tag{5.6b}$$

Here we used numerical values for M, m_p, l_p, and L given in Table 5.1, to get the approximate results

$$M/m_p \approx 0.5 \times 10^{63} \sim 10^{60} = m^{3/4}, \tag{5.7}$$

the latter by Eq. (5.1), and

$$l_p/L \approx 0.8 \times 10^{-62} \sim 10^{-60} = m^{-3/4}, \tag{5.8}$$

again by Eq. (5.1). The result (5.6b) is what we set out to show, namely that the attribute (5.5) of relative mass density is a cosmological attribute corresponding to $n/4 = -3/4$. We emphasize that quantity m is here—and in all these cosmological cases $n/4$—the unitless mass m of the universe, as defined at Eq. (5.1).

The mass M of the universe relative to its area (rather than to a volume) seems at first a disturbingly nonintuitive attribute. However, as we saw, this is not unreasonable and in fact agrees with accepted ideas about BHs. Furthermore, as discussed below, it lends credence to, and predicts the correctness of, recent models [18–23] of cosmological evolution.

We showed at Eqs. (5.7) and (5.8) that ratios m_p/M and $l_p/L \sim m^{-3/4}$ as well. These are not explicitly *numbers/area* as are postulated for the physical interpretation (1st column Table 5.2). However, our aim is to show the converse, that cosmological numbers/area, when taken relative to appropriate attributes, do represent attributes that obey the $m^{-3/4}$ law. From the result (5.7), quantity $(m_p/l_p^2)/(M/l_p^2)$ is another number of this type. This is the mass per surface area of a Planck particle relative to the universe's mass per surface area of a Planck particle.

Power $n/4 = -1/4$

Here the biological attributes in column two are small or *incremental rates*—heart rate, number of births/year, etc. We seek the same kinds of attribute in cosmology. Going down the list of basic constants in Table 5.1, we note that the mass M_{star} of a star is an increment in comparison with the mass M of the universe. Likewise the radius r of a BH of mass equal to one stellar mass is an increment in comparison with the Hubble length L of the universe. This suggests the use of alternative attributes

$$y \equiv \left(\frac{\text{mass of star}}{\text{mass of universe}} \right) \equiv \frac{M_{star}}{M}$$

or

$$y' \equiv \left(\frac{\text{radius of BH of stellar mass}}{\text{Hubble length}} \right) \equiv \frac{r}{L}. \tag{5.9}$$

As an added benefit these attributes y, y' are already unitless, as required.

The values for M_{star}, M, r, and L in Table 5.1, and Eq. (5.1), give respectively

$$\frac{M_{star}}{M} \approx 0.4 \times 10^{-24} \sim 10^{-20} = m^{-1/4} \tag{5.10}$$

and

$$\frac{r}{L} \approx 1.6 \times 10^{-24} \sim 10^{-20} = m^{-1/4} \tag{5.11}$$

for each of the two attributes. Thus, the cosmological attributes (5.9) of relative mass or relative extension correspond to certain "incremental rates" in biology.

They are determined by the same power $n/4 = -1/4$ as are the *biological* incremental rates in the power-law (5.4).

Power $n/4 = 0$

Here the biological attributes in column 2 set certain background reference scales (hemoglobin level, body temperature). Our aim is to find corresponding reference attributes in cosmology. Also, the requirement $n/4 = 0$ means that the attributes must be of order unity in size. There are many such dimensionless cosmological attributes [2]. By Table 5.1, among these are the fine structure constant $\alpha = 2\pi e^2/(\hbar c) \approx 1/137$, the ratio $m_n/m_e = 1836$ of a nucleon mass to the electron mass, and the ratio of the gravitational length $GM/c^2 \sim 10^{30}$ of the universe to the Hubble length $L = 1.9 \times 10^{29}$. Another physical constant of this type is the temperature T of a blackbody relative to its mean optical frequency v, that is $kT/(h\langle v \rangle)$, where k is Boltzmann's constant. Calculation of $\langle v \rangle$ using the blackbody radiation curve shows that the constant has a value of approximately 7.0. This is of order unity, as required. Note that the cosmological background temperature of about 3 K is, in fact, nearly constant [2], and has been shown to be that of a blackbody to an accuracy of better than 1 part in 10^4. Hence the cosmological background temperature is likewise an order-unity constant, i.e., of the type $n/4 = 0$ indicated.

Power $n/4 = 1/4$

Here the biological attributes in column 2 set certain space- or timescales (life span, time between heartbeats, etc.). This power $1/4$ is often said to define a "biological time" scale for organisms. The corresponding cosmological constants must, then, likewise define total scales, and vary as $m^{1/4}$. One such constant [2] is the Hubble expansion time L/c in units of the BH time r/c, the constant $(L/c)/(r/c)$. By the values in Table 5.1 this is approximately $0.6 \times 10^{24} \sim 10^{20} = m^{1/4}$. Another is [2] the mass m of the universe relative to the mass M_{star} of a star, the constant M/M_{star}. By the values in Table 5.1, this is approximately $2.5 \times 10^{24} \sim 10^{20} = m^{1/4}$ by (5.1).

Power $n/4 = 2/4$

The rate of heat flow from a mammal or bird (mW/°C) varies as $m^{2/4}$. This empirically observed rate depends upon the *mass or energy density* of the organism, as well as its surface area and fur/feather insulation effects. A corresponding cosmological quantity [2] that likewise varies as $m^{2/4}$ is the mass density (m_n/a_n^3) of a nucleon, where a_n is its extension, in units of the density (M/L^3) of the universe. This is then the unitless attribute $y = (m_n L^3)/(M a_n^3)$. By the numbers in Table 5.1 and by (5.1), this has value $1.6 \times 10^{47} \sim 10^{40} = m^{2/4}$. Other density quantities are the Hubble length in units of the nucleon size, or $L/a_n \approx 1.0 \times 10^{43} \sim 10^{40} = m^{2/4}$; and the electrical force per particle between an electron and a proton relative to their mutual gravitational force per $e^2/(G m_n m_e) \approx 0.22 \times 10^{40} \sim 10^{40} = m^{2/4}$.

Power $n/4 = 3/4$

The biological attributes of metabolic rate, brain size, power (energy/time) use of a tree in a forest, etc., follow an $m^{3/4}$ dependence. They measure *total power* use.

For example, the large brain of a human is a major source of radiated heat power. Note that the word "use" in quotes (first column) is meant to simply imply "function" and, of course, not necessarily "intent" as in biological cases. Corresponding cosmological attributes that measure power are the total mass–energy contained in a star relative to that in a nucleon, that is $(M_{star}c^2)/(m_n c^2) \approx 2.4 \times 10^{57} \sim 10^{60} = m^{3/4}$; and the mass–energy of the universe relative to that in the Planck particle, or $(Mc^2)/(m_p c^2) \approx 0.5 \times 10^{63} \sim 10^{60} = m^{3/4}$. Both attributes go as $m^{3/4}$ as required.

The preceding are purely ratios of energies. Constants of the $m^{3/4}$ type that mix energies and lengths are, by Table 5.1, ratios $(Mc^2/a_n)/(M_{star}c^2/L) \sim m^{3/4}$, and $(Mc^2/L)/(m_n c^2/r) \sim m^{3/4}$.

Likewise, by Table 5.1, the ratio $(L/l_p) \approx 1.2 \times 10^{62} \sim 10^{60} = m^{3/4}$ as well, and therefore is listed. Thus, our theory predicts that this pure ratio of *lengths* likewise measures total power (energy/time), albeit in some as yet *unknown* way.

Power $n/4 = 4/4$

All these attributes, by definition, are to vary as $m^{4/4}$, i.e., linearly with the mass m. Biological attributes of this type are the weight of a heart and the home-range area of a mammal; and, by Eq. (5.1), the cosmological attribute of the mass m of the universe, in units of the mass m_n of a nucleon. Also, by the values for m_p, l_p, and a_n given in Table 5.1, the ratio $(m_p/l_p^3)/(m_n/a_n^3)$ of the Planck mass density to the nucleon mass density has the approximate value $0.3 \times 10^{77} \approx 10^{80} = m$ (by Eq. (5.1)) once again. The common attribute over these biological and cosmological systems is then a total mass range.

Power $n/4 = 5/4$

The energy expended by an adult mammal in birthing and raising a litter varies as $m^{5/4}$. A corresponding cosmological attribute is the effective total number of "daughter" nucleons, i.e. (M/m_n), relative to the fraction (M_{star}/M) of all energy that is contained with a typical star. This is the attribute $(M/m_n)/(M_{star}/M)$. It has a value $\sim m^{5/4} = 10^{100}$ by Table 5.1 and Eq. (5.2). The division by the typical star energy fraction M_{star}/M regards a typical star as the medium or "family" for creating the M/m_n effective daughter nucleons in the universe. This also means that the mass m of the universe is the geometric mean between the mass m_n of a nucleon and $10^{100} M_{star}$.

Power $n/4 = 6/4$

From Table 5.1, the Hubble length L in units of the extension a_n of a nucleon obeys

$$\frac{L}{a_n} \approx 1.0 \times 10^{43} \sim 10^{40} = m^{1/2} \tag{5.12}$$

by (5.1). Cubing both sides of this equation shows that the number of nucleons that can fit within a Hubble sphere (the volume of the Universe) varies as $m^{3/2} = m^{6/4}$ as required by the $n/4$ rule.

Another cosmic number of this type is the Planck density m_p/l_p^3 expressed in units of the microwave back ground radiation density ρ_{bk}. By the numbers in Table 5.1, we get $m_p/(l_p^3 \rho_{bk}) \approx 1.0 \times 10^{127} \sim 10^{120} = m^{3/2} = m^{6/4}$ as required.

A final such cosmic number is, by Table 5.1, $(M/m_n)/(r^2/L^2)$, the effective number of nucleons per surface area of a BH relative to that of the universe.

There is no currently reported *biological* attribute varying as this power of the species mass m. As before, Eq. (5.4) is therefore taken to predict that one or more such attributes *will* be found (an example of aim (iii)). We could get higher-order powers than 6/4 by merely cubing length ratios such as (5.12) to get volume ratios. However, these would not be fundamentally new constants, and so we ignore these. There appear to be no fundamental biological or cosmological attributes at this time corresponding to powers of m beyond value 6/4, although there is no a priori reason to rule these out as eventual possibilities.

5.1.7. Discussion

We showed that *existing data* define a power-law relation (5.4) that describes *in common* many attributes of biology and cosmology. The correspondences are often approximate, but they definitely exist. We also showed that all biological and cosmic attributes of a given order n tend to be of a similar *physical* type (as listed in italics in the first column of Table 5.2). Thus, the index n has physical significance, and is not merely a numerical curiosity. The approach has also served to predict the existence of unknown effects that correspond to *missing items* in Table 5.2 (aim (iii), Section 5.1.4). In summary, the latter effects are as follows:

(a) A physical effect $n/4 = 3/4$ whose power use varies as the ratio (L/l_p), the length L of the universe in units of the Planck length.
(b) A biological effect $n/4 = 6/4$ that corresponds to the unitless Hubble length L/a_n, or to the Planck density expressed in units of the microwave background radiation density.

That some cosmic and biological attributes have common power-law behavior has also been noted before [24], although not for specifically quarter powers, nor for the Dirac numbers in particular.

The common behavior suggests the following hypothesis: That they arise out of systems whose macroscopic statistical behavior is similar. (We do not address microscopic behavior in detail in this chapter.) The statistical behavior (5.4) is, in fact, characteristic of a class of *complex systems*, in particular those obeying fractal growth. Fractal growth is power-law growth, precisely the form of Eq. (5.4). Power-law behavior is known to arise out of self-organized criticality [25].

More recently, power-law behavior was found to arise alternatively as an *optimal flow of information* to an observer [26], via EPI. (See the derivation in Section 8.3 of Chapter 8.) That derivation does not make any assumption as to the absolute scale sizes of the attributes y. Thus, it potentially applies from the microscopic scale sizes of elementary particle interactions, to the intermediary scale sizes of biological systems, to the cosmic scale sizes of intergalactic interactions. These are also properties of complex systems in states of equilibrium or *nonequilibrium*. Hence we propose that *both living systems and astronomical systems evolve, in the large, as complex systems* of this type. We call this the *hypothesis of macrosimilarity*. Supportive material for this hypothesis follows.

Both bio- and cosmo-system types have evolved as stochastic effects. Biological organisms have evolved quasi-randomly out of Darwinian evolution; and galaxies likewise "evolve" randomly, with many aspects of their structure following power laws. These are properties that characterize complex stochastic systems. Regarding cosmology, it has long been hypothesized that the universe has a "fractal nature" [20, 21]. Fractal processes obey power laws and this is precisely the form of Eq. (5.4). Other examples of cosmological power laws also exist. For example, observations of young and populous star clusters have shown that the stellar initial mass function (IMF) obeys a power law with an exponent of about 2.35 for stars more massive than a few $M_\odot$ (solar mass). Also, a consensus has emerged that the IMF of young star clusters in the solar neighborhood and in interacting galaxies can be described, over the mass range of a few 10 to 107 M, as a power law with an exponent of 2 [27]. (Note that these are not *unitless* cosmic constants, and so are not included in Table 5.2.) Such power laws are characteristic of self-organized criticality, or of optimal information flow in complex systems [26].

Some *visual* support for the hypothesis of macro-similarity is offered by the very recently discovered [28] "double helix nebula" shown in Figure 5.1. This nebula has the appearance of a biological DNA molecule! Considering that its length is about 80 light-years, is this macro-similarity to an extreme? The image has not yet been confirmed as having fractal structure, although its branching nature makes it a good candidate. The nebula also is interesting in that it is located relatively near (about 300 light-years from) the supermassive BH *at the center* of our Milky Way. (By comparison, the Earth is more than 25,000 light years from the BH.) This is of course reminiscent of the "baby universe" model of Smolin (Appendix). The correspondence would be compelling if its helical structure contained evidence of "ladder rungs" that somehow encoded for a particular set of cosmological constants. This possibility, although remote, cannot be ruled out since the nebula has only been imaged over a limited range of wavelengths (centered on 24 μ).

Aside from these macroscopic effects, many *microscopic* cosmic processes obey, as well, power-law (called "scaling law") behavior. This is on the level of elementary particle interactions, particularly in their cross-sectional rates of change $d\sigma/dt$. A classic work on this subject is, for example, [29]. However, our emphasis in this chapter is on macro-similarity, so that we do not further discuss such microscale behavior.

There is further supporting evidence for the hypothesis of macro-similarity. *Both* living and nonliving systems are known to obey dynamics governed by *transport equations* [24, 30]

$$\frac{\partial p(i|t)}{\partial t} = p(i|t)[g(i, t, p) - d(i, t, p)], \quad i = 1, 2, \ldots \qquad (5.13)$$

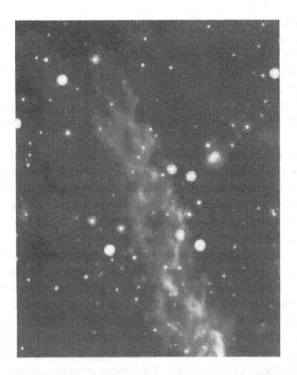

FIGURE 5.1. The "double helix nebula" image, taken with the Spitzer Space Telescope with 6-arcsec resolution [2]. The nebula is located near the BH at the galactic center of the Milky Way. (Original color-coded picture on cover.) Reprinted by permission of Macmillan Publishers Ltd, [28], copyright 2006.

Here $p(i|t)$ denotes the occurrence rate, at a general time t, of a particle of type i, $i = 1, 2, \ldots$. The rate $p(i|t)$ also represents for large enough population levels the probability that a randomly selected particle from the system will be of type i. Equation (5.13) is called the "equation of genetic change" in population biology. Since the time t is general, eq. (5.13) generally describes *nonequilibrium* states, or equilibrium states as t becomes large. The coefficients g, d are, respectively, growth and depletion effects, and in general depend upon (i, t, p), with the notation p denoting all $p(i|t)$, $i = 1, 2, \ldots$. According to the choice of coefficients g, d, Eq. (5.13) describes many nonliving or living systems [30]. For example, with g and d selected on a continuum, (5.13) becomes the Boltzmann transport equation [30]. Other examples of nonliving systems described by (5.13) are laser-resonator energy level dynamics, Poiseuille flow, neutron transport in nuclear fission reactors, Navier–Stokes equations of hydrodynamical flow, and the growth of polymer chains. Examples of living system effects (5.13) are Lotka–Volterra growth of biological populations, RNA replication growth, and phenome growth in ecological systems.

The transport Eq. (5.13) is, as with the quarter-power laws in Chapter 8, derived on the basis of optimal information flow [30]. In turn, this requires the existence of a complex system in a non-equilibrium state. Again, this supports the hypothesis of macro-similarity.

Further, and independent, evidence for the hypothesis of macro-similarity is provided by recent *inflationary theories* of the formation of universes [18–23]. In brief, multiple universes obey macro-similarity because they act individually and collectively like a biological system of reproducing organisms. Aspects of these inflationary theories that have a bearing on this hypothesis are discussed in the Appendix.

One important candidate explanation for why the cosmological constants in particular obey the power law form (5.4) is the *anthropic principle* [2, 31–33]. (We limit attention to the so-called "weak form" of the principle.) The anthropic principle serves as a practical guide to selecting among contending hypotheses in favor of those that are consistent with *our existence* in the Universe [2]. We likewise use it that way.

Consider first a general state of the universe, where intelligent observers *do not* necessarily exist. In this scenario, the law (5.4) no longer holds as a unique expression [2]. Rather, *it is known to split into two laws,* where two different bases m_1 and m_2 replace m according to which constants y_n are represented.

On the other hand, the anthropic principle assumes *the existence* of intelligent, carbon-based observers: It arises as an epistemic nature of physical laws—that scientific laws are not only attempts at formulating ontologies, but also are epistemologies, i.e., bodies of knowledge. Knowledge requires intelligent observers. Such observers, and indeed all of life, cannot exist until the universe is at least as old as the lifetime of the first generation of stars. The connection with the power law (5.4) is that at that time and beyond [2] $m_1 = m_2 \equiv m$. There is then a unique base m in (5.4), as required. In this way, the presence of human observers implies the validity of a unique law (5.4) for the cosmological constants.

Thus, the validity of the law (5.4) is a necessary condition for the presence of human observers (*if* human observers, *then* the law). This is what we wanted to show. However, it is of course not sufficient—the validity of (5.4) does not imply with certainty the existence of human observers. This is, however, irrelevant to our argument, which is based on the irrefutable *existence* of human observers.

Interestingly, the *biological* laws obeying (5.4), which hold for carbon-based life, follow as well from the anthropic principle. This is in fact on two levels. First, they follow from the simple need for carbon-based observers, as in the cosmological case (i.e., laws of biology, like laws of physics, have epistemological content). Second, these biological laws are of course premised upon the existence of *complex biological* systems. The observers who activate the anthropic principle are, of course, such systems. Thus, the anthropic principle implies both the biological and cosmological aspects of (5.4).

The observer has a key role to play, as well, in derivation of the biological power laws (5.4) on the basis of information flow and EPI [25]. This derivation is given in Section 8.3 of Chapter 8.

In summary, theory and experimental data imply that cosmic evolution on (at least) the macroscale has paralleled Darwinian evolution in being stochastic and characteristic of nonequilibrium, complex systems. The theoretical evidence is fourfold: the common power-law behavior (5.4) of both biological and astronomical systems, the common growth effects (5.13), the recent inflationary models of multiple universes, and the weak anthropic principle.

The experimental evidence we show is a great deal of astronomical data showing power-law behavior.

Predictions of the theory are as follows:

(1) The cosmological magic numbers of *offspring* BH *universes* of a parent BH obey the same power law (5.4) as do attributes of *biological systems* on Earth (Sections 5.1.3–5.1.5, with examples in Section 5.1.6).

(2) In turn, this correspondence implies that certain gaps in these biological and cosmological attributes, as indicated in Table 5.2, will eventually be filled by particular effects. These are (a) and (b) mentioned at the beginning of this section.

(3) Correspondence (5.3) can be accounted for by the premise of the *inflationary model* of multiple universes (Appendix), that the cosmic magic numbers y_n act as genetic traits that are passed on by universes to "offspring" universes. Thus, the inflationary model and the correspondence (5.4) support one another.

(4) The population of offspring universes obeys the same Lotka–Volterra growth law (5.13) as do biological populations (Appendix). Again, this supports the inflationary model.

(5) We saw in the preceding that the cosmological law (5.4) loses its validity at earlier epochs, splitting into two laws. Only after sufficient time has passed do the two laws coalesce into one. The fact that we observe the one law means that sufficient time has passed to form observers, and therefore life, in our universe. Then likewise since baby universes follow the law (5.4) they must be roughly our age. Therefore (by de Sitter expansion) they must be roughly our size. This is consistent with a recent analysis on the elementary particle level [34] which estimates the size of our "mother" universe (the one that presently contains us) to be roughly a factor $\sqrt{2}$ the size of ours.

The main objection [2] to such inflationary theory is that its "fittest" black-hole universes, i.e., those that evolve in the direction of maximum production of offspring universes, do not necessarily contain stars and living creatures. That is, the multiverse would not necessarily lead to our existence. However, this is a moot point since living creatures *do* exist in our universe and we can create theory *conditional upon that fact*. Thus, an inflationary multiverse *could* have been the origin of our particular universe. In effect, the multiverse can be regarded as necessary, but not sufficient, to *our existence*.

5.1.8. Concluding Remarks

Just as our universe and its multiverse obey fractal laws, so do our own manifestations as complex biological systems. We share the fractal behavior of those grander universal complex systems. Hence, life as we experience it mimics the larger reality that surrounds us. Indeed, this was the point of the numerous examples in Section 5.1.6, which demonstrate that the quarter-power law applies *both* to biological attributes, which is derived on the basis of EPI in Chapter 8, *and* to *corresponding* cosmological attributes. In this regard the quarter-power law obeys the most basic requirement of a fractal process, invariance to change of scale. Our lives and environments are scaled versions of these grander realities. Figure 5.1 offers perhaps a visual confirmation of this statement.

5.2. Quantum Basis for Systems Exhibiting Population Growth

5.2.1. Summary

The laws of population growth of living and nonliving systems may be derived out of quantum mechanics. This is as a limiting case of the Schroedinger wave equation (SWE). The derivation is based upon use of the Hartree approximation. A time-dependent, complex potential function is required, defining a non-Hermetian problem. The imaginary part of the potential directly gives rise to the "fitness" properties governing growth and depletion of the populations that comprise the system. In this way, biological life can arise out of quantum mechanics. Speculations are made on how the required complex potential function might be realized in practice.

5.2.2. Introduction

Growth equations typically have a form (5.13), which we rewrite more generically as

$$\frac{dp_k(t)}{dt} = p_k(t)G_k(\mathbf{r}, t), \quad k = 1, \dots, K, \tag{5.14a}$$

where

$$G_k(r, t) = g_k(\mathbf{r}, t) + \sum_{j \neq k} p_j(t) \int d\mathbf{r}_j g_{jk}(\mathbf{r}_{jk}, t). \tag{5.14b}$$

In Eq. (5.14a), $p_k(t)$ is the PDF for finding a particle of class k anywhere in space $\mathbf{r}$ at the time t, in terms of its overall "fitness" G_k, the latter defined as the relative increase in p_k per unit time at time t. Enough time is assumed to have passed such that the medium is essentially homogeneous in the spatial occurrence of particle type k,

$$p_k(\mathbf{r}, t) \approx p_k(t). \tag{5.15}$$

Effectively, the *spatial* dependence of the fitness $G_k(\mathbf{r},t)$ negligibly affects the particle densities at these times.

Turning to Eq. (5.14b), the total fitness G_k of class k is assumed to be formed out of two additive effects: an intrinsic fitness g_k (independent of other particles), and an interactive fitness demarked by the sum over all other particles j. The sum represents the mean interactive fitness g_{jk} over all other particles j. These are located at general distances $\mathbf{r}_{jk} \equiv \mathbf{r}_k - \mathbf{r}_j$ from particle k.

It may be noted that the interactive fitnesses $g_{jk} \equiv g_{jk}(\mathbf{r}_{jk}, t)$ in (5.14b) are expressed on the microscale, i.e., using microscopic coordinates $\mathbf{r}_{jk}, t$. They can equivalently be expressed macroscopically in terms of all other population densities p_j. With such a replacement, each Eq. (5.14a) becomes a highly nonlinear first-order differential equation in all the PDFs $p_j, j = 1, \dots, K$.

The equations of growth (5.14a, b) describe many living and nonliving systems, including genetic growth, polymer growth, DNA growth, ecological growth, population levels of two-level atoms in a lasing gas, etc. Each particular problem is defined by a different choice of growth coefficients $g_k(\mathbf{r},t)$ and $g_{jk}(\mathbf{r}_{jk}, t)$. We next consider how these growth equations arise as a natural phenomenon of quantum mechanics, and specialize it to the case of living systems by a particular choice of coefficients $g_k(\mathbf{r},t)$ and $g_{jk}(\mathbf{r}_{jk}, t)$.

5.2.3. Hartree Approximation

Consider a scenario where N particles of classes $k = 1, \dots, K$ are moving under the influence of an exterior potential plus interaction potentials among them. Let $\mathbf{r}_j = (x, y, z)_j$ denote the position of the jth particle relative to a fixed origin, and let $\mathbf{r}_{jk} \equiv \mathbf{r}_k - \mathbf{r}_j$ denote the position of the kth particle relative to the jth. We want to establish the quantum dynamics of the system.

We use Hartree's self-consistent field approach [35], assuming that the kth particle moves in a net potential field that is the sum of its own local potential $v_k(\mathbf{r},t)$ plus its interaction potentials $V_{jk}(\mathbf{r}_{jk}, t)$ with all the other particles $j \neq k$. Each of the latter potentials is taken to be the mean of the interaction potential $v_{jk}(\mathbf{r}_{jk}, t)$ between the jth and kth particles as weighted by the jth particle's position probability density $|\psi_j(x, t)|^2$. Hence the kth particle "sees" a total potential

$$V_k(\mathbf{r}, t) = v_k(\mathbf{r}, t) + \sum_{j \neq k} V_{jk}(\mathbf{r}_{jk}, t), \qquad (5.16a)$$

where

$$V_{jk}(\mathbf{r}_{jk}, t) \equiv \langle v_{jk}(\mathbf{r}_{jk}, t) \rangle = \int d\mathbf{r}_j |\psi_j(\mathbf{r}_j, t)|^2 v_{jk}(\mathbf{r}_{jk}, t). \qquad (5.16b)$$

Note that the indicated average is over all space but not the time, since we are interested in the system behavior as a function of the time.

5.2.4. Schrodinger Equation

The Hartree approach separates out the many-body problem, allowing a distinct SWE for the kth particle to be formed as

$$-\frac{\hbar^2}{2m_k} \nabla^2 \psi_k + V_k(\mathbf{r},t)\psi_k = i\hbar \frac{\partial \psi_k}{\partial t}, \quad \psi_k \equiv \psi_k(\mathbf{r}_k, t), \qquad (5.17a)$$

where $V_k(\mathbf{r},t)$ is given by (5.16b). The net SWE is then

$$-\frac{\hbar^2}{2m_k}\nabla^2\psi_k + \left[v_k(\mathbf{r},t) + \sum_{j\neq k}\int d\mathbf{r}_j|\psi_j(\mathbf{r}_j,t)|^2 v_{jk}(\mathbf{r}_{jk},t).\right]\psi_k = i\hbar\frac{\partial\psi_k}{\partial t}.$$
(5.17b)

Note that the SWE derives out of EPI [30], and is also used in Chapters 2, 4 and 6. In the latter applications the Planck constant $\hbar$ does not enter in, so that the effects are classical, and yet follow the mathematics of quantum theory. This analysis will likewise lead to a classical system, namely, one exhibiting classical growth and depletion. However, system requirements will be on the quantum level.

It is to be noted that the interpretation of the SWE as derived by EPI [30] allows for growth or depletion, i.e., nonconservation of particle number, as required here. In general, such SWE–EPI solutions do not obey normalization over space at *each* time but, rather, normalization over all space and *all* time. Therefore at each time a generally different number of particles can exist in the system, as required in this growth problem.

5.2.5. Force-Free Medium

Next, assume that a general particle of the medium experiences zero net force, so that the potential has no space-dependence,

$$V_k(\mathbf{r}, t) \equiv V_k(t).$$
(5.18a)

By Eqs. (5.16a,b), consistently,

$$v_k(\mathbf{r},t) \equiv v_k(t), \quad v_{jk}(\mathbf{r}_{jk}, t) \equiv v_{jk}(t).$$
(5.18b)

The SWE (5.17a) becomes

$$-\frac{\hbar^2}{2m_k}\nabla^2\psi_k + V_k(t)\psi_k = i\hbar\frac{\partial\psi_k}{\partial t}, \quad \psi_k \equiv \psi_k(\mathbf{r}_k, t).$$
(5.19)

This is obviously non-Hermetian.

By separation of variables $\psi_k(\mathbf{r}_k, t) \equiv R(\mathbf{r}_k)T(t)$, this SWE separates into a *free-particle* wave equation

$$\left(\nabla^2 + \frac{2m_k\beta}{\hbar^2}\right)R(\mathbf{r}_k) = 0$$
(5.20a)

in $R(\mathbf{r}_k)$, and a first-order equation

$$i\hbar\frac{dT(t)}{dt} = (\beta + V_k(t))T(t).$$
(5.20b)

in $T(t)$. The constant β is the separation constant of the separation of variables approach, and is taken to be real. Assume that the medium is a box with a very large side L. The potentials $V_k(t) \equiv$ const. within the box and infinite at its walls.

Then the solution to the SWE (5.20a) is that for the free particle in a box,

$$R(\mathbf{r}_k) = (2/L)^{3/2} \sin(l\pi x_k/L) \, \sin(m\pi y_k/L) \, \sin(n\pi z_k/L), \quad l, m, n$$
$$= 1, 2, \ldots, \quad L \to \infty. \tag{5.21}$$

Then, for a position $\mathbf{r}_k$ any moderate distance away from the sides of the box, and with finite solution orders l, m, n, effectively

$$R(\mathbf{r}_k) \approx \text{const.} = A. \tag{5.22}$$

This spatial uniformity is not too surprising, considering that the potential likewise has no spatial variation.

Thus, the medium is homogeneous or "well mixed." Anticipating our application, a macroscopic example would be a pond, where on average a given species of fish (say) is equally likely to be everywhere. However, by the same token the likelihood can vary with the time, waxing and waning with the presence of predators, preys, weather, etc. This is accomodated by $T(t)$.

From material preceding Eq. (5.20a), the total wave function is now

$$\psi_k(\mathbf{r}_k, t) = AT(t) \equiv \psi_k(t) \tag{5.23}$$

as an obvious extension of the notation.

We next turn to the time-dependence $T(t)$. This obeys the differential equation (5.20b). By notation (5.23) this becomes

$$i\hbar \frac{d\psi_k(t)}{dt} = (\beta + V_k(t))\psi_k(t). \tag{5.24}$$

Note that this would also result from (5.17a) if, instead of the force-free assumption (5.18a), it was assumed that the mass m_k is sufficiently macroscopic that $\hbar^2/2m_k \nabla^2 \psi_k$ is negligible in size compared with the other terms of (5.17a). However, we stick with assumption (5.18a) because we are not interested in the spatial dependence of the ψ_k. The "growth equations" we seek are conventionally of the form (5.13), expressing growth in *time alone*.

5.2.6. Case of a Complex Potential

The potential function $V_k(t)$ introduced in (5.18a) is as yet, undefined. Consider the particular case where it is complex [36–38], and of the form

$$V_k(t) \equiv i\frac{\hbar}{2}G_k(t) - \beta, \tag{5.25}$$

where function $G_k(t)$ is real (later representing a fitness function). The other potentials of the Hartree approach (5.16 a, b) must correspondingly be

$$v_k(t) \equiv i\frac{\hbar}{2}g_k(t) - \beta,$$

$$v_{jk}(t) \equiv i\frac{\hbar}{2}g_{jk}(t). \tag{5.26}$$

Using Eqs. (5.25) and (5.26) in (5.16 a, b) shows that these new functions $G_k(t)$, $g_k(t)$, and $g_{jk}(t)$ relate to one another in the same way that the original Hartree potentials co-relate in (5.16 a, b). That is

$$G_k(t) = g_k(t) + \sum_{j \neq k} G_{jk}(t) \qquad (5.27)$$

and

$$G_{jk}(t) = \int d\mathbf{r}_j \, p_j(t) g_{jk}(t) = L^3 p_j(t) g_{jk}(t),$$

where

$$p_j(t) = |\psi_j(t)|^2. \qquad (5.28)$$

The latter is the usual probability (1.57) in terms of its squared amplitude function.

Note from (5.25) and (5.26) that these special potentials $V_k(t)$, $v_k(t)$, and $v_{jk}(t)$ go as $\hbar$, which suggests *weak* temporal effects.

5.2.7. SWE Without Planck Constant

Putting (5.25) into the SWE (5.24) gives

$$\frac{1}{2} G_k(t) \psi_k = \frac{d\psi_k}{dt}, \qquad (5.29)$$

the factor $\hbar$ having cancelled out! This is a key step in the derivation, since it will give rise to the macroscopic growth equations (5.14a, b) *without* having to evaluate them in the usual approximation $\hbar \to 0$. Thus, the derivation will be for, simultaneously, quantum- *or* classical growth effects. This is in fact a prerequisite of the derivation, since growth equations hold for both living and nonliving systems [24, 30], and some nonliving systems operate distinctly on the quantum level, such as the bilevel atoms of a lasing medium.

5.2.8. Growth Equation

Multiply (5.29) by ψ_k^*, giving

$$\frac{1}{2} G_k(t) \psi_k^* \psi_k = \psi_k^* \frac{d\psi_k}{dt}. \qquad (5.30)$$

The complex conjugate is

$$\frac{1}{2} G_k(t) \psi_k \psi_k^* = \psi_k \frac{d\psi_k^*}{dt}. \qquad (5.31)$$

Add the two last equations, using (5.27) and (5.28). The result is of the form Eqs. (5.14a, b),

$$\frac{dp_k(t)}{dt} = p_k(t) \left[g_k(t) + \sum_{j \neq k} \alpha_{jk}(t) p_j(t) \right], \quad \alpha_{jk}(t) \equiv L^3 g_{jk}(t). \qquad (5.32)$$

These define growth equations, and in particular Lotka–Volterra growth equations [5.39]. Here the $g_k(t)$ represent the intrinsic fitness of species k at the time t, while the summation represents contributions $g_{jk}(t)$ to fitness due to interactions with all other species j of the ecology. To obey L–V equations is a necessary, although not sufficient, condition for a system to be *living*. The correspondence of this total growth factor (in the square bracket) with the Hartree terms (in square bracket) in Eq. (5.17b) is striking. In essence, the growth of populations of classical systems follows the same kind of shielding and interaction effects as do the interactive potentials of the many-body scenario of quantum mechanics.

5.2.9. How Could Such a System Be Realized?

Complex potentials, such as we require at Eq. (5.25), characterize dissipative systems, and also can occur in scattering processes [37, 38]. Thus, our required potential could conceivably be physically realized in a scattering experiment. It is possible that a special particle could provide the complex potential. Of course, even then it might not result in a viable ecological system. For example, ecological systems tend to be periodic, so that only a special subset that allows for periodic solutions could result in such a system. But, there is of course no guarantee that a periodic system is "alive." Living systems have other properties as well, such as biological attributes (metabolic rate, age to maturity, etc.), that obey power laws, as we saw previously. Nevertheless the possibility of such a "life-enabling" particle is an intriguing possibility.

A scattering experiment characterized by a complex potential was recently reported in [40]. An atomic beam of metastable argon in a certain state is subjected to the imaginary potential exerted by an optical field. This potential is provided by a circularly polarized standing light wave produced by the use of a laser beam. The beam is retroreflected upon the atoms, and is in a state that is resonant with their state. Here the "special particle" alluded to above is merely a photon.

However, in these scattering experiments the complex potential that was accomplished was not, additionally, proportional to Planck's constant $\hbar$ as required in Eq. (5.25). A recent experiment [41] has actually accomplished this proportionality. However, the experiment seems limited in practice to laboratory implementation, rather than occurring naturally as, e.g., in a scattering experiment.

These realizations of the required complex potential are consistent with theories of panspermia. These propose that life-enabling particles permeate the universe and randomly give rise to life when they interact with matter under proper conditions. From this perspective the proper conditions would be the above Hartree conditions, in the special case of a complex potential obeying (5.25).

Equations (5.32) describe a dynamically growing system of arbitrary spatial properties. Since spatial scale is avoided by the approach, the scale of a system obeying (5.32) is presently unknown. It could conceivably be microscopic or macroscopic, depending upon the scale at which the effect (5.25) holds. However, the thrust of this work is that, no matter what the spatial scale of the system might be, its dynamics arise in the Hartree model coupled with a special potential function.

5.2.10. Current Evidence for Nanolife

Many life forms that originated, as in the preceding, on the quantum level, would have presumably evolved into the macroscopic forms we presently observe. However, some living systems might presently still exist on the quantum scale, remnants of the original life forms. Is there evidence for the occurrence of such life (so-called "nanolife")? Given that the wavelength of light is from 400 to 700 nm, such life would tend to be this size or less to be affected by quantum effects. At the lower end of this range, microbes of the order of 400 nm in width were recently found in volcanic waters off Iceland (*Daily Telegraph*, UK, Feb. 5, 2002). These evidently belong to a new group of archaea. Preceding downward in size, viruses are of the order of 100 nm in width and, so, likewise qualify as nanolife. However, many people do not consider them fully alive because they are parasitic and need a host for reproduction. On the same level of size (10 to 150 nm), so-called "nanobes" have been found in ancient sandstone about 3 mi below the seabed of Western Australia. These are smaller than cells, fungi, or bacteria. Also, tiny fossil microbes of size 20 to 200 nm have been found in a 4.5-billion-year-old Martian meteorite found in Antarctica. Many scientists believe it possible that primitive life forms of 50 nm have also existed. By comparison, 10 DNA molecules have a total width of about 20 nm, so that the complexity of 50-nm life would be very low and it would be quite primitive. [*New York Times* Web site www.nytimes.com/library/national/science/011800-sci-space-nanobes.html]. As to whether these life forms originated in the manner described above remains a fascinating question.

A preliminary version of this approach to the L–V equations may also be found in [42].

Appendix: Inflationary Models

A very young universe is thought to exert sufficient compressive pressure on its mass particles to produce *mini black holes* [2]. These are the size of atoms ($\sim 10^{-8}$ cm) or even the Planck length ($\sim 10^{-35}$ m). Smolin [2, 18, 19] posits that every universe arises, in fact, as a "baby" BH, that expands outward and begets other BHs, which likewise become universes. The ensemble of such universes is called a "multiverse." Linde [20, 21], Guth [22], and Hawking [23] have also pioneered this idea of an inflationary, self-reproducing multiverse. For brevity we therefore call it the SLGH model.

A key assumption of the SLGH model is that each such outburst of a BH is effectively a time reversal of the collapse of a massive object into a BH. (This was anticipated over 30 years ago by R. Penrose and S. Hawking.) Another assumption is that each such baby universe is not a perfect replica of its parent but a slightly changed or "mutated" form. Its mutations are in the form of slightly altered values of the universal physical constants y_n. It is reasonable that certain sets of these will be better than others for the purpose of producing offspring. Therefore the "fittest" universes (concept due to David Hume [43, 44]) will tend to dominate after many generations.

It is useful to compare this process with the famous definition of *natural selection* as provided by Darwin [45]:

Natural selection can be expressed as the following general law:

(i) If there are organisms that reproduce, and
(ii) if offspring inherit traits from their progenitor(s), and
(iii) if there is variability of traits, and
(iv) if the environment cannot support all members of a growing population, then
(v) those members of the population with less-adaptive traits (determined by the environment) will die out, and then those members with more-adaptive traits (determined by the environment) will thrive. The result is the evolution of species.

Viewed in this light, the SLGH model describes a system undergoing natural selection. In particular its cosmological attributes y_n act like mutation-prone, inheritable genes in the organisms (universes) of a biological system (multiverse). Overall, the SLGH model amounts to a biological model for the creation and propagation of universes. On this basis, one could predict that the population dynamics of such universes should obey the "biological" laws of growth (5.13), for certain choices of growth g and depletion d coefficients.

Of course, many aspects of the SLGH model can be questioned. For example, a seemingly paradoxical result is that the "natural" size for a black-hole universe is down in the subatomic region, on the scale of the Planck length 10^{-35} m, which is of course much smaller than our universe. Thus, the one piece of data we have on sizes of universes seems to violate the model.

However, SLGH theory predicts that the net number $(g - d)$ of new universes that are produced per generation is roughly *proportional to the volume V* of the parent universe,

$$(g - d) \propto V. \tag{5.A1}$$

This guarantees that larger universes reproduce the most successfully by leaving the largest number of progeny after many generations. It results that *tiny* universes would become *extinct* over time. Thus, those that do survive would be comparable in size to our own (see below), which seems a valid prediction since it is consistent with the evidence of our existence.

SLGH theory also assumes that a stupendous number of BHs exist, each a potential universe. (Our universe alone is estimated to contain 10^{240} "virtual" quantum BHs [2].) Such a number can be accounted for by the model assumption that each such potential universe actually contains almost no mass. For example, in our universe it is thought that the amount of positive mass–energy is nearly cancelled by the total amount of negative energy due to gravitational attraction. The net difference can be the order of but "a few ounces" [22, 23].

SLGH theory is also questioned on the grounds that the "fittest" BH universes, i.e., those that evolved in the direction of maximum production of offspring universes, would not necessarily contain stars and living creatures, and therefore would not necessarily lead to our existence. This question is addressed at the end of Section 5.1.7.

As we saw, the commonality of the power law Eq. (5.4) to biology and cosmology offers support for the SLGH model. The principal distinction between biological and cosmological uses of Eqs. (5.4) is that in biological uses m is the mass of a randomly observed organism from a population, whereas in cosmological uses m is, by (5.1) and (5.2), a *single* mass, i.e., the mass of the known universe. However, if Eq. (5.4) is to describe both biological and cosmological *systems* in the same way, the mass m should be selected *in the same way* for *either* type of system. This can be satisfied if the cosmological mass m is taken to be the Eddington mass of a *randomly selected universe* from the SLGH multiverse. The multiverse then acts in this regard as an ecology of universes, some of which are "fitter" than others to produce viable offspring.

Equation (5.13) offers further support for the SLGH model, in describing cosmological growth by the same law that describes biological population growth. Further consistency of the power-law effect (5.4) with the SLGH model is that the model was also used to justify the cosmological example $n/4 = -3/4$ in Table 5.2. See Section 5.1.6, beginning.

Acknowledgments. We thank Professor Edward Harrison for reading, and providing valuable comments on, a preliminary version of Section 5.1. We found his book [2] to provide a particularly fascinating introduction to cosmology, including modern inflationary theories. The book, in fact, inspired us to do the research reported in Section 5.1. Of course we take responsibility for all errors of commission and omission in carrying through on it.

6
Encryption of Covert Information Through a Fisher Game

RAVI. C. VENKATESAN

Securing covert information has immense ramifications in a variety of strategically important disciplines. A novel and promising method for encrypting covert information in a host distribution is presented here. This is based upon the use of Fisher information, in particular its quantum mechanical connotations. This use of Fisher information nicely supplements its use in other chapters of the book, where it is shown how Fisher information can be used to discover scientific knowledge. Here we show how to use Fisher information to secure that knowledge.

6.1. Summary

Encryption and decryption are important problems of the information age. A novel procedure is developed for encrypting a given code (covert information) in a host statistical distribution. The procedure is developed from the perspective of a problem of probability density reconstruction from incomplete constraints (physical observables). This is attacked within the framework of a Fisher game. Information encryption is achieved through unitary projections of the code onto the null space of the ill-conditioned eigensystem of the host statistical distribution. This is described by a time-independent Schrödinger-like wave equation. This Schrödinger wave equation (SWE) contains an empirical pseudopotential defined by undetermined Lagrange multipliers. The latter are accurately obtained by employing a modified game corollary. A secret key to the covert information is found by exploiting the extreme sensitivity of the eigensystem of the statistical system to perturbations. Favorable comparisons are made with an analogous maximum entropy formulation. Numerical examples of the reconstructed probability densities and code exemplify the efficacy of the model.

6.1.1. Fisher Information and Extreme Physical Information

Sections 6.1.1 to 6.1.14 briefly review, and expand upon, some of the introductory material on Fisher information set out in Chapter 1. Also, new terminology and material are presented. Lastly, the overall problem is defined, the game corollary is discussed, and the objectives of the approach to be taken in Section 6.2 and beyond are mapped out.

The Fisher information measure (FIM) is the subject of intense research in areas as diverse as quantum physics, financial mathematics, carcinogenesis, statistics, and information geometry [1, 2]. The FIM governs the amount of information that is obtained in an ensemble of data. The measure is related to the Fubini–Study metric defined in Riemannian space, which has recently received much attention in quantum information processing [3]. Extreme physical information (EPI) [1] utilizes the difference $I - J$ of two FIM measures (Chapter 1 and below). EPI is a self-consistent and principled methodology to elicit physical laws on the basis of observation (measurements). EPI theory involves the interaction of a gedanken observer and the system under observation. The observer measures the system, which responds as a result of this interaction. The data collected by the observer are assumed to be identically and independently distributed and have an FIM which is, in particular, the trace of the Fisher information matrix. This is known henceforth as the Fisher channel capacity (FCC) [1] and is denoted by I. The FCC is the Fisher-equivalent of the Shannon quantal entropy (e.g. [4]). The Shannon quantal entropy $S_Q = -Tr(\rho_d \ln \rho_d)$ is defined in terms of the *trace of the density operator*.

If I is the information in the measurements that are initiated by the observer, the system's response to the measurement activity is the bound information J. In this chapter, the space inhabited by the observer is referred to as measurement space, and the space occupied by the physical or empirical system subjected to measurements is known as system space.

6.1.2. Fisher Game

The above interplay between the observer and the system effectively results in a competitive game [5] of information acquisition. This is referred to as the *Fisher game* (called the information game in Chapter 1). The game is largely a mnemonic device for picturing the Fisher measurement process, but also has practical consequences (as in this chapter). A simplified description of the game is as follows. The observer would like to maximize her/his level of acquired information about the system. However, the system contains but a finite amount of information, so that any gain in information by the observer is at the expense of the system. The system is reluctant to part with its information. Therefore, the observer and system effectively play a zero-sum game, with information as the prize. The system is characterized as a demon since it (selfishly) seeks to limit transfer of information to the observer. A play of the game gives rise to the law of science governing formation of the observation. Thus, the observer (the "participatory observer" of Chapter 1) and the demon together form laws of nature as the outcome of a game. The demon will likewise play a key role in the coding approach, in particular through the "game corollary" (Chapter 1).

6.1.3. Extreme Physical Information vs Minimum Fisher Information

The EPI theory of Frieden is built upon the mathematical foundations laid by the Dutch mathematician Stam [6], and, the Szilard–Brillouin theory of measurements [7]. A superficially similar theory to EPI is the minimum Fisher information (MFI) principle of Hüber [8]. Both EPI and MFI commence with a Lagrangian functional that contains the FCC as the measure of uncertainty. EPI theory guarantees a solution if the amplitude $\psi(x)$ of the FCC space is related to the amplitude $\varphi(\tilde{x})$ of the bound information J space by a unitary transform $\psi[x] = U\varphi[\tilde{x}]$ (the amplitudes defined in Section 6.1.5). Here U is a unitary operator, and coordinates x and $\tilde{x}$ are those of the conjugate measurement space and system space, respectively. In quantum mechanics, the Cartesian coordinate x and the linear momentum $\tilde{x} = p_x$ are the simplest examples of mutually conjugate spaces.

By comparison, MFI utilizes ad hoc constraint terms in place of the bound information J. In the basic MFI model, no unitary space is used. Its solutions are regarded as smooth approximations to ground truth (see Section 1.5, Chapter 1).

6.1.4. Time-Independent Schrödinger Equation

In time independent scenarios, information theoretic investigations based on Lagrangians (cost functions) containing the FIM as the measure of uncertainty yield on variational extremization an equation similar to the time-independent Schrödinger equation (TISE) [9] (see also Chapter's 2 and 4 in this context). The Schrödinger equation is a fundamental equation of physics, which describes the behavior of a particle under the influence of an external potential $V(x)$

$$\underbrace{-\frac{\hbar^2}{2m}\frac{d^2\psi(x)}{dx^2} + V(x)\,\psi(x)}_{H^{QM}\psi(x)} = E\psi(x), \tag{6.1}$$

where E is the total energy, and, H^{QM} is the time independent Schrödinger Hamiltonian operator. For *translational invariant* families of distributions, $p(y\,|\theta) = p(y - \theta) = p(x)$ [1, 10]. Here, y is a measured random variable parameterized by θ (the "true" value), and $x = y - \theta$ is a *fluctuation*, i.e., a random variable. In 1-D, the Fisher channel capacity (FCC) is

$$I^{FCC} = \int dy p(y\,|\theta)\left(\frac{\partial \ln p(y\,|\theta)}{\partial \theta}\right)^2 = \int dx p(x)\left(\frac{\partial \ln p(x)}{\partial x}\right)^2$$

under *shift invariance*. We emphasize that these interpretations of the operators and eigenvalues of the TISE (6.1) as physical quantities are only motivational in nature. We use (6.1) as a numerical tool, not a physical effect. For example, we later describe a "pseudo-potential" rather than an actual potential.

6.1.5. Real Probability Amplitudes and Lagrangians

Introducing the probability amplitude (wave function), which is related to the probability density function (PDF) as $\psi(x) = \sqrt{p(x)}$, the FCC acquires the compact form

$$I^{FCC} = 4 \int dx \left(\frac{\partial \psi(x)}{\partial x} \right)^2. \tag{6.2}$$

Next, the MFI, EPI, and MaxEnt approach to the problem of reconstructing $\psi(x)$ are discussed. The MFI approach extremizes I^{FCC} subject to a set of known constraints:

$$L^{FIM} = 4 \int dx \left(\frac{\partial \psi(x)}{\partial x} \right)^2 + \int dx \sum_{i=1}^{M} \lambda_i \Theta_i(x) \psi^2(x) - \lambda_o \int dx \psi^2(x), \tag{6.3}$$

where the Lagrange multiplier λ_o corresponds to the PDF normalization condition $\int dx \psi^2(x) = 1$. The Lagrange multipliers $\lambda_i; i = 1, \ldots, M$ correspond to actual (physical) constraints of the form $\int dx \Theta_i(x) \psi^2(x) = \langle \Theta_i(x) \rangle = d_i$. Here, $\Theta_i(x)$ are operators and d_i are the constraints (physical observables). It is important to note that in (6.3), the FCC acquires the role of a kinetic energy, and the *empirical bound information* effectively becomes a potential energy. We emphasize that these roles are but useful mnemonic devices rather than being physical, since our TISE is descriptive of a general problem of knowledge acquisition (6.3) rather than a physical action phenomenon.

EPI deals with a *physical information* defined as $K = I^{FCC} - J$. Hence, to provide an EPI connotation to the Lagrangian (6.3), we assume that the empirical terms correspond to the EPI bound information J. Specifically, we define $J[x] = -\int dx \sum_{i=1}^{M} \lambda_i \Theta_i(x) \psi^2(x) + \lambda_o \int dx \psi^2(x)$. Note that in (6.3) a negative sign has been ascribed to the PDF normalization term in (6.3) in order to achieve consistency with the TISE in physics.

In EPI theory, the bound information J is self-consistently derived on the basis of an invariance principle [1], rather than empirically as described herein. Thus, (6.3) may be construed as an empirical manifestation of $K = I^{FCC} - J$. See also Chapter 1 in this regard.

An equivalent MaxEnt cost function [11] is

$$L^{ME} = - \int dx p(x) \ln p(x) + \sum_{i=0}^{M} \mu_i \langle \Theta_i(x) \rangle. \tag{6.4}$$

Note that the Lagrange multipliers λ_i and μ_i generally have different values.

In inference problems, the case of incomplete constraints corresponds to scenarios, where $M < N$. Here, M is the number of constraints and N is the dimensionality of the statistical distribution. When expressed in the form of a matrix operator (Section 6.2), the case of incomplete constraints usually results in the operator being ill conditioned.

6.1.6. MFI Output and the Time-Independent Schrödinger-Like Equation

Variational extremization of (6.3) yields the MFI result

$$\underbrace{-\frac{d^2\psi(x)}{dx^2} + \frac{1}{4}\sum_{i=1}^{M}\lambda_i\Theta_i(x)\psi(x)}_{H^{FI}\psi(x)} = \frac{\lambda_o}{4}\psi(x),\qquad(6.5)$$

where H^{FI} is an empirical Hamiltonian operator. Here, (6.5) is referred to as a time independent Schrödinger-*like* equation (TISLE). The familiar MaxEnt solution is

$$\psi(x) = \exp\left[\frac{-1}{2}\sum_{i=1}^{M}\mu_i\Theta_i(x)\right] = \exp[\Phi(x)].\qquad(6.6)$$

Comparing (6.5) with the TISE in physics (6.1), it is immediately obvious that the constraint terms that describe the physical observables constitute an empirical pseudo-potential. The normalization Lagrange multiplier and the total energy eigenvalue relate as $\lambda_o = 4E$, and $\hbar^2/2m = 1$, where $\hbar$ is the Planck constant and m is the particle mass. Again, these are effective relations enforced for the sake of operational convenience, and not physical correspondences.

6.1.7. Quantum Mechanical Methods in Pattern Recognition Problems

The above discussion establishes the basis to describe the leitmotiv for utilizing the FIM as a measure of uncertainty in problems of inference. A primary advantage of utilizing the FIM in problems of inference and knowledge discovery stems from the fact that it facilitates a quantum mechanical connotation to statistical problems. This feature was first effectively utilized in pattern recognition, for the case of quantum clustering, in the seminal works of Horn and Gottlieb [12].

Later work by Venkatesan [13] provided an information theoretic basis for the cost functions of possibilistic c-means (Pc-M, *fuzzy*) clustering. This was accomplished with the aid of the EPI/MFI principles, and a discrete variational complex [14] for continuum-free variational models based on the seminal work of T.D. Lee [15]. A significant result of this work was that the constraint terms in the cost functions of Pc-M clustering satisfied the Heisenberg uncertainty principle. Also, the FIM provides a means of solving inference problems through the solution of differential equations, instead of an algebraic problem encountered in the MaxEnt solution procedure.

6.1.8. Solutions of the TISE and TISLE

The solution of (6.5) is a superposition of Hermite–Gauss (H–G) expansions, which are the general solutions of the Boltzmann transport equation [1, 16]. The

H–G solutions are obtained as a consequence of solving a differential equation. These solutions are a lowest-order *equilibrium state* and higher-order effectively *excited states*.

The basic MaxEnt solution (6.6) is not capable of reproducing H–G solutions [1, 17, 18]. Instead, deviations from equilibrium are accounted for by addition of constraint terms and their concomitant Lagrange multipliers. For constraints of the geometric moment type $\Theta_i(x) = x^i; i = 0, \ldots, M$, it is easily demonstrated that the solutions of (6.5) and (6.6) exactly coincide for the case of Maxwell–Boltzmann PDF's $\Theta_i(x) = x^i; i = 0, 1, 2$, contingent to the Lagrange multipliers λ_i and μ_i satisfying certain relations (Appendix A in [1]). This equivalence relation is further expanded upon in this chapter, and its relevance to the encryption model is highlighted.

Substituting (6.6) and a pseudo-potential in the form of a polynomial *ansatz* into the TISLE (6.5), a family of pseudo-potentials and amplitudes/PDF's that minimize the FIM and maximize the Shannon entropy is derived. Specifically, the resulting pseudo-potentials and amplitudes simultaneously satisfy both Hüber's MFI principle and the MaxEnt principle of Jaynes. Details of the calculation are provided in Section 6.3.2.

The significance of the above procedure is that it provides the minimum input information, in the form of the expectation of the coordinates $\langle x^i \rangle; i = 1, \ldots, M$, required to accurately reconstruct a physical TISE potential, approximated by a polynomial ansatz, and the concomitant amplitudes through the process of inference.

The TISE in physics (6.1) relates the number of energy states of the particle in a potential well to the wave function (and consequentially the PDF), through the solution of an eigenvalue problem. Solutions exist for the TISE only for certain values of energy, called the *energy eigenvalues*. Corresponding to each eigenvalue is an eigenfunction. The *zero-energy (ground) state*, which is the lowest energy level, corresponds to exponential PDF's that constitute the leading-order term of the H–G expansion. Study of the *zero-energy state* permits important comparisons to be drawn on the relative abilities of the FIM and Shannon entropy in problems of inference.

In inference problems, the Hamiltonian operator H^{FI} is unknown because the form of the pseudo-potential is undetermined. The pseudo-potential corresponding to a *physical* TISE potential is approximated from the constraints. Subsequently, the amplitude and associated PDF are inferred by solving (6.5) as an eigenvalue problem [19].

6.1.9. Information Theory and Securing Covert Information

The needs for security and privacy protection are rapidly gaining prominence as important problems. These needs are of course basic to knowledge acquisition and knowledge discovery. The encryption of covert information, in particular, constitutes a critical issue in security and privacy protection. Cryptography, steganography, and digital watermarking constitute some of the most prominent and vigorously researched areas in security and privacy protection. Steganograply and digital watermarking are areas of study closely related to cryptography, which are used to address digital rights management, protect information, and conceal covert information [20]. While steganography and digital watermarking share a similar goal with cryptography—the hiding of covert information—there are certain significant differences between these tools, as described in the following discussion.

Steganography. By definition, steganography involves hiding data inside other (host) data. The formats of the covert data and that of the host could differ substantially. For example, a text file could be hidden "inside" an image or even a sound file. By looking at the image, or listening to the sound, an attacker/intruder would not know that there is extra information present. Specifically, in steganography, the covert information is concealed in a manner such that no one apart from the intended recipient knows of the existence of the message. This is in contrast to cryptography, where the existence of the message is obvious, but the meaning is obscured.

Digital watermarking. Digital watermarking involves the insertion of a pattern of bits into a digital image, audio or video file that identifies the file's copyright information (author, rights, etc.). The purpose of digital watermarks is to provide copyright protection for intellectual property that is in digital format.

Concept of keys. Before proceeding further, we briefly explain the concept of a *key* in the encryption of covert information. The distribution of keys is an issue of primary concern in cryptography and allied disciplines. A key may be a program, a number, or a string of numbers that enables the legitimate recipient of the message (decrypter) to access the covert information.

In cryptography, a secret, shared, or private key is an encryption/decryption key known only to the entities that exchange secret messages. In traditional secret key cryptography, a key would be shared by the communicators so that each could encrypt and decrypt messages. The risk in this system is that if either party loses the key or it is stolen, the system is broken.

A more recent alternative is to use a combination of public and private keys. In this system, a public key is used together with a secret key. The RSA protocol[1] is a prominent example of a *public key infrastructure* (PKI). A PKI often employs a *key ring strategy*. Specifically, one key is kept secret while the others are made public. PKI is the preferred approach on the Internet. The secret key system is sometimes known as *symmetric cryptography* and the public key system as *asymmetric cryptography*. In general, contemporary approaches in securing keys are based on the degree of difficulty in computing (e.g., integer factorizing) certain functions [21, 22].

Information Theory. Since the pioneering work of Claude Shannon who discovered the Shannon entropy as a consequence of his wartime work on cryptography,

---

[1] RSA is an acronym for Rivest–Shamir–Adelman. The first reference on the RSA cryptosystem algorithm is R.L. Rivest, A. Shamir, and L. Adelman. On digital signatures and public key cryptosystems. MIT Laboratory for Computer Science Technical Memorandum 82 (April 1977).

information theoretic measures of uncertainty have often been employed in securing covert and proprietary information. The Shannon entropy has been utilized in cryptography to study public key distribution [21, 23]. The Renyi entropy has been employed by Cachin [24] to provide an information theoretic foundation for steganography. The Fisher information has been used by Moulin and Ivanovic [25] to provide an optimal design for synchronization in blind watermarking. Encryption of covert information in a host distribution has recently been studied within the framework of the MaxEnt theory by Rebollo-Neira and Plastino [26].

6.1.10. Rationale for Encrypting Covert Information in a Statistical Distribution

This chapter presents a novel and self-consistent information theoretic formulation to encrypt covert information into a host statistical distribution. The latter is inferred from an incomplete set of constraints (physical observations) in a Fisher game. The game possesses quantum mechanical connotations.

Embedding covert information in a host distribution has a number of advantages, the most prominent of which being that the amount of information that can be encrypted increases with the dimension of the statistical distribution. There are three other distinct advantages of immense practical importance, which accrue as a consequence of embedding covert information in a statistical distribution. First, embedding covert information in a statistical distribution allows for a single unified model that possesses critical characteristics of *both* cryptography and steganography.

The example presented in Section 6.4 is a demonstration of the cryptographic applications of this model. Also, an example in steganography follows from the statistical representation of image feature space [27]. Thus, image features may be encrypted into a given statistical distribution without significantly altering the nature of the host distribution. Work along these lines is presently being carried out by the author.

Next, as is demonstrated in Section 6.4 and described in Section 6.5, the model described in this chapter allows for both secret key and public key distribution strategies. The formulation of keys and their subsequent distribution is a distinguished feature of *quantum cryptography* [28]. Key distribution in quantum cryptography is performed with the aid of either the Heisenberg uncertainty principle, or entanglement.

Quantum key distribution protocols were pioneered by the Bennett–Brassard model [29–31]. Quantum transmissions presently do not facilitate the transfer of large amounts of information. By comparison, *the encryption strategy presented in this chapter may be efficiently employed for large amounts of data*. The key is a single number denoting the perturbation of the values of an element of an operator describing the eigenstructure, or a string of numbers. This is a part of a statistical encryption model or procedure, with a quantum mechanical connotation, that may be effectively employed to complement existing key distribution protocols in quantum cryptography.

6.1.11. Reconstruction Problem

This chapter studies a problem of density reconstruction, also known as *the inverse problem in statistics*, within the framework of a Fisher game. To provide a practical application that is of immense contemporary relevance, the reconstruction model is studied as part of an attempt to embed covert information in a host statistical distribution. The latter is inferred from an incomplete set of constraints (physical observations).

The specific problem is as follows: Given physical observables $d_1, \ldots, d_M$, the zero-energy amplitudes and PDF's are obtained by solving (6.5) as an eigenvalue problem. This is accomplished by inferring a TISLE pseudo-potential $V(x) = \sum_{i=1}^{M} \lambda_i \Theta_i(x) = \sum_{i=1}^{M} \lambda_i x^i$ corresponding to a TISLE potential in physics. Here, (6.5) is solved iteratively. After each iteration, "optimal" values of the Lagrange multipliers are obtained that are consistent with (i) the quantum mechanical virial theorem and (ii) the *modified game corollary* (described in Section 6.1.13). The solution is by a constrained optimization procedure. Details of the procedure are provided in Section 6.3.3, and numerical examples that demonstrate the efficacy of the Fisher game are presented and analyzed in Section 6.4. The inferred PDF constitutes the host for the covert information.

6.1.12. Ill-Conditioned Nature of Encryption Problem

The chief effect behind the proposed approach is the ill-conditioned nature of the given statistical system (see Section 6.2.2). The eigenstructures of statistical systems inferred from incomplete constraints are usually ill conditioned. The ill-conditioned eigenstructure of the inferred PDF is decomposed into its respective *range* and *null spaces* [32]. By definition, given an operator A, the null space is the set of all solutions $|b\rangle$ corresponding to $A|b\rangle = 0$. All information pertinent to the distribution is instead contained in the *range space*.

The null space is data-independent. We use it as a convenient "reservoir" for encrypting the covert data d_i. Thus, what previously was regarded generically as input data is now specifically made the covert message to be encrypted. This covert information (code) is embedded into the null space with the aid of a unitary operator (encryption operator) A. This is derived in Section 6.2.3.

The eigenstructure of the ill-conditioned system is extremely sensitive to perturbations. The condition number of a matrix operator may be evaluated using the MATLAB® function $cond(\bullet)$. Given a matrix operator A, values of $cond(A) > 1$ imply that the operator is sensitive to perturbations. This sensitivity is exploited in order to obtain a secret key. Specifically, a strategy to formulate a secret key is introduced by perturbing a single element of the eigenstructure of the operator that represents a combination of the range space of the host PDF and the embedded code.

A PKI may be obtained by perturbing more than one element of the eigenstructure operator. This results in a string of keys being derived. Next, a decryption operator, which is the adjoint of the encryption operator, is derived. The decryption operator enables recovery of the code. It is important to note that the unitary encryption and decryption operators are formulated by determining the normalized eigenvectors of the *null space of an operator*. These eigenvectors have eigenvalues of zero. This utilizes a matrix operator that represents the constraints, and is accomplished with the aid of singular value decomposition (SVD) [32]. Details of the implementation of the above procedure are described in Section 6.4.

6.1.13. Use of the Game Corollary

The *game corollary* introduced by Frieden [1] seeks to obtain undetermined coefficients by imposing the constraint that the demon plays the last move in the competitive game. This is accomplished by maximizing the uncertainty of the *observer* by specifying that the FCC defined in terms of the undetermined coefficient satisfies $dI^{\mathrm{FCC}}(c)/dc = 0$. Specifically, the coefficient c extremizes the FCC. Usually the extremum is a minimum.

This notion of the game corollary is reasonable for simple problems, and notably does not take into account the bound information. This chapter introduces a modification of practical significance to the original notion of the game corollary, by including the contribution of the bound information, on the intuitive grounds that it is a manifestation of the FCC/FIM. For complex problems possessing undetermined Lagrange multipliers, the modified game corollary may be defined within an iterative setting in terms of a constrained optimization problem as

$$\frac{\partial \left[I^{\mathrm{FCC}} - \Im(\lambda_i) \right]}{\partial \lambda_i} = \frac{\partial R(\lambda_i)}{\partial \lambda_i} \rightarrow 0; \quad i = 1, \ldots, M. \tag{6.7}$$

Here, $R(\lambda_i)$ is a residual "mismatch" between the bound information at a specific iteration level and the FCC. This permits the Lagrange multiplier to be formed by a regularized procedure, during simultaneous iterative reconstruction of PDF's from observational data. This calculation is based on information theoretic and game theoretic criteria.

In summary, the encryption formulation described in this chapter is physically tantamount to embedding the covert information into the pseudo-potential of a TISLE. The secret key is the perturbation. Note in particular that *it is not the projection unitary operator that encrypts the covert information into the null space of the eigenstructure.*

6.1.14. Objectives to be Accomplished

This chapter is concluded by satisfying a threefold objective:

(1) The implications of the measure of uncertainty on the encryption process are briefly elucidated. Specifically, the effect of the choice of the measure of

uncertainty on the sensitivity of the eigenstructure of the host distribution is briefly examined.

(2) Arguments are advanced indicating that the encryption model described herein is immune to *plaintext attacks*.[2] By definition, a *plaintext attack* is one where the prior messages have intercepted and decrypted prior messages in order to decrypt others [21, 22].

(3) Finally, a qualitative extension of the encryption model presented herein is briefly discussed. Specifically, the unitary projection of select portions of the code into different *eigenstates* of the TISLE is suggested.

6.2. Embedding Covert Information in a Statistical Distribution

6.2.1. Operators and the Dirac Notation

Consider M constraints (physical observables) $d_1, \ldots, d_M$ that are expectation values of a random variable $x_{i,n}; n = 1, \ldots, N$ according to the probability p_n as

$$d_i = \sum_{n=1}^{N} p_n x_{i,n}, \quad i = 1, \ldots, M; \quad \sum_{n=1}^{N} p_n = 1. \tag{6.8}$$

The number of available measurements M is taken to be less than the dimension of the probability space N. The primary stage in the theory of embedding in a statistical distribution is to describe the constraints in a matrix form. This is accomplished in a compact form by evoking the Dirac *bra-ket* notation [33, 34] to describe linear algebraic operations.[3] In the Dirac notation, a *bra* $\langle \bullet |$ signifies a row vector, and, a *ket* $| \bullet \rangle$ denotes a column vector. The probability density $|p\rangle \in R^N$ is a column vector (ket), where $|n\rangle, n = 1, \ldots, N$ is the standard basis in R^N, is expressed as

$$|p\rangle = \sum_{n=1}^{N} |n\rangle\langle n|p\rangle \sum_{n=1}^{N} p_n |n\rangle. \tag{6.9}$$

Here, $\langle \bullet | \bullet \rangle$ denotes the *inner product* between two vectors. Given basis vectors $n, m, \langle n|m \rangle = \delta_{nm}$ defines the Krönecker delta. Within the context of (6.9), the operation $\langle n|p \rangle$. extracts the nth entry of $|p\rangle$. In the Dirac notation, projections are defined by the *outer product* $|\bullet\rangle\langle\bullet|$.

It is convenient to restate (6.8) in a more compact manner. Defining the column vector of observations as $|d\rangle \in R^{M+1}$ with components $d_1, \ldots, d_M, 1$ and an

---

[2] In cryptography, the information to be encrypted is known as a plaintext.
[3] While physicists prefer the compact representation of the Dirac notation, mathematicians loathe it because it ignores certain exceptions!

operator $A: \mathbf{R}^N \to \mathbf{R}^{M+1}$ given by

$$A = \sum_{n=1}^{N} |x_n\rangle\langle n|, \tag{6.10}$$

define vectors $|x_n\rangle \in \mathbf{R}^{M+1}, n = 1, \ldots, N$ as the expansion

$$|x_n\rangle = \sum_{i=1}^{M+1} |i\rangle\langle i|x_n\rangle = \sum_{i=1}^{M+1} x_{i,n}|i\rangle, \tag{6.11}$$

where i is a basis vector in $\mathbf{R}^{M+1}$, (6.8) may be expressed simply as

$$|d\rangle = A|p\rangle. \tag{6.12}$$

6.2.2. Eigenstructure of the Constraint Operator

It is evident from (6.10) and (6.12) that the unity element in $|d\rangle \in \mathbf{R}^{M+1}$ enforces the normalization constraint of the probability density. Here the operator $A : \mathbf{R}^N \to \mathbf{R}^{M+1}$ is ill-conditioned and rectangular. Thus, (6.12) may be expressed as

$$|p\rangle = \underbrace{A'^{-1}|d\rangle}_{\text{Range space}} + \underbrace{|p'\rangle}_{\text{Null space}}, \tag{6.13}$$

where A'^{-1} is the pseudoinverse [32] of A defined in $range(A)$, and $A'^{-1}|d\rangle$ contains all the data-related information.

The *null space* term in (6.13) is of particular importance in the encryption strategy introduced in this chapter, since the code is embedded into it via a unitary projection. By definition, given a column vector $|b\rangle$, the null space of A is the set of all solutions corresponding to $A|b\rangle = 0$. Thus, (6.13) states that $|p'\rangle$ lies in the null space of A, i.e., $|p'\rangle \in null(A)$ is data independent. However, it is critically dependent on the solution methodology employed to solve (6.13).

The system (6.13) does not have a unique solution $|p\rangle$. The solution is dependent upon a number of factors, such as the measure of uncertainty (or disorder) and the inference process. *However, the process of embedding information into a statistical distribution via unitary projections into $|p'\rangle \in$ null(A) is independent of the measure of uncertainty.* At this stage it is assumed that the inference of the PDF provides us with the values of the operator A and the column vector $|p\rangle$. A procedure for systematically categorizing the operator A on the basis of physical potentials admitted by the TISE (6.1) is presented in Section 6.4.

To define the unitary projections of the embedded code, we introduce the operator $G = A^\dagger A$. Here, $A^\dagger$ is the conjugate transpose of A. For real matrices, $A^\dagger = A^T$.[4] The SVD is effectively utilized to evaluate the range and null spaces of the operator G.[5] The unitary encryption and decryption operators are obtained by

---

[4] The conjugate transpose is sometimes referred to as the adjoint. This is not to be confused with the adjugate of a matrix involved in matrix inversion, which is also sometimes referred to as the adjoint.

[5] The SVD of A is equivalent to the eigenvalue decomposition of the operator G.

determining the normalized eigenvectors that correspond to the zero eigenvalues of the operator.

It is of course legitimate to evaluate the unitary encryption and decryption operators by considering the operator A instead of G. However, our strategy is to use G in Section 6.2.3 and beyond.

As one might expect, the "loss of information" caused by floating point errors could make the evaluation of $null(G)$ using SVD prohibitively unstable for many applications. Specifically, this operation squares the condition number, resulting in the large singular values being increased and the small singular values decreased.[6] In fact *the unique distinction of the encryption model presented herein is that this instability is used to the advantage of the designer, in increasing the security of the embedded covert information.*

6.2.3. The Encryption and Decryption Operators

Assuming the availability of the operator A and the probability vector $|p\rangle$, whose inference is described in Section 6.3, the normalized eigenvectors corresponding to the eigenvalues in the null space of G having value zero are defined as $|\eta_n\rangle$; $n = 1, \ldots, N - (M + 1)$. In practice, as will be described in Section 6.4, the values of $|\eta_n\rangle$ are efficiently obtained by using the MATLAB® function $null(\bullet)$. Using these eigenvectors, the decryption operator $\hat{U}_{\text{dec}} : R^N \to R^{N-(M+1)}$ is defined as [26]

$$\hat{U}_{\text{dec}} = \sum_{n=1}^{N-M-1} |n\rangle\langle\eta_n|. \tag{6.14}$$

The adjoint of the decryption operator is the encryption operator defined by

$$\hat{U}_{\text{enc}} = \hat{U}_{\text{dec}}^{\dagger} = \sum_{n=1}^{N-M-1} |\eta_n\rangle\langle n|. \tag{6.15}$$

Note that $\hat{U}_{dec} \bullet \hat{U}_{enc} = I \in R^{N-(M+1)}$, where I is the identity operator. This section concludes with a description of how a code $|q\rangle$ may be embedded into a host statistical distribution defined by the vector $|p\rangle$.

6.2.4. Generic Strategy for Encryption and Decryption

Given a code $|q\rangle \in R^{N-(M+1)}$, the $N - (M + 1)$ components are given by $\langle n| \rangle = q_n; n = 1, \ldots, N - (M + 1)$. A simple example of such a code is a random string generated by the MATLAB® random number generator $rand(\bullet)$. The PDF of the

---

[6] It may be demonstrated that by considering the operator G instead of A, the instability/sensitivity to perturbations is substantially enhanced (more than quadrupled). This may be easily evaluated using the MATLAB® function $cond(\bullet)$.

embedded code is defined as

$$|p_c\rangle = \hat{U}_{\text{enc}}|q\rangle = \sum_{n=1}^{N-M-1} |\eta_n\rangle\langle n|q\rangle. \tag{6.16}$$

Given a distribution $|p\rangle$ extremized by a FCC I^{FCC} and $M+1$ constraints (6.3), the code $|q\rangle$ is embedded into $|p\rangle$ as

$$|\tilde{p}\rangle = |p\rangle + |p_c\rangle. \tag{6.17}$$

It is important to note that since $|p_c\rangle$ inhabits the null space of the operator A, $A|p_c\rangle = 0$. The constraint operator A and the column vector $|\tilde{p}\rangle$ are transmitted to the receiver. The constraint vector (physical observations) are recovered by

$$|d\rangle = A|\tilde{p}\rangle. \tag{6.18}$$

The constraint vector $|d\rangle$ is employed to reconstruct the host PDF $|p\rangle$ through extremization of I^{FCC} via the iterative reconstruction procedure described in Section 6.3.3. Since $|p_c\rangle$ inhabits the null space, identically $A|p_c\rangle = 0$. The PDF of the embedded code is obtained from

$$|p_c\rangle = |\tilde{p}\rangle - |p\rangle. \tag{6.19}$$

The encrypted code is recovered by the operation

$$|q\rangle = \hat{U}_{\text{dec}}|p_c\rangle = \sum_{n=1}^{N-M-1} |n\rangle\langle\eta_n|\rho_c\rangle. \tag{6.20}$$

Note that the above procedure describes only the fixed procedure of the encryption/decryption strategy. The derivation of a key by perturbing the eigenstructure of the operator G is empirical in nature, and is discussed in detail in Section 6.4.

6.3. Inference of the Host Distribution Using Fisher Information

This section serves a threefold objective. First, in Sections 6.3.1 and 6.3.2 conditions are derived for obtaining a family of TISLE pseudo-potentials having a polynomial form. These are constructed to yield amplitudes that simultaneously minimize the FIM (FCC) and maximize the Shannon quantal entropy.

Next, an iterative scheme is introduced for reconstructing amplitudes and PDF's from empirical observations. This is accomplished in Section 6.3.3 by evoking the quantum mechanical virial theorem and the modified game corollary. As a consequence, optimal values of the undetermined Lagrange multipliers are obtained via constrained optimization (i.e. steepest descent).

Finally, in Section 6.3.4 a numerical experiment is described, which validates the approach. This is accomplished by comparing the results of (i) the inference of the amplitude from a priori provided physical observables using the Fisher game, (ii) the exact solution of the TISE, and (iii) an equivalent MaxEnt procedure.

6.3.1. Correspondence Between MFI and MaxEnt

The determination of amplitudes that simultaneously satisfy Hüber's MFI principle and the MaxEnt theory of Jaynes involves establishing a relation between the FIM Lagrange multipliers λ_i and the Shannon–Jaynes multipliers μ_i (see Appendix A of [1]). To extend this analysis, the pseudo-potential for a PDF described by (6.6) is a priori specified to be of the form $V(x) = \sum_{i=1}^{N} c_i x^i$, where the coefficient c_i is a combination of Shannon–Jaynes Lagrange multipliers. Substituting (6.6) into the TISE (6.1) yields

$$\left(\frac{d\Phi(x)}{dx}\right)^2 = V(x) - E + \frac{d^2\Phi(x)}{dx^2}; \quad \Phi(x) = \frac{-1}{2}\sum_{i=1}^{M}\mu_i x^i \qquad (6.21)$$

Substituting $V(x) = \sum_{i=1}^{N} c_i x^i$ into (6.21) gives

$$\left(-\sum_{i=1}^{M}\frac{i}{2}\mu_i x^{i-1}\right)^2 = \sum_{i=1}^{N} c_i x^i - E + \sum_{i=1}^{M}\frac{i(i-1)}{2}\mu_i x^{i-2}, \qquad (6.22)$$

where $E = \lambda_o/4$ by comparison of the physical TISE (6.1) and the TISLE (6.5). Broadly speaking, empirical representations of the Schrödinger potential fall into two categories—asymmetric and symmetric. Asymmetric potentials possess both odd and even moment terms. Symmetric potentials have only even moment terms. From (6.22), the FIM normalization Lagrange multiplier relates to the Shannon-Jaynes Lagrange multipliers as $\lambda_o = 4\mu_2 - \mu_1^2$.

For asymmetric potentials to have a physical meaning, $\lambda_o = 4\mu_2 - \mu_1^2 \neq 0$. Violation of this criterion will not yield a TISLE pseudo-potential that corresponds to a physical TISE potential. Consequently, the inference of amplitudes which satisfies the operator relation $H^{QM}\psi(x) = E\psi(x)$ will be inhibited. Likewise, for symmetric potentials, $\lambda_o = 4\mu_2 \Rightarrow \mu_2 \neq 0$. In order that the PDF's $p(x) = \psi^2(x) \in L_2$, M is an even number. Further, $N = 2M - 2$.

6.3.2. Amplitudes and Pseudo-Potentials Satisfying MFI and MaxEnt

The principle that underlies the inference of a polynomial TISLE pseudo-potential of the form $V(x) = \sum_{i=1}^{M}\lambda_i\Theta_i(x) = \sum_{i=1}^{M}\lambda_i x^i$, and its accompanying amplitude, is that the Mth degree polynomial can be exactly reconstructed from M expectation values. However, our problem by definition involves incomplete constraints and hence, it is possible to provide *only a few low-order expectations* as the input values. The methodology described in Section 6.3.1 provides a useful tool to estimate the minimum order of the moments that are required for this purpose.

TABLE 6.1. Expectations required for select Schrödinger potentials.

Name	Potential	Expectation values
Symmetric harmonic oscillator	$V(x) = \frac{kx^2}{2}$	$\langle x^2 \rangle$
Asymmetric harmonic oscillator	$V(x) = \frac{kx^2}{2} + x$	$\langle x \rangle$ and $\langle x^2 \rangle$
Symmetric sextic	$V(x) = x^2 + x^4 + kx^6$	$\langle x^2 \rangle$, $\langle x^4 \rangle$ and $\langle x^6 \rangle$
Morse	$V(x) = k\left(1 - e^{-x}\right)^2$	$\langle x \rangle$, $\langle x^2 \rangle$, $\langle x^3 \rangle$ and $\langle x^4 \rangle$

Note that k is a constant, which differs for each potential.

The asymmetric harmonic oscillator and the symmetric sextic potentials are the simplest forms of physical TISE potentials corresponding to asymmetric and symmetric TISLE pseudo-potentials, respectively. The asymmetric harmonic oscillator amplitude and potential expressed in terms of the MaxEnt Lagrange multipliers, and the relation between the FIM and Shannon entropy Lagrange multipliers are

$$\psi(x) = -\frac{\mu_o}{2} - \frac{\mu_1 x}{2} - \frac{\mu_2 x^2}{2}, \quad V(x) = \mu_2(\mu_1 x + \mu_2 x^2), \quad (6.23)$$

$$\lambda_o = 4\mu_2 - \mu_1^2 \neq 0, \quad \lambda_1 = \mu_1 \mu_2, \quad \text{and} \quad \lambda_2 = \mu_2^2.$$

The corresponding values for the symmetric sextic potential are

$$\psi(x) = -\frac{\mu_o}{2} - \frac{\mu_2 x^2}{2} - \frac{\mu_4 x^4}{2}, \quad V(x) = \left(\mu_2^2 - 6\mu_4\right) x^2 + 4\mu_2\mu_4 x^4 + 4\mu_4^2 x^6$$

$$\lambda_o = 4\mu_2, \quad \lambda_2 = \left(\mu_2^2 - 6\mu_4\right), \lambda_4 = 4\mu_2\mu_4, \quad \text{and} \quad \lambda_6 = 4\mu_4^2. \quad (6.24)$$

Table 6.1 provides a sample of some of the commonly studied Schrödinger potentials, and the expectation values required as input information.

6.3.3. Modified Game Corollary

Left-multiplying (6.5) by the amplitude and integrating over the region of interest, we obtain[7]

$$L^{\text{FIM}} = 4 \int dx \left(\frac{\partial \psi(x)}{\partial x}\right)^2 + \int dx \sum_{i=1}^{M} \lambda_i \Theta_i(x) \psi^2(x) - \lambda_o \int dx \psi^2(x). \quad (6.25)$$

It is obvious that the first term on the LHS of (6.25) is the FCC (6.2) subjected to a single integration by parts, followed by application of the boundary conditions. In accordance with the usual practice for solving the SWE [35], (6.5) is expressed as a tridiagonal matrix. The input values for the pseudo-potential are provided by known moments of the coordinate $\langle x^i \rangle = d_i$; $i = 1, \ldots, M$, corresponding to a physical TISE potential chosen a priori.

The number of moments to be employed is determined in accordance with the procedure described in Sections 6.3.1 and 6.3.2. Arbitrary values for the Lagrange

---

[7] In this analysis, the amplitudes of the TISLE are considered as being real quantities. This is tenable within a 1-D framework.

multipliers are provided as input information to start the calculation. The values of the Lagrange multipliers are iteratively corrected to attain optimal values using the following procedure: The tridiagonal matrix is solved as an eigenvalue problem employing the MATLAB® function $eigs(\bullet)$. For the Nth iteration level, the corresponding wave functions, Lagrange multipliers, and total energy eigenvalue are ψ_N, λ_{iN}, and λ_{oN}, respectively.

It is important to note that the values of ψ_N and the concomitant Lagrange multipliers at some intermediate iteration level N do not satisfy the normalization condition. The Lagrange multipliers are corrected by evoking an iterative representation of the *modified game corollary* (described below). The process is repeated till the desired values of the amplitudes and Lagrange multipliers are obtained.

To recapitulate from Section 6.1.5, the FCC takes on the role of the kinetic energy in the EPI/MFI principle. Evoking the quantum mechanical virial theorem, the FCC relates to the expectation of the linear momentum p_x (Appendix D in [1]) and the expectation of the TISLE pseudo-potential as

$$I^{\text{FCC}} = -4 \int dx \psi(x) \frac{\partial^2 \psi(x)}{\partial x^2} = \frac{4}{\hbar} \langle p_x^2 \rangle = 4 \left\langle \frac{p_x^2}{2m} \right\rangle = \frac{1}{2} \left\langle x \frac{dV(x)}{dx} \right\rangle. \quad (6.26)$$

For polynomial pseudo-potentials of the form $V(x) = \sum_{i=1}^{M} \lambda_i x^i$, we obtain

$$I^{\text{FCC}} = \frac{1}{2} \left\langle x \frac{dV(x)}{dx} \right\rangle = \sum_{i=1}^{M} \frac{i}{2} \langle V(x) \rangle = \sum_{i=1}^{M} \frac{i}{2} \lambda_i d_i. \quad (6.27)$$

The bound information at the Nth iteration level calculated with the true values of $\psi(x)$(the desired values) is

$$J(\lambda_{iN}) = I^{\text{FCC}}(\psi_N) = \lambda_{oN} \langle \psi_N \psi_N \rangle - \sum_{i=1}^{M} \lambda_{iN} \langle \psi_N | \Theta(x) | \psi_N \rangle. \quad (6.28)$$

Here, (6.28) is the *empirical bound information* in (6.3), evaluated at some intermediate Nth iteration level. The Lagrange multipliers in (6.27), λ_i, are the true/desired values to be obtained as a consequence of the constrained optimization procedure, and are undetermined. To make (6.27) consistent with the iteration process, we define the relation between the true/desired Lagrange multipliers, and the Lagrange multipliers at some intermediate Nth iteration level as

$$\sum_{i=1}^{M} \frac{i}{2} \lambda_i \langle \psi(x) | \psi(x) \rangle = \sum_{i=1}^{M} \frac{i}{2} \lambda_i = \sum_{i=1}^{M} \lambda_{iN} \langle \psi_N | \psi_N \rangle. \quad (6.29)$$

The residue (or "mismatch") between the bound information at the Nth iteration level and the FCC calculated with the true values of $\psi(x)$ (the desired values) is

$$R(\lambda_{iN}) = -\lambda_{oN} \langle \psi_N | \psi_N \rangle + \sum_{i=1}^{M} \lambda_{iN} \langle \psi_N | \Theta(x) | \psi_N \rangle + \sum_{i=1}^{M} \lambda_{iN} d_i \langle \psi_N | \psi_N \rangle \quad (6.30)$$

$$\Rightarrow \frac{R(\lambda_{iN})}{\langle \psi_N | \psi_N \rangle} = \tilde{R}(\lambda_{iN}) = -\lambda_{oN} + \sum_{i=1}^{M} \lambda_{iN} \left\{ \frac{\langle \psi_N | \Theta(x) | \psi_N \rangle}{\langle \psi_N | \psi_N \rangle} + d_i \right\};$$

$$i = 1, \ldots, M.$$

Here, $\tilde{R}(\lambda_{iN})$ may be construed as being the EPI $K = I^{\text{FCC}} - J$ evaluated at a specific iteration level N, *normalized* with respect to $\langle \psi_N | \psi_N \rangle$. In EPI theory, the bound information is often an expectation of a probability law. Evoking (6.29), the normalized FCC $\tilde{I}_N^{\text{FCC}}(\lambda_{iN}, d_i)$ and the normalized empirical bound information $\Im(\lambda_{oN}, \lambda_{iN})$ at a given iteration level N are defined as

$$\tilde{I}_N^{\text{FCC}}(\lambda_{iN}, d_i) = \frac{I_N^{\text{FCC}}(\lambda_{iN}, d_i)}{\langle \psi_N | \psi_N \rangle} = \sum_{i=1}^{M} \lambda_{iN} d_i = \sum_{i=1}^{M} \frac{i}{2} \frac{\lambda_i d_i}{\langle \psi_N | \psi_N \rangle} = \frac{I^{\text{FCC}}(\lambda_i, d_i)}{\langle \psi_N | \psi_N \rangle},$$
(6.31)

and

$$\Im(\lambda_{oN}, \lambda_{iN}) = \frac{J(\lambda_{oN}, \lambda_{iN})}{\langle \psi_N | \psi_N \rangle} = \lambda_{oN} - \sum_{i=1}^{M} \lambda_{iN} \frac{\langle \psi_N | \Theta(x) | \psi_N \rangle}{\langle \psi_N | \psi_N \rangle}.$$

It is immediately obvious from (6.31) that (6.30) is a manifestation of the modified game corollary expressed in normalized form

$$\frac{\partial \tilde{R}(\lambda_{iN})}{\partial \lambda_{iN}} \to 0 \Rightarrow \frac{\partial \left[\tilde{I}_N^{\text{FCC}}(\lambda_{iN}, d_i) - \Im(\lambda_{oN}, \lambda_{iN}) \right]}{\partial \lambda_{iN}} \to 0; \quad i = 1, \dots, M.$$
(6.32)

Performing a steepest descent procedure $\tilde{R}(\lambda_{iN})$ along the gradient of the Lagrange multipliers, $\partial \tilde{R}(\lambda_{iN})/\partial \lambda_{iN} \to 0$ yields optimal values of the Lagrange multipliers, which are consistent with the modified game corollary, and the quantum mechanical virial theorem.

At this juncture, attention is drawn to an important criterion that is crucial to the modified game corollary being applied in an iterative manner. Specifically, the constraints (physical observables) d_i; $i = 1, \dots, M$ correspond to the expectations of the operator $\Theta(x)$ evaluated with the true value of $\psi(x)$, which obeys the PDF normalization condition (6.29). These expectation values are either evaluated with a priori knowledge of the physical TISE potential by solving the forward SWE eigenvalue problem, or are provided as *input information*.

Constrained optimization techniques (i.e., steepest descent) require the gradient of the function to be extremized. Here, (6.5) readily yields $\partial \lambda_o / \partial \lambda_i = d_i$; $i = 1, \dots, M$. This is consistent with the Legendre transform structure for the FIM [36].

It is noteworthy to point out that an expression similar to (6.30) has been determined within the context of inference of amplitudes and PDF's from empirical data [19]. *However, the work presented in this section is the first of its kind to describe the process of iterative regularization of Lagrange multipliers within an information theoretic and game theoretic framework.*

The iterative modified game corollary has a threefold significance:

(1) It includes the contribution of the bound information.
(2) Unlike the game corollary [1], which is performed at the end of the EPI solution procedure to obtain undetermined coefficients, (6.32) is applied at each iteration

level N. This has the following physical significance. At the completion of each iteration level, the Fisher game is played, and $I^{FCC} - J$ is minimized at some intermediate Nth iteration level. As with the game corollary [1], this increases the uncertainty of the observer by ensuring the demon plays the last move. However, within the context of EPI, the implication is that *each iteration level signifies a given instant at which a new measurement is initiated by the gedanken observer.* Effectively, many Fisher games are now played. At present, it is premature to speculate about the relative magnitudes of the uncertainties obtained from the modified game corollary (after the completion of the final iteration level) and from the game corollary of Frieden [1]. This is left to future investigation.

(3) Finally, the iterative implementation of the modified game corollary holds forth the prospect of ascribing an energy landscape to the process of inference. This energy landscape provides an impetus to describe the Fisher game as an example of semisupervised learning.[8] Work along these lines has been pursued by the author and will be presented in a forthcoming publication.

This sub-section is concluded by drawing attention to the fact that the pseudo-potential in (6.3) may be expressed in terms of *zero-mean operators*, $(\Theta_i(x) - d_i)$, obeying

$$V^*(x) = \left(V(x) - \sum_{i=1}^{M} \lambda_i d_i\right) \Rightarrow \langle V^*(x)\rangle = \left\langle V(x) - \sum_{i=1}^{M} \lambda_i d_i \right\rangle = 0.$$

This minor modification is a fiduciary shift resulting in the kinetic energy being the total energy [19]. While this procedure does demonstrate operational convenience in the iterative regularization process, it modifies the operators described in Section 6.2.1, and hence will not be utilized here. The encryption model for the case of zero-mean operators will be presented in a separate publication.

6.3.4. Fisher Game vs MaxEnt

The above subsections have introduced an approach to inferring the eigenstructure and accompanying wave functions of statistical systems from incomplete constraints. The systems are of a quantum mechanical nature. A key step is to specify criteria that guarantee solutions that satisfy both the MFI and MaxEnt principles; see Sections 6.3.1 and 6.3.2. This is a useful tool for benchmarking results. In Section 6.3.3 a concept is introduced for obtaining the undetermined Lagrange multipliers through an iterative application of the modified game corollary.

For purposes of interpreting the process of iteratively inferring amplitudes and PDF's from empirical data, as above, the significance of the results presented in

---

[8] The term *information landscape* is more apt in capturing the spirit of the analysis. However, in deference to learning theory terminology, the term *energy landscape* is employed to describe the "evolution" of the FCC as a consequence of iterative application of the modified game corollary. (Comment by editor B. Roy Frieden: The term *fitness landscape* is used analogously in studies of biological population change.)

Sections 6.3.1 and 6.3.2 is largely theoretical. The reason is that the computational inference of amplitudes and PDF's from empirical data first requires that the TISLE (6.5) be solved as an eigenvalue problem. Specifically, it is practically impossible to manipulate an eigenvalue solver to yield a total energy eigenvalue $E = \lambda_o/4$ that exactly satisfies (6.22) for a given form of the pseudo-potential.

These results are, however, useful in providing initial "starting" values for the undetermined Lagrange multipliers, as will be demonstrated in this subsection. *Further, since the relations in Sections 6.3.1 and 6.3.2 are exact, they may be legitimately used to relate the final optimal values of the FIM and Shannon–Jaynes Lagrange multipliers.* These are obtained after the completion of the iterative process.

Before proceeding to the next section, it is judicious to demonstrate the results of numerical simulation that validates the above-introduced theory for the Fisher game by comparison with an equivalent MaxEnt solution. For this purpose, the case of the asymmetric harmonic oscillator is chosen. The TISE physical potential is $V^{\mathrm{AHO}}(x) = x^2/2 + x$. The exact solution is obtained by solving the TISE (6.1) as an eigenvalue problem for the potential $V^{\mathrm{AHO}}(x)$.

The TISLE (6.5) is solved specifying the pseudo-potential to be $V(x) = \tilde{\lambda}_1 x + \tilde{\lambda}_2 x^2$. For the TISE and TISLE cases, the eigenvalue problem is solved for 201 equally spaced points defined in $[-1, 1]$. The boundary conditions for the wave function/amplitude are given as $\psi(-1) = \psi(1) = 0$.

Solution of the TISE (6.1) yields the values of the expectations $\langle x \rangle = -0.0344$, and $\langle x^2 \rangle = 0.1302$. The value of the total energy eigenvalue is $E_{\mathrm{exact}} = 2.5152$. The Lagrange multipliers described by (6.23) are normalized with respect to the computationally obtained TISE total energy eigenvalue E_{exact}. Applying the theory presented in Sections 6.3.1 and 6.3.2 to the TISLE problem (6.5) with an asymmetric harmonic oscillator potential, the "starting" values of the Lagrange multipliers are given by $\lambda_1^{\mathrm{initial}} = 0.3975$, and $\lambda_2^{\mathrm{initial}} = 0.1987$. Since in (6.5), the FIM Lagrange multipliers are divided by a factor of 4, the effective input FIM Lagrange multipliers are $\tilde{\lambda}_1^{\mathrm{initial}} = 0.09939$ and $\tilde{\lambda}_2^{\mathrm{initial}} = 0.00496$. The corresponding value of the total energy eigenvalue is $E_{\mathrm{init}} = 2.4737$.

Given the starting values, (6.5) is solved as an eigenvalue problem, and the FIM Lagrange multipliers are regularized by iteratively evoking the modified game corollary (6.32). The final values of the FIM Lagrange multipliers are $\tilde{\lambda}_1^{\mathrm{final}} = 0.5043$, and $\tilde{\lambda}_2^{\mathrm{final}} = 1.0718$. The final total energy eigenvalue is $E_{\mathrm{final}}^{\mathrm{FIM}} = 2.6018$. The relative error of the Fisher game vis-à-vis the exact solution in terms of the total energy eigenvalue is $\left| \frac{E_{\mathrm{exact}} - E_{\mathrm{final}}^{\mathrm{FIM}}}{E_{\mathrm{exact}}} \right| = 0.0344$. *It is important to explicitly state that the procedure to obtain starting values of the FIM Lagrange multipliers substantially reduces the computational overhead.*

The relative error in terms of the total energy eigenvalue is provided as a measure of the iterative "evolution" of the Lagrange multipliers as a consequence of the modified game corollary. The true measure of the quality of the reconstruction is only possible using statistical estimation techniques, such as the χ^2 test. It is worthwhile to mention that numerical simulations have revealed certain cases of reconstructed PDF's possessing a relative error of the total eigenvalue less than

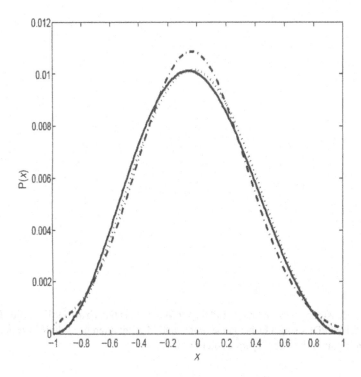

FIGURE 6.1. Numerical simulation of PDFs for asymmetric harmonic oscillator potential. Exact TISE solution (*solid line*), TISLE solution inferred using the Fisher game (*dots*), and inferred MaxEnt solution (*dash-dots*). Note that the Fisher-game solution only differs from the exact solution by a *marginal shift* to the right.

0.0344, but having a more pronounced distortion from the exact solution than the example depicted in Figure 6.1.

The equivalent MaxEnt solution is obtained by solving (6.4) by employing a Newton–Raphson procedure. This utilizes as input information the expectation values of $\langle x \rangle$ and $\langle x^2 \rangle$ obtained from the forward TISE problem. The final values of the Shannon–Jaynes Lagrange multipliers are $\mu_1^{\text{final}} = 0.2664$ and, $\mu_2^{\text{final}} = 3.6611$. These values are not in accord with (6.23) and correspond to a total energy eigenvalue of $E_{\text{final}}^{\text{ME}} = 3.6433$. Further, the relative error of the MaxEnt solution in terms of the total energy eigenvalue is $\left| \frac{E_{\text{exact}} - E_{\text{final}}^{\text{ME}}}{E_{\text{exact}}} \right| = 0.4484$. A number of factors may be attributed to this discrepancy, an analysis of which is beyond the scope of this work.

The results of the simulation are depicted in Figure 6.1. As is observed, the PDF obtained from the exact solution and the PDF inferred from incomplete constraints using the Fisher game have similar shapes. The Fisher game solution only errs marginally through a *small rigid shift* to the right. However, the PDF inferred using the MaxEnt theory differs significantly in shape (more peaked) from that of the exact solution and from the inferred PDF using the Fisher game.

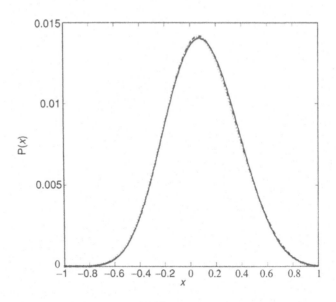

FIGURE 6.2. Numerical simulation of PDFs for Morse potential. Exact TISE solution (*solid line*), inferred PDF using a Fisher game (*dots*), and MaxEnt inference solution (*dash-dots*). The agreement of the three separate solutions is evident.

The process of encryption–decryption of covert information in a statistical distribution uses a Fisher game that is described in Section 6.4. Here, the *Morse potential* is employed to provide a statistical host. Figure 6.2 depicts the results of the inferred PDF using a Fisher game, and compares it with the exact solution. The expression for the Morse potential (in Table 6.1) has the chosen form $V(x) = k[e^{-2x} - 2e^{-x}]; k = 40$. The value of the constant $k = 40$ was chosen because it yields inferred PDF's that coincide with the exact TISE solution, using either the Fisher game or the MaxEnt theory. The aim is to encourage reproduction of the results by the interested reader, using either of the inference strategies. Coincidently, this value of the constant in the Morse potential has also been employed in [19].

This potential is substituted into the TISLE (6.5), which is solved as a forward eigenvalue problem. The values of the expectations are $\langle x \rangle = 0.0931$, $\langle x^2 \rangle = 0.0841$, $\langle x^3 \rangle = 0.0234$, and $\langle x^4 \rangle = 0.0203$. The resulting total energy eigenvalue of $E = -33.7086$. The theory presented in Sections 6.3.1 and 6.3.2 is utilized to obtain starting values, by expanding the Morse potential as a Taylor series. The procedure employed for the example with the asymmetric harmonic oscillator potential is extended to the case of the Morse potential. Specifically, the FIM Lagrange multipliers are optimized using the modified game corollary.

The final values of the FIM Lagrange multipliers are $\tilde{\lambda}_1^{final} = -0.0356$, $\tilde{\lambda}_2^{final} = 39.9909$, $\tilde{\lambda}_3^{final} = -40.6768$, and $\tilde{\lambda}_4^{final} = 15.6448$. These values correspond to the total energy eigenvalue $E = -33.8221$.

As is observed in Figure 6.2, the inferred PDF using a Fisher game virtually coincides with the exact solution. It is interesting to note that the PDF inferred using the MaxEnt theory also virtually coincides with the exact solution, for this case of a Morse potential. We state the values of the Shannon–Jaynes Lagrange multipliers for the sake of completeness and benchmarking: $\mu_1 = -0.8966$, $\mu_2 = 6.0446$, $\mu_3 = -1.7651$, and $\mu_4 = 2.2983$. These values correspond to a total energy eigenvalue of $E = -34.7592$.

6.4. Implementation of Encryption and Decryption Procedures

This section treats the implementation of encryption into, and decryption from, a host statistical distribution, as introduced in Section 6.2 through the use of a Fisher game. The *raison d'etré* for employing the FIM as a measure of uncertainty is two fold:

(1) To systematically categorize the constraint operator A. This work provides a credible physical basis for constructing the operator A by relating it to physical potentials admitted by a fundamental equation of physics, the TISE (6.1). The operator A is of dimension $(M + 1) \times N$ and consists of the powers of the coordinate x^i; $i = 1, \ldots, M$, evaluated at N different points on the computational lattice. As indicated by (6.8), (6.10), and (6.12), the vector of incomplete constraints $|d\rangle$, which is the *input* to the inference procedure, is formed by evaluating the expectation of A^9. As indicated in Section 6.2.2, the $(M + 1)$ th row of the operator A is populated by *unity* elements. This enforces the PDF normalization constraint. The value of the index M is contingent upon the physical potential that is modeled by the TISE (6.1).

Table 6.1 details, for select physical potentials, the expectations required to infer the TISE solutions from incomplete constraints (physical observations). The physical distributions required to obtain the column vector of incomplete constraints $|d\rangle$, as described by (6.12), are the PDF's corresponding to certain amplitudes (wave functions). These amplitudes are obtained by solving the TISE (6.1) as a forward eigenvalue problem, for an a priori specified physical potential.

Within the framework of the encryption model, the successful reconstruction of a physical distribution requires that both the constraints and the measure of uncertainty be properly chosen, and complement each other.

Within the basic MaxEnt framework [26], the operator A, and consequently the vector of incomplete constraints $|d\rangle$, do not possess a physical rationale. Specifically, they may be constructed based on an ad hoc choice of constraints. Consequently the Shannon–Jaynes Lagrange multipliers and the inferred PDF's are not bound by principles that correspond to physical systems.

---

[9] The operator A is independent of the measure of uncertainty.

As an example, in the case of the asymmetric harmonic oscillator potential studied in Section 6.3.4, the basic MaxEnt solution substantially differs from the exact physical distribution obtained from the TISE (6.1) for the case of the input expectations $\langle x \rangle$ and $\langle x^2 \rangle$.

In contrast, the TISLE (6.5), derived from the EPI Lagrangian (6.3), uses the FIM as the measure of uncertainty, and the empirical constraint terms are replaced with an extremized empirical bound information J. In comparison with the preceding MaxEnt framework, this *does* correspond to a physical system. This is why, in Section 6.3, solutions were found that are consistent with the TISE (6.1). Furthermore, the modified game corollary (see Section 6.3.3) yields FIM Lagrange multipliers corresponding to the incomplete constraints. This is by satisfying the quantum mechanical virial theorem[10].

Expectations obtained by solving the TISE (6.1) as a forward eigenvalue problem are henceforth referred to as *correct constraints*. It is of course possible for an arbitrary number of constraints to be generated as a consequence of solving the TISE. However, correct constraints are those that fill the requirements (in terms of required values of moments of the coordinates) of the TISE approach.

Constraints that differ from those generated as a consequence of solving the TISE are hereafter referred to as *incorrect constraints*. In essence, inferring a PDF using incorrect constraints is tantamount to modeling a physical system incorrectly. This will, of course, generally give incorrect results.

It is within the realm of possibility that a MaxEnt solution can be obtained that precisely matches the exact distribution corresponding to the asymmetric harmonic oscillator potential. However, this hypothetical MaxEnt solution would necessarily have to be inferred from a set of expectations that *differs* from that legitimately required to reconstruct a given physical potential and amplitudes/PDF's that satisfy the TISE. Specifically, for this specific example, this amounts to modeling a physical system as *other than* the asymmetric harmonic oscillator.

To provide an actual example, let *six expectation values* (order of moments) be generated by solving the TISE for the asymmetric harmonic oscillator potential. These six values are provided as inputs to a MaxEnt inference routine. A MaxEnt solution that actually coincides with the exact solution is found by a process of trial and error. This result is depicted in Figure 6.3. Again, this depended upon six inputs.

By comparison, from Table 6.1, the expectations required to reconstruct the TISE solution for the asymmetric harmonic oscillator potential are just $\langle x \rangle$ and $\langle x^2 \rangle$. These are *only two* in number.

In effect, the MaxEnt solution that exactly coincides with the exact solution of the asymmetric harmonic oscillator uses incorrect constraints. These in fact correspond to the six constraints needed for inferring an asymmetric sextic potential, since that would require six moments as input expectation values.

---

[10] The demon, on playing the last move, determines the value of the FIM Lagrange multiplier corresponding to the set of expectations that are generated from the exact solution of the TISE, in order that the gedanken observer's uncertainty be *maximized*.

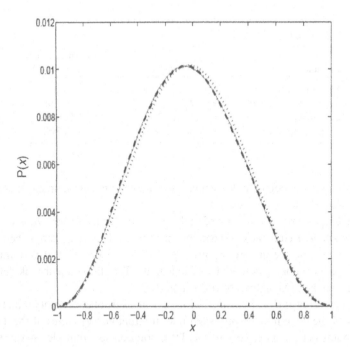

FIGURE 6.3. Numerical simulation of PDFs for asymmetric harmonic oscillator potential. Exact TISE solution (*solid line*), TISLE solution inferred using the Fisher game (*dots*) using correct constraints, and inferred MaxEnt solution (*dash-dots*) using incorrect constraints. In effect, the MaxEnt solution infers the asymmetric sextic potential.

The values of the expectations are $\langle x \rangle = -0.0344$, $\langle x^2 \rangle = 0.1302$, $\langle x^3 \rangle = -0.0103$, $\langle x^4 \rangle = 0.0408$, $\langle x^5 \rangle = -0.0043$, and $\langle x^6 \rangle = 0.0178$. The values of the Shannon–Jaynes Lagrange multipliers are $\mu_1 = 0.2799$, $\mu_2 = 2.8145$, $\mu_3 = 0.0212$, $\mu_4 = -1.0781$, $\mu_5 = -0.1502$, and $\mu_6 = 4.1689$.

It is generally observed that MaxEnt solutions coincide with the exact TISE solution and/or the inferred Fisher game solution when a larger number of expectation values are employed. This is to be expected, since the problem then becomes progressively more well defined, with decreasing dependence upon regularization. An example of this is for the case of the Morse potential.

Keeping the stated objective in mind, the constraint operator A is categorized on the basis of expectation values corresponding to select generic TISE physical potentials. Table 6.2 displays the condition numbers of A evaluated for 201 data points in $[-1, 1]$. The sensitivity of A is obtained by utilizing the MATLAB® function $cond(\bullet)$, which evaluates the condition number using SVD. A L_2 norm is specified.

As is noted from Table 6.2, the value of condition numbers of the operator A is greater than *unity*, even for the simplest TISE physical potential, i.e., the quantum mechanical harmonic oscillator. It is interesting to note that the values of the condition numbers for the first two potentials in Table 6.2 coincide, and those

TABLE 6.2. Condition numbers of constraint operator A for select Schrödinger potentials.

Potential	Expectations	Condition number of operator A
Symmetric quadratic	$\langle x^2 \rangle$	3.7305
Asymmetric quadratic	$\langle x \rangle$ and $\langle x^2 \rangle$	3.7305
Symmetric sextic	$\langle x^2 \rangle$ and $\langle x^4 \rangle$	18.6851
Asymmetric quartic	$\langle x \rangle$, $\langle x^2 \rangle$, $\langle x^3 \rangle$ and $\langle x^4 \rangle$	18.6851

for the last two also coincide. Why this is so, and its implication on the encryption process are tasks for future research.

The encryption process described in this study performs for both *rank deficient* and *rank efficient* (full rank) constraint operator A. This is accomplished without compromising the underlying principles of the theory. Extensive numerical simulations have not encountered solutions to the TISLE that incur rank deficient operators, so the FIM approach works in any event.

(2) The second reason for choosing the FIM as a measure of uncertainty is that select portions of the code may be projected into different *energy states* of the TISLE, having equivalent counterparts in the TISE solution for multiple energy states. This would entail a modification of the definition of the decryption and encryption operators described by (6.14) and (6.15), resulting in a qualitative enhancement of the theory of encrypting covert information in a host statistical distribution. Section 6.5 contains a brief description of embedding covert information into multiple energy states of a TISLE.

The process of encryption of the covert information is treated in Section 6.4.1. This includes salient implementation details of the process of encryption and the derivation of a key by introducing perturbations to the eigenstructure of the operator G. The process of decryption is described in Section 6.4.2. A concise summary of the encryption–decryption process, catering to a prospective implementer, is provided in Section 6.4.3. The ability of the model to employ both a *symmetric* and an *asymmetric cryptographic strategy* and the ability of the encryption model to withstand a *plaintext attack* are discussed in Section 6.4.4.

The terminology in cryptography and allied disciplines refers to two communicating parties as *Alice* and *Bob*, and an eavesdropper as *Eve* [21, 22]. In this study, the author plays the role of both *Alice* and *Bob* by implementing the encryption on an IBM RS-6000 workstation cluster and the decryption on an IBM Thinkpad running MATLAB® v 7.01

6.4.1. Encryption Process

The methods developed in this subsection are meant to augment a PDF reconstruction from incomplete constraints described in Section 6.3. Such a PDF and an empirical pseudo-potential corresponding to the Morse potential have been reconstructed for 201 equally spaced points in $[-1, 1]$.

Thus, it is assumed that a constraint operator A having dimensions $(M + 1) \times N$, a column PDF vector $|p\rangle \in {}^N$, and the concomitant FCC I^{FCC} are available. The operator A has dimensions of 5×201, where $N = 201$, $M = 4$.

This case is examined in detail next.

Embedding. Given A, the operator G having dimensions $N \times N$ is evaluated. The normalized eigenvectors $|\eta_n\rangle \in N - (N + 1)$ of the eigenvalues of the *null space* of G possessing value *zero* are evaluated using the MATLAB® function $null(\bullet)$.

This $null(\bullet)$ function is specific to the task at hand and provides the normalized eigenvectors using SVD. The encryption operator (6.15) may now be formed. The code $|q\rangle \in N - (M + 1) = 196$ is created using the MATLAB® random number generator $rand(\bullet)$ to generate $N - (M + 1)$ numbers in $[0, 1]$. A seed value (of 7) is specified as an input option to $rand(\bullet)$.

The encryption operator $\hat{U}_{\text{enc}}$ operates on $|q\rangle \in R^{N-(M+1)=196}$ to yield the pdf of the code $|p_c\rangle \in R^{N-(M+1)=196}$. The null space of the operator G having dimensions $N \times N$ has $N - (M + 1)$ zero eigenvalues. It is into these zero eigenvalues that the code is now projected. The resulting vector $|p_c\rangle$ is added to $|p\rangle$ as described in (6.17) to yield $|\tilde{p}\rangle$.

It is instructive to provide the interested reader with a numerical representation of the operations in the *null space* of an operator. For this purpose, the values of $|\tilde{p}\rangle$ are plotted in Figure 6.4. The net PDF appears chaotic, and quite unlike the smooth curves encountered in Section 6.3.4. The chaotic behavior is potentially deceptive because the *null space* of the operator G into which the code has been projected does not possess any information concerning the PDF inferred, using the Fisher game (depicted in Figure 6.2). Information concerning the inferred PDF resides in the *range space* of the operator A.

The oscillatory behavior of the combined PDF of the range and null spaces of the operators is exploited to develop a secret key. It is important to reemphasize that the oscillatory nature of the PDF $|\tilde{p}\rangle$, which is representative of the range space of A and the null space of G, is natural for ill-conditioned eigensystems.

As a prelude to the following, which discusses perturbation of the eigenstructure of the operator G in order to obtain a secret key, a threshold value for the perturbation is evaluated. The criterion that is chosen to ascertain the magnitude of the perturbation is of critical importance. This criterion may be decided by calculating the L_2 norm of the error of reconstruction of the code *without perturbations*. For this purpose the designer has to first perform the encryption procedure described above and then perform the decryption of the code using the procedure detailed in Section 6.4.2, without the perturbation. For our example, the error of reconstruction is found to be $\varepsilon = 2.2599e - 014$. This is then the threshold value for perturbations. Thus any perturbation to the operator G should be greater than ε. Hence the size of ε is the criterion that was sought.

Perturbations. To review, the above discussion described the embedding procedure. The highly oscillatory nature of ill-conditioned eigenstructures was highlighted with the aid of a numerical example. Finally, the procedure for obtaining a threshold value for perturbations was determined. We next consider the all-important issue of formulating a key.

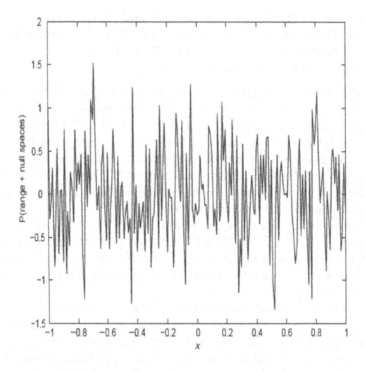

FIGURE 6.4. Numerical simulation of $|\tilde{p}\rangle$ for the Morse potential. The oscillatory behavior is representative of eigenstructures of ill-conditioned systems.

At this stage a single element, say the element $G_{1,3}$ at the first row and third column of the operator G, is perturbed by the addition of a small number $\delta G_{1,3} = 3.0e - 013 > \varepsilon$, where ε is the threshold value. The perturbation $\delta G_{1,3}$ comprises the secret key. Thus, perturbing the ith row and jth column of the operator G results in the operator $\tilde{G} = G + \delta G_{i,j}$.

The encryption procedure is repeated and the combined PDF $|\tilde{p}\rangle$ is reevaluated using this operator $\tilde{G}$ instead of the operator G. Specifically, first the normalized eigenvectors of the eigenvectors of the *zero* eigenvalues of the *null space* of $\tilde{G}$ are evaluated using the MATLAB® function $null(\bullet)$. These are denoted as $|\tilde{\eta}_n\rangle$. Then the encrypted code is embedded into the *null space* of $\tilde{G}$. Using (6.16), the code PDF is obtained as $|\tilde{p}_c\rangle = \tilde{\tilde{U}}_{\text{enc}} |q\rangle = \sum_{n=1}^{N-M-1} |\tilde{\eta}_n\rangle \langle n|q\rangle$. The column vector $|\tilde{p}_c\rangle$ is added to the host PDF to give the combined PDF $|\tilde{p}_{\text{pert}}\rangle$.

The effect of perturbing a single element of the operator is illustrated in Figure 6.5. Despite the highly oscillatory nature of Figures 6.4 and 6.5, the difference between them is evident.

As will be discussed in Section 6.4.4, a number of keys may be constructed in this manner in order to thwart plaintext attacks.

Transmission. The data is transmitted from the designer to the legitimate receiver by sending across $|\tilde{p}_{\text{pert}}\rangle$, and the values of the FIM Lagrange multipliers. Note that

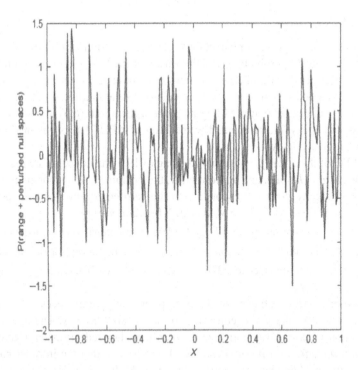

FIGURE 6.5. Numerical simulation of the combined PDF with perturbations,$|\tilde{p}_{\text{pert}}\rangle$, for the Morse potential. A single element of the operator G is perturbed by $3.0e - 013$, the secret key.

the $|\tilde{p}_{\text{pert}}\rangle$ has been computed after the addition of the perturbation/perturbations to select elements of the operator G. These quantities are transmitted through a *public channel*. The key/keys that represent the perturbations $\delta G_{i,j}$ are to be transmitted using a *secure channel*, such as a quantum key distribution channel. *The key/keys are given a label to indicate the element/elements of the operator G whose perturbations they represent.*

Care is to be taken that both the sender and the receiver use compatible software for reconstructing the pdfs (especially eigenvalue solvers and routines to evaluate the normalized eigenvectors of *null spaces* of operators).

The quantities that are required to be transmitted over the public channel constitute an important qualitative distinction between the encryption model using a Fisher game and a corresponding MaxEnt model.

When the MaxEnt model is used, the quantities that are to be transmitted are the operator A and $|\tilde{p}\rangle$. The constraints are then obtained through the operation $d_i = \langle i| A |\tilde{p}\rangle$, where i is a basis vector in R^{M+1}. The host PDF $|p\rangle$ is then inferred via the MaxEnt procedure using the expectations $|d\rangle$.

By comparison *when the Fisher game is used, the operator A need not be transferred.* This drastically reduces the amount of information that needs to be transmitted through a public channel.

6.4.2. Decryption

The process of decryption commences with the receiver solving the TISLE (6.5) as a forward eigenvalue problem, using the values of the FIM Lagrange multipliers that are transmitted. The operator A and as a consequence the operator G are readily calculated as a consequence of calculating the amplitude $\psi(x)$. This is by substituting the transmitted FIM Lagrange multipliers into (6.5) and solving the TISLE as an eigenvalue problem. The resulting host PDF $|p\rangle$ is obtained in this way with high accuracy.

The PDF of the code $|\tilde{p}_c\rangle$ is retrieved using (6.19). The key/keys are recovered from the *secure channel*, and appropriately, the operator $\tilde{G}$ is reconstructed. The normalized eigenvectors corresponding to the *zero* eigenvalues of the operator $\tilde{G}$, $|\tilde{\eta}_n\rangle$, are evaluated using the MATLAB® function *null* (•).

Finally, the decryption operator $\tilde{U}_{dec}$ (6.14) is formulated from $\tilde{G}$ in an analogous manner as that employed in Section 6.4.1 to formulate the encryption operator $\tilde{U}_{enc}$. The covert information is retrieved by the application of (6.20), $|q_{r1}\rangle = \tilde{U}_{dec}|p_c\rangle = \sum_{n=1}^{N-M-1} |n\rangle\langle\tilde{\eta}_n|\tilde{p}_c\rangle$.

To demonstrate the efficacy of the encryption-decryption process, the reconstructed code with the key is compared to that without the key. The case of *reconstruction without a key* corresponds to an unauthorized intrusion. (Note that *this is not the case of a plaintext attack*.) It is assumed that the intruder has acquired sufficient knowledge to retrieve the code PDF $|\tilde{p}_c\rangle$, using the procedure detailed above. However, the intruder does not know the secret key that effects the transformation $G \rightarrow \tilde{G}$, or is unaware of its existence. The intruder thus cannot evaluate the eigenvectors $|\tilde{\eta}_n\rangle$, and hence is unable to form the correct decryption operator.

The case of reconstruction with and without a key is described in Table 6.3 for select cases. Two column vectors of dimension 196 representing the reconstruction with the key ($|q_{r1}\rangle$) and reconstruction without the key ($|q_{r2}\rangle$) are obtained. The values provided for *decryption without a key* correspond to the intruder attempting an incorrect decryption with the eigenvectors of the unperturbed system $|\eta_n\rangle$. Using (6.20), this corresponds to $|\tilde{q}_{r2}\rangle = \tilde{U}_{dec}|p_c\rangle = \sum_{n=1}^{N-M-1} |n\rangle\langle\eta_n|\tilde{p}_c\rangle$.

The error of reconstruction *without the key* is the modulus of the L_2 norm $\|(|\tilde{q}\rangle - |\tilde{q}_{r2}\rangle)\| = 11.01112$, where $|\tilde{q}\rangle$ represents the original code embedded into $\tilde{G}$. Defining the column vector $|\text{err}\rangle = |\tilde{q}\rangle - |\tilde{q}_{r2}\rangle$, and the number of elements in $|\text{err}\rangle$ as length (err) = 196, the corresponding RMS error is found to be $\||\text{err}\rangle/\|\sqrt{\text{length(err)}} = 0.786508$. The standard deviation of the column vector $|\text{err}\rangle$ is $\sigma(\text{err}) = 0.6168067$. Therefore the relative RMS error is $RMS(\text{err})/\sigma(\text{err}) = 1.27513$ or 128%. By comparison, the RMS error in the reconstruction *using* the key *is zero* to 14 decimal places.

When interpreting the RMS error and the relative RMS error, it should be borne in mind that the code has been generated by $rand(•)$, which randomly generates numbers, having a *uniform distribution* in [0,1]. But the thrust of the results is independent of distribution: *The covert information cannot be obtained from the reconstructed code without possessing the secret key.*

TABLE 6.3. Reconstruction of the covert information with and without the key.

Original value $|q\rangle$	Reconstruction with key $|q_{r1}\rangle$	Reconstruction without key $|q_{r2}\rangle$
0.23813682639005	0.23813682639005	0.39070924582122
0.69913526160795	0.69913526160795	−0.38520555664833
0.27379424177629	0.27379424177629	−0.91260190810576
0.90226539453884	0.90226539453884	−0.64390540882811

Table 6.3 displays some representative elements $\{1, 75, 177,$ and $196\}$ of the column vector of the original code $|\tilde{q}\rangle$, the reconstructed code with the key $|\tilde{q}_{r1}\rangle$, and the reconstructed code without the key $|\tilde{q}_{r2}\rangle$. As is observed, the reconstructed code without the key does not bear any resemblance to the original code.

6.4.3. Summary of the Encryption–Decryption Strategy

This subsection summarizes the encryption–decryption strategy detailed in Sections 6.4.1 and 6.4.2 to facilitate its use.

Encryption. The following stages comprise the encryption process:

(1) The host PDF $|p\rangle$ is inferred from incomplete constraints using a Fisher game. The FIM Lagrange multipliers are derived using the modified game corollary (see Sections 6.3.3 and 6.3.4).

(2) The ill-conditioned operator A is formed from the values of the constraints that constitute the TISLE pseudo-potential (6.5) by taking the discrete values of the expectations defined on the computational lattice. The operator $G = A^{\dagger}A$ is formed.

(3) The code (covert information) $|q\rangle$ to be embedded into the *null space* of G is obtained using a random number generator (in a simulation). In a real use of the procedure it would be simply the message to be coded.

(4) The key/keys are determined by perturbing one or more elements of the operator G. A threshold for the perturbations is obtained by evaluating the reconstruction error ε for the unperturbed operator G. The normalized eigenvectors corresponding to the *zero*-valued eigenvalues of G, $|\eta_n\rangle$; $n = 1, \ldots, N - (M + 1)$, are obtained using the MATLAB® function $null(\bullet)$, or an equivalent routine. Here, N is the dimension of the physical distribution and M is the number of incomplete constraints used in the inference of the host PDF (Step (1)).

(5) The encryption of the code into the host distribution is effected using the encryption operator (6.15), and the code PDF $|p_c\rangle$ is determined by (6.16). The combined PDF $|\tilde{p}\rangle$ is determined from (6.17).

(6) The threshold for perturbations is obtained by the designer solving the TISLE (6.5) as a forward eigenvalue problem, using the FIM Lagrange multipliers. This provides the reconstructed host PDF $|p_r\rangle$. The reconstructed code PDF is obtained from the combined PDF $|\tilde{p}\rangle$ using (6.19). The reconstructed code $|q_r\rangle$ is obtained

using (6.20). The threshold for perturbations is obtained by evaluating the L_2 norm $\||q\rangle - |q_r\rangle\| = \varepsilon$. Any perturbation to an element/elements of G should be greater than ε.

(7) Applying a perturbation $\delta G_{i,j}$ to the element $G_{i,j}$ of the operator G results in the operator $\tilde{G} = G + \delta G_{i,j}$, $\delta G_{i,j} > \varepsilon$.

(8) The code $|\tilde{q}\rangle$ to be projected into $null(\tilde{G})$ is obtained from a random number generator.

(9) Step (4) is repeated for the operator $\tilde{G}$, yielding the eigenvectors $|\tilde{\eta}_n\rangle$; $n = 1, \ldots, N - (M + 1)$.

(10) Step (5) is repeated for the code $|\tilde{q}\rangle$ resulting in a code PDF $|\tilde{p}_c\rangle$ and a combined PDF $|\tilde{p}_{\text{pert}}\rangle$.

Transmission. The FIM Lagrange multipliers λ_i; $i = 1, \ldots, M$ and the combined PDF $|\tilde{p}_{\text{pert}}\rangle$, obtained in Step (10), are transmitted by the sender to the receiver through a public channel. The secret key $\delta G_{i,j}$ is transmitted via a *secure channel*.

Decryption. The following stages comprise the decryption process:

(11) The legitimate receiver recovers the FIM Lagrange multipliers λ_i; $i = 1, \ldots, M$ and the combined PDF $|\tilde{p}_{\text{pert}}\rangle$ from the public channel. The secret key $\delta G_{i,j}$ is recovered from the secure channel.

(12) The receiver and the sender have a prearranged understanding about the nature of the TISLE pseudo-potential (6.5) and the boundaries $[a, b]$ over which the TISLE is to be solved. In the chapter, this is $[-1, 1]$.

(13) The receiver solves the TISLE using the FIM Lagrange multipliers as a forward eigenvalue problem and obtains the host PDF $|p\rangle$ from the amplitudes $\psi(x)$. The constraint operator A and the operator $G = A^\dagger A$ are subsequently constructed.

(14) The code PDF $|\tilde{p}_c\rangle$ is obtained using (6.19), $|\tilde{p}_c\rangle = |\tilde{p}_{\text{pert}}\rangle - |p\rangle$.

(15) The key $\delta G_{i,j}$, recovered from the secure channel, is added to the element/elements of the operator G. This yields the perturbed operator $\tilde{G}$.

(16) Step (4) is repeated. Specifically, the normalized eigenvectors of the eigenvalues of *zero* value in *null space* of $\tilde{G}$, $|\tilde{\eta}_n\rangle$; $n = 1, \ldots, N - (M + 1)$, are evaluated. This is by use of the MATLAB function $null()$, or an equivalent routine.

(17) The decryption operator (6.14) is obtained using the eigenvectors $|\tilde{\eta}_n\rangle$, $i = 1, \ldots, N - (M + 1)$.

(18) The reconstructed code is obtained using (6.20). Specifically,
$$|\tilde{q}_r\rangle = \tilde{U}_{\text{dec}}|\tilde{p}_c\rangle = \sum_{n=1}^{N-M-1} |n\rangle\langle\tilde{\eta}_n|\tilde{p}_c\rangle. \text{ The algorithm ends.}$$

6.4.4. Security Against Malicious Attacks

This subsection briefly describes security by the FIM encryption strategy against malicious attacks. An example is a plaintext attack. As discussed in Section 6.1.9, the secret key or symmetric cryptographic strategy is not immune to plaintext attacks.

To thwart such an attack, a *key ring strategy* is to be employed. Within the context of the encryption model presented in this chapter, an asymmetric cryptographic strategy is readily formulated by randomly perturbing more than one element of the operator G. *The key is thus a string of numbers* representing perturbations to multiple elements of G. Each perturbation has to satisfy the criterion described in Section 6.4.1, that is $\delta G_{i,j} > \varepsilon$, where $\delta G_{i,j}$ is the perturbation to the element in the ith row and jth column of G and ε is the error of reconstruction without perturbations.

To thwart the plaintext attack using the key ring strategy, a string of key numbers is generated. One of these is kept secret and the remainder are declared public. A separate key is used for each different message. Thus, the successful interception of a single key, and the subsequent decoding of a single message by a malicious attacker, will not compromise the security of the other messages.

Figure 6.6 depicts the combined PDF $|\tilde{p}_{\text{pert}}\rangle$ with three elements of the operator G being perturbed. The keys are $\delta G_{1,3} = 3.0e - 013 > \varepsilon$, $\delta G_{2,7} = 3.3e - 013 > \varepsilon$,

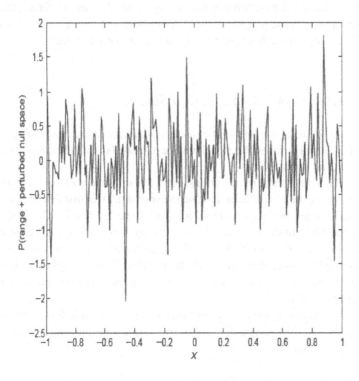

FIGURE 6.6. Numerical simulation of the combined PDF with perturbations, $|\tilde{p}_{\text{pert}}\rangle$, for the Morse potential. Three elements of the operator G are perturbed resulting in *three* keys.

TABLE 6.4. Reconstruction of the covert information with and without the key. Three elements of the operator G are perturbed.

Original value $|q\rangle$	Reconstruction with key $|q_{r1}\rangle$	Reconstruction without key $|q_{r2}\rangle$
0.23813682639005	0.23813682639005	−0.22939187230541
0.69913526160795	0.69913526160795	0.42610010630556
0.27379424177629	0.27379424177629	0.27187680327071
0.90226539453884	0.90226539453884	0.61955980427953

and $\delta G_{5,9} = 3.5e - 013 > \varepsilon$. The keys form part of a key ring strategy, where one of the keys is kept secret and the remainder are declared public. The differences among Figures 6.4 to 6.6 are evident.

The error of reconstruction without the keys is the modulus of the L_2 norm $\|err\| = \|(|\tilde{q}\rangle - |\tilde{q}_{r2}\rangle)\| = 11.03564$. The RMS error is $\|\|err\rangle\|/\sqrt{\text{length(err)}} = 0.788260$. The standard deviation of the column vector $|err\rangle$ is σ (err) $= 0.658659$. The relative RMS error is RMS(err)$/\sigma$(err) $= 1.19676$ or 120%.

Table 6.4 displays a select number of elements $\{1, 75, 177, \text{and } 196\}$ of the column vector of the original code $|\tilde{q}\rangle$, the reconstructed code with the key $|\tilde{q}_{r1}\rangle$, and the reconstructed code without the key $|\tilde{q}_{r2}\rangle$, for the case of three elements of the operator G being perturbed. As is *again* observed, the reconstructed code without the key does not bear any resemblance to the original code.

6.5. Extension of the Encryption Strategy

The work presented above treats the case of *ground state (equilibrium)* amplitudes of a TISLE (6.5). However, the model may be extended to include the case of both *equilibrium* and *excited states*.

This extension has three ramifications. First, the amplitudes of the excited states are described by the H–G (Hermite-Gauss) expansion. The first few terms of the H–G expansion are analytically tractable. Thus, the exact relation (6.22) relating the MFI and MaxEnt theories may be modified by deriving exact representations for the amplitudes of the excited states. Next, each *energy state* will have its own set of incomplete constraints (physical observables). This implies that each state will have its own constraint operator A and the operator into which the code is embedded $G = A^{\dagger}A$.

Thus, the encryption and decryption operators would need to be suitably modified. Specifically, the decryption operator (6.14) in a multiple energy state model would become for a state k

$$\hat{U}_{\text{dec}}^k = \sum_{n=1}^{N-M-1} |n^k\rangle \langle \eta_n^k |; \quad k = 0, 1, \ldots, \tag{6.33}$$

and the encryption operator (6.15) would acquire the form for a state k

$$\hat{U}_{\text{enc}}^k = \sum_{n=1}^{N-M-1} |\eta_n^k\rangle \langle n^k|; \quad k = 0, 1, \ldots . \qquad (6.34)$$

Here, $|\eta_n^k\rangle$ are the normalized eigenvectors for the *zero*-valued eigenvalues of the operator $G^k = A^{k\dagger} A^k$, where A^k is the constraint operator for the state k. The modified game corollary (6.31) would remain unchanged in form and meaning, and would be applied to obtain the optimal FIM Lagrange multipliers that characterize the empirical bound information for each energy state.

Finally, a multiple excited state encryption strategy would enable different portions of the same code, or of different codes, to be encrypted through unitary projections into different states that satisfy the TISLE (6.5). Further, it would extend the concept of a key ring strategy to a multiple key ring strategy.

Preliminary studies indicate that such an encryption strategy would provide a significantly greater degree of security to covert information than an encryption strategy based on an equilibrium state Fisher game model, or on the MaxEnt approach. Furthermore, it would more fully exploit the true quantum mechanical nature of the FIM and the TISLE. These results will be presented in a forthcoming publication.

6.6. Summary of Concepts

We have introduced in this chapter a variety of concepts:

(1) A modified game corollary was derived, and used, for iteratively evaluating the undetermined Lagrange multipliers of the problem, while inferring PDF's from empirical data.

(2) A set of exact relations were derived for describing amplitudes that simultaneously satisfy the MFI principle and the MaxEnt theory. These are utilized to computationally infer the PDF's of the Fisher game. Numerical examples of the inference of PDF's from empirical data demonstrate the superiority of the Fisher game versus the MaxEnt principle. This is especially true for the case of a limited number of constraints (physical observables).

(3) A strategy was described for encrypting covert information via unitary projections into the *null space* of a matrix operator G. The *null space* is representative of the constraint A. A methodology is introduced to systematically categorize the constraint operator A based on the physical potentials satisfying the TISE. The advantages of solving the encryption model within the ambit of a Fisher game, rather than by MaxEnt theory, were described and exemplified.

(4) In the encryption model, the key is obtained by perturbing select elements of an operator that is representative of the constraints. Numerical results using a single secret key (symmetric cryptography) are demonstrated. The results demonstrate

the efficacy of the encryption strategy by presenting results of decryption of a code (covert information) both with and without a key. The extension of the encryption model to incorporate an asymmetric cryptographic strategy is discussed, for the purpose of acquiring immunity to plaintext attacks.

(5) Finally, an approach for extending the encryption-decryption strategy to the use of excited energy levels is briefly discussed.

Acknowledgments. Gratitude is expressed toward B. Roy Frieden and L. Rebollo–Neira for helpful and informative discussions. This work was carried out under the auspices of *MSR contract CSM-DI&M-QIT-101107-2005.*

7
Applications of Fisher Information to the Management of Sustainable Environmental Systems

AUDREY L. MAYER, CHRISTOPHER W. PAWLOWSKI, BRIAN D. FATH, AND HERIBERTO CABEZAS

All organisms alter their surroundings, and humans now have the ability to affect environments at increasingly larger temporal and spatial scales. Indeed, mechanical and engineering advances of the twentieth century greatly enhanced the scale of human activities. Among these are the use and redistribution of natural resources. Unfortunately, these activities can have unexpected and unintended consequences. Environmental systems often respond to these activities with diminished or lost capacity of natural function. Fortunately, environmental management can play an important role in ameliorating these negative effects. The aim is to promote sustainable development, i.e., enrichment of the lives of the majority of people without seriously degrading the diversity and richness of the environment. However, the management tools themselves often fall prey to the same narrow levels of perspective that generated the negative conditions. The challenge is to develop a system-level index, one that indicates the organization and direction of ecological system dynamics. This index could detect when the system is changing its configuration to a new, perhaps less desirable, dynamic regime and may be incorporated into a sustainable management plan for the system. In this chapter, we demonstrate the use of Fisher information (FI) as such an environmental system index.

7.1. Summary

We derive an expression for FI based on sampling of the system trajectory as it evolves in the phase space defined by the state variables of the system. *This FI index is derived as a measure of system dynamic order, as defined by its speed and acceleration along periodic steady-state trajectories.* We illustrate the concepts on data collected from both computer model simulations and real-world environmental systems. FI is found to provide a valuable tool to identify impending and in-progress shifts in regime, as distinguished from normal cycles, fluctuations, and noise in the systems.

7.2. Introduction

"Sustainability" is often used in a qualitative sense. However, there is at present a great need to quantitatively measure (and monitor) its many qualitative aspects in real systems. Real systems are regarded as sustainable if they can maintain their current, desirable productivity and character without creating unfavorable conditions elsewhere or in the future [1–4]. Sustainability therefore incorporates both concern for the future of the current system (temporal sustainability) and concern about the degree to which some areas and cultures of the planet are improved at the expense of other areas and cultures (spatial sustainability). That is, sustainability is to hold over both space and time.

Sustainability encompasses many disciplines. For example, economic systems are not sustainable if they degrade their natural resource base and impoverish some sectors of the human population [5, 6]. Indices are needed that will measure sustainability through time, and over space, at several scales. These indices must also have the ability to aggregate the many disciplinary facets of sustainability, often incorporated through a large number of environmental, social, and economic variables. Such a multidisciplinary dynamic system can be regarded as sustainable if it maintains a desirable steady state or regime[1], including fluctuations that are desirable (such as those that respond to natural disturbances [8]).

Generally dynamic regimes can be identified observationally by a characteristic variability and behavior. Systems maintain themselves in these regimes through negative feedback loops and interactions between system compartments, until internal or external perturbations cause these interactions to break down. Then the system shifts (often catastrophically) to a new regime. Catastrophic regime shifts have been observed in systems ranging from the global climate [9], aquatic and terrestrial ecosystems at all scales [7], and even nation states [10,11]. (Catastrophic population shifts are treated in Chapter 8.)

Particularly for ecological and environmental systems, a local focus on the transient behavior between stable regimes could make important contributions to understanding and managing human impacts on ecosystems, and restoring systems to more desirable regimes [12–14]. Indices that can identify, and help foreshadow, these regime shifts are critical to managing dynamic systems and desirable regimes. However, environmental management actions must not aim to stop or interfere with the natural and often complex cycles that exist within any dynamic regime, and which are integral to the functioning of biological systems. Interference with these cycles would likely destroy the biological system. There is, therefore, a need for scientific criteria that are passive, i.e., would accommodate system fluctuations and cycling while the regime is being *preserved*. Such criteria are needed for the successful management of the interactions and interdependencies between human and environmental systems.

---

[1] Similar concepts to dynamic regimes have also been called multiple equilibria, alternative stable states, and stable attractors. We follow the dynamic regimes terminology of [7].

Indices derived from information theory offer particular promise for identifying catastrophic shifts in ecological systems [15,16]. Information theory has commonly been used to monitor and investigate the structure and behavior of ecosystems in several ways [17], such as measuring ecosystem complexity (e.g., [18,19]) or stability (e.g., [15, 20]). Shannon information is most well known to ecologists, and has been used for decades as a measure of species diversity, which can be compared across communities and ecosystems. However, Shannon information is a global measure of order, and while is it appropriate for comparing complexity between systems, it cannot measure changes in order as a particular system travels through time. A *local* measure, such as FI, is therefore needed [15].

7.3. Fisher Information Theory

We next summarize those properties of FI that will be used in the development. The emphasis is upon those aspects of the information that are not already covered in Chapter 1. This is mainly the formation of an appropriate scalar Fisher variable s out of a vector of degrees of freedom that define the system. The material is largely drawn from [21] and [17], with some additions and modifications.

7.3.1. Definition

Ronald Fisher [22] devised a measure of indeterminacy, now called FI, that measures the level of indeterminacy or disorder in data. The FI I in the observation y of a parameter of true value θ is defined as [23, 24]

$$I(\theta) \equiv \int \frac{dy}{p_0(y|\theta)} \left[\frac{dp_0(y|\theta)}{d\theta} \right]^2. \tag{7.1a}$$

Here $p_0(y|\theta)$ is the probability density, or likelihood law, for observing a particular measured value of y in the presence of a value of θ. The FI is a measure of the amount or the quality of the information obtainable from the measurement of y.

To apply FI to the complex dynamic systems treated further in this chapter, let $y = s$, where s is an observable state of the system as further discussed in Section 7.4. Also let $\theta = \langle s \rangle$, where $\langle s \rangle$ is the mean of s over the period of time T defined by

$$\langle s \rangle \equiv \frac{1}{T} \int_0^T dt s(t), \tag{7.1b}$$

where T is the period of time included in all observations, and $\langle s \rangle$ is now a constant. The FI expression of Eq. (7.1a) now becomes

$$I(\langle s \rangle) = \int \frac{ds}{p_0(s|\langle s \rangle)} \left[\frac{dp_0(s|\langle s \rangle)}{d\langle s \rangle} \right]^2. \tag{7.1c}$$

7.3.2. Shift-Invariant Cases

Certain systems obey shift invariance [23],

$$p_0(s|\langle s \rangle) \equiv p(s - \langle s \rangle | \langle s \rangle) = p(s - \langle s \rangle), \tag{7.2}$$

of their likelihood laws. The first equality defines the new probability law p on the fluctuations $s - \langle s \rangle$ in measurement conditional upon $\langle s \rangle$ The second equality states that for a shift-invariant system these fluctuations are independent of the value of $\langle s \rangle$. Since $\langle s \rangle$ is a constant computed from integrating s over all time, it seems reasonable that subtracting a constant from s will not change the shape of p_0, and it will not change the information. For purposes of providing further illustrative examples, consider that when observing the time on an analog clock dial in an effort to estimate the actual time, the *distribution* p_0 of times shown on the dial has the same shape, independent of the actual time, provided the observations occur over a time bracket spanning at least one complete 12-hour cycle. For the case of time series obtained from observing a cyclic biological system, this implies that the distribution of the values of the observed variable s is independent of the time bracket used for the observations, provided the time bracket is at least one-cycle wide.

With these assumptions in mind, we now define a new variable $\hat{s}$ as $\hat{s} = s - \langle s \rangle$ so that $d\hat{s} = ds - d\langle s \rangle = ds$, since $\langle s \rangle$ is a constant. These together with (7.2) make Eq. (7.1c) simplify to

$$I = \int \frac{d\hat{s}}{p(\hat{s})} \left[\frac{dp(\hat{s})}{d\hat{s}} \right]^2. \tag{7.3a}$$

Here, $p(\hat{s})$ simply represents the probability density for observing a particular value of the variable $\hat{s}$. The expression of Eq. (7.3a) is important because the derivative of $p(\hat{s})$ with respect to $\hat{s}$ can be readily evaluated from either data (by finite differences) or models (by analytical differentiation). In contrast, in the absence of shift invariance, there would be no straightforward way of evaluating the derivative with respect to the parameter θ in Eq. (7.1a) or $\langle s \rangle$ in Eq. (7.1c).

Since the subtraction of $\langle s \rangle$ from s does not affect the calculation under the assumptions already discussed, we now tacitly use $\hat{s}$ in place of s through the remainder of this Chapter 7. Thus Eq. (7.3a) becomes (cf. Eq. (1.50))

$$I = \int \frac{ds}{p(s)} \left[\frac{dp(s)}{ds} \right]^2, \tag{7.3b}$$

where $p(s)$ is the probability density for s. (In fact (7.3b) may also be proven rigorously.) A form of information (7.3b) that can prove handy is its re-expression in terms of a probability *amplitude* function $q(s)$ (cf. Eq. (1.55)) ,

$$I = 4 \int ds \left[\frac{dq(s)}{ds} \right]^2, \quad p(s) \equiv q^2(s). \tag{7.3c}$$

FIGURE 7.1. Two-dimensional cyclic dynamic system, with one period showing the system orbit in phase space as a function of time. Changes in the shape of the orbit constitute dynamic regime changes. The observational uncertainty (Δx_1 and Δx_2) for the two measurable variables effectively quantizes the phase space into discrete states of the system.

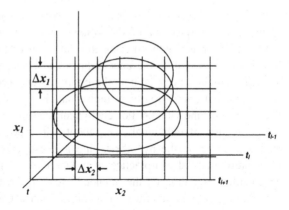

7.3.3. *Phase States s of Dynamic Systems*

By adapting a standard approach of statistical mechanics, we represent an ecological system by its *phase space*. This space is defined by the measurable variables x_i of the system and the time t. A system phase region that is distinguishable from other regions either by observation or by measurement is called a state of the system. Hence we let *the Fisher variable s represent a particular state of phase space*. Consider, for illustrative purposes, a cyclic dynamic system that has two measurable state variables x_1 and x_2. This system exists, or has a time trajectory, in a space defined by two dimensions and time. The phase-space coordinate s is a known function $s(x_1, x_2)$ of the two variables.

A trajectory for such a hypothetical system is shown in Figure 7.1 at three different times t_{i-1}, t_i, and t_{i+1}. Note that there are uncertainties Δx_1, Δx_2 inherent in the measurements of the state variables, just as there is in any observation of reality. For example, if the variables represent the populations of individuals belonging to specific biological species, there are ultimate uncertainties present of size unity ($\Delta x_1 = \Delta x_2 = 1$), since there is no such thing as a half-individual. This also implies that two values ($x_i^{(1)}$ and $x_i^{(2)}$) of the same variable are indistinguishable from one another if they differ by less than the uncertainty in the measurement ($|x_i^{(1)} - x_i^{(2)}| < \Delta x_i$), and they therefore are said to be within the same state of the system.

We represent this uncertainty in the observation and measurement of reality as a discretization of phase space. This is shown for the present example in Figure 7.1 as a series of squares (or rectangles) where the sides are Δx_1 and Δx_2, respectively. It should be noted that while both Δx_1 and Δx_2 are shown as equal, there is no particular requirement that they be the same, since the uncertainty involved in different types of measurements is generally not the same. We call each box or rectangle a state of the system. Simultaneous or time-series measurements of x_1 and x_2 will generate a series of two-component vectors ($x_1|_{ti}$ and $x_2|_{ti}$), one pair at each measurement time ti. Inserted into the discretized phase space, these measurements get distributed among the states of the system.

In Figure 7.1, each vector pair of variables (x_1, x_2) is represented as a point in the circular trajectory of the system. For illustrative purposes, Figure 7.1 (and 7.2) is drawn with unreasonably large uncertainties and correspondingly broad states. In reality, the uncertainties and corresponding sizes of the states are much smaller. For example, the minimum uncertainty in measuring the size of a biological population of N individuals is $\pm 1/N$, e.g., 0.01% for a population of 10,000 individuals. For ecological measurements, an error of 0.05% in the population of individuals, i.e., about 5 in a population of 10,000, is tolerated depending on the hypothesis under investigation and the accuracy of the observation methods. In practice, even larger errors in population counts can be acceptable, depending on the intended use of the data. However, perceived errors may be due to natural cycling rather than actual errors in the population count.

The calculation is simplified by the fact that the changes in state are very small. In the limit of high-accuracy measurements ($\Delta x_i \to 0$), the size of a state of the system approaches zero, and the differences in the path lengths l_S through different states (see Figure 7.1) of the system approach a common small number Δs, i.e.,

$$\lim_{\Delta x_i \to 0} l_S = \Delta s. \tag{7.4}$$

In practice, for purposes of calculation, we then tacitly replace the square states of the system depicted in Figures 7.1 and 7.2 with a sequence of segments of fixed length Δs along the system path as shown in Figure 7.4. That is, we treat states of the system as path segments of length Δs.

For many data sets there will likely be states containing *multiple* data vectors and states with no observations. The data sets within a state simply represent multiple observations of the same state of the system since they are indistinguishable from one another, with the observer seeing the same state more than once. Another interesting result of the argument is that when uncertainty is considered, from the perspective of an observer, the system trajectory is a series of jumps from one state to another. Generalizing this result to a multidimensional system with m state variables is difficult to visualize, but mathematically straightforward. Figure 7.2 presents our attempt at visually representing the trajectory of a system in m dimensions plus time.

7.4. Probability Law on State Variable s

Since different states of the system are not in general observed with the same frequency, an important quantity from the perspective of calculating the FI is the probability of observing a particular state of the system. Let the system be approximately cyclic, of period T, and let it obey a definite trajectory $s(t)$ over the period. For such a system, the probability $p(s)$ of observing the system in any particular state s should increase with the amount of time Δt_S that the system

FIGURE 7.2. Illustration of the path through time in phase space for a cyclic multidimensional system with m measurable variables. Note that the uncertainty in the observation of the variables (Δx_1, Δx_2, and Δx_m) discretizes the phase space into states of the system.

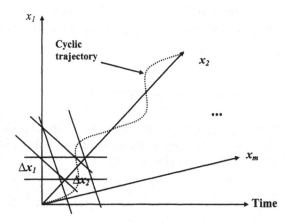

spends in the state, i.e.,

$$p(s) \propto \Delta t_S. \tag{7.5}$$

This intuitive effect will follow from the probabilistic analysis to follow. Another intuitive effect which should follow is that $p(s)$ be inversely proportional to the local velocity ds/dt through space, i.e.,

$$p(s) = \frac{K}{ds/dt} \tag{7.6}$$

for some constant K.

The analysis is an outgrowth of the simple fact that the system follows a definite trajectory $s(t)$ through phase space. Then each trajectory event $(s, s + ds)$ should occur one-to-one as often as a corresponding time event $(t, t + dt)$, or they occur equally often. Then, by definition of the probability densities $p(s)$ and $p(t)$

$$p(s)ds = p(t)dt. \tag{7.7}$$

Hence $p(s)$ increases with dt, as was conjectured in (7.5). However, the increase is not simple proportionality to dt, owing to the multiplication by ds in (7.7).

Of course any time value is equally probable over the given interval $(0, T)$ so that

$$p(t) = \frac{1}{T}. \tag{7.8}$$

Using this relation in (7.7) gives

$$p(s) = \frac{1}{T} \frac{1}{ds/dt}. \tag{7.9}$$

This confirms the intuitive result (7.6), that the probability of a phase state being observed should be inversely proportional to the velocity through the state. The constant $K = 1/T$.

7.5. Evaluating the Information

We now proceed to use the definition (7.3b) and the preceding elementary results to derive the expression for information that will be used in the test cases. Differentiating (7.9) gives directly

$$\frac{dp}{ds} = -\frac{1}{T}\frac{1}{(ds/dt)^2}\frac{d}{ds}\left(\frac{ds}{dt}\right). \qquad (7.10a)$$

By the chain rule of differentiation,

$$\frac{d}{ds}\left(\frac{ds}{dt}\right) = \frac{d}{dt}\left(\frac{ds}{dt}\right)\frac{dt}{ds} = \frac{s''}{s'}, \qquad (7.10b)$$

where each prime denotes a derivative d/dt.

Substituting (7.10b) into (7.10a) gives

$$\frac{dp}{ds} = -\frac{1}{T}\frac{1}{s'^2}\frac{s''}{s'} = -\frac{1}{T}\frac{s''}{s'^3}. \qquad (7.10c)$$

Finally, substituting Eqs. (7.9) and (7.10c) into definition (7.3b) gives

$$I = \frac{T}{T^2}\int ds\, s'\frac{s''^2}{s'^6} = \frac{1}{T}\int ds\frac{1}{s'}\frac{s''^2}{s'^4}. \qquad (7.11)$$

Or, since $ds = s'dt$,

$$I = \frac{1}{T}\int_0^T dt\frac{s''^2}{s'^4}. \qquad (7.12)$$

To review, T is a characteristic time that can be interpreted as the period for one cycle of the system; $s'(t)$ is the speed tangential to the system path in phase space; $s''(t)$ is the scalar acceleration tangential to the system path in phase space; and the integration is conducted over a time bracket from 0 to T. That is, the integration is done over at least one cycle of the system.

The phase-state tangential velocity and acceleration $s'(t)$ and $s''(t)$ are defined in terms of the m system state variables x_i as

$$s'(t) = \sqrt{\sum_i^m \left(\frac{dx_i}{dt}\right)^2}, \qquad (7.13)$$

$$s''(t) = \frac{1}{s'(t)}\sum_i^m \frac{dx_i}{dt}\frac{d^2x_i}{dt^2}. \qquad (7.14)$$

Equation (7.12) together with the expressions in Eqs. (7.13) and (7.14) form our final expression for computing the FI. Note, from the form of Eq. (7.12), that the function $s(t)$ per se is not needed for calculating the information. (Of course $s(t)$ is formally the integral of Eq. (7.13).) The inputs $s'(t)$ and $s''(t)$ to Eq. (7.12) can be calculated from their definitions (7.13) and (7.14) either by directly differentiating the time series data x_i or from model equations.

7.6. Dynamic Order

To start the discussion on dynamic order, we first consider the concepts of perfect order, perfect disorder, and real dynamic systems. A perfectly *ordered* system is one that does not change from one observation to the next, i.e., each measurement is exactly like any other so one measurement gives you all of the information that is obtainable about the system. For static systems, this translates into the observable variables having the same values within measurement error from observation to observation. A perfectly ordered system spends all of its time in one particular state of its phase space. In contrast, a perfectly *disordered* system is one that unpredictably changes from one observation to the next, i.e., each measurement is completely uncorrelated to every other measurement, so no amount of measuring can give you any information about the system. For dynamic systems, this translates into the observable variables having completely unpredictable and uncorrelated values from one observation to the next. For a perfectly disordered system, the likelihood of observing any particular state is the same as that of observing any other state. For real dynamic systems, however, order is a more complex and subtle matter falling between the two idealizations of perfect order and perfect disorder. By their very nature, real dynamic systems will travel through, or exist in, different states of their phase space at different times, and the dynamics may be cyclical or noncyclical.

To relate the aforementioned concepts of order to FI, we start by considering Eq. (7.3c). For a perfectly ordered system, the system exists in only one state. Therefore the probability density $p(s)$ for observing a particular state s is a tall spike, as is its amplitude function $q(s)$ (see Eq. (7.3c)). Then the derivatives $dq(s)/ds$ within the spike are infinite, so that the FI (7.3c) becomes *infinite*. This means that a great deal of information can be obtained from observing the states of the system. For a perfectly disordered system, the system exists in many equally probable states, the probability density $p(s)$ and its amplitude $q(s)$ for observing any particular state s are the same as that for all other states, the derivative $dq(s)/ds$ is zero, and the FI (7.3c) is zero as well. This means that no information can be obtained from observing states s of this system.

Since FI is a measure of the information obtainable from observation, in practice these tendencies again simply mean that (1) very little useful information can be obtained from observing constantly changing, never repeating systems that show no behavior patterns, and (2) a lot of useful information can be extracted from observations of systems that have behavior patterns. For systems, particularly dynamic systems such as biological systems that are between perfect order and perfect disorder, the FI will vary between zero and infinity depending on the degree of dynamic order. Starting with a system in perfect *disorder*, the FI will rise as the system shows preference ($dq(s)/ds$ rising above zero) for particular states of the system, and it will reach a maximum if the system shows preference for only one state ($dq(s)/ds \to \infty$). Or, starting with a system in perfect *order*, the FI will decrease as the system's preference for its one state decreases ($dq(s)/ds \to 0$), and it will reach zero if the system loses all preference for any particular state ($dq(s)/ds = 0$). Thus, the FI measure (7.3) when calculated from observations of dynamic systems correlates with their levels of order or disorder.

7.7. Dynamic Regimes and Fisher Information

One thesis being proposed here is that a change in the value of FI can signal a change in regime for a dynamic system. This thesis is important because it offers a generalized methodology for detecting regime changes. The logic is as follows: (1) if a change in dynamic regime is observable then there must be a corresponding change in the measurable variables of the system, i.e., anything that has no manifestation in the observable variables of the system is unobservable; (2) an observable change in the measurable variables implies a corresponding change in the distribution of states of the system; and (3) a corresponding change in the distribution of system states implies a change in the FI. While we have found these arguments useful under many circumstances, we have no formal mathematical proof of their rigor and for that reason offer them here as a thesis rather than a law of nature. However, since normal system cycles also generate changes in the FI, *one confounding issue is sorting out the FI signal of regime change from that originating from natural cycles.* The strategy to mitigate the problem is to draw out the range of the integral in Eq. (7.12) wide enough to average over at least a few system cycles but not so wide as to average out phenomena, and then look for changes in this averaged FI. It is particularly effective when an estimate of the characteristic time for the longest natural period of the system is available. While this method will not resolve the problem under all circumstances, particularly for real systems that show complex cyclic behavior with multiple periodicities, we have found it generally effective.

A second related thesis that follows from the above is that the FI associated with a particular regime of the system, when averaged over sufficient system cycles, will remain constant as long as the system does not leave the particular regime. Note, however, that the measurable variables of the system x_i are free to vary and fluctuate. To further expand on the concept, consider Eq. (7.12) and note that I will not vary with time as long as the speed s' and the scalar acceleration s'' remain constant. Note, as well, that s' and s'' can be constant, while the components (dx_i/dt and d^2x_i/dt^2) of the sum in Eqs. (7.13) and (7.14) *vary* according to the system dynamics. In other words, the only real requirement for I to be constant is that the sums in Eqs. (7.13) and (7.14) remain constant. These would allow the measurable variables x_i to increase, decrease, and fluctuate even to zero while the FI remains *constant*. When calculating I by integrating over a period, I will be constant if s' and s'' have the same temporal profile from one period of integration to the next, i.e., s' and s'' are not temporally constant, but change consistently from one period of integration to the next. This thesis is important because it represents a criterion that would allow the measurable variables of the system to fluctuate, while system dynamic order and regime are being *conserved*, i.e., it allows for the system dynamic regime to be preserved without requiring that the system variables be held constant. This is critical because biological systems have cycles, e.g., high and low growth seasons, and fluctuations that cannot be stopped without destroying the system. Note again that consistent cycles also are acceptable fluctuations, i.e., s' and s'' may not be constant, but cycle consistently as already discussed.

A third and, perhaps, more speculative thesis is that a trend with a steady decrease in the FI with time signifies a loss of dynamic order. Functioning biological systems universally exhibit *dynamic order*, i.e., the measurable variables of the system have observable regularities. A trend where dynamic order is being steadily lost is important because dynamic order is a fundamental property of all biological systems, and it is something that must be preserved for the system to survive (see also Chapter 3). Hence, we speculate that *loss of dynamic order is likely to be indicative of some sort of loss of organization or loss of function*. For example, living biological systems that are steadily moving toward death typically suffer a degradation of internal feedbacks, causing the loss of patterns and regular behaviors observed in their measurable variables. This thesis presents one interpretation of the significance of a steady decrease in FI. This thesis is, however, not applicable to dead biological systems where all measurable variables might be fixed and the FI infinite.

7.8. Evaluation of Fisher Information

Equation (7.12) for the FI requires first and second derivatives of the measurable variables x_i of the system with respect to the time t. The derivatives can be obtained analytically if a mathematical model for the system in the form of differential equations is available. However, when working with time series obtained from field or experimental data, it is necessary to evaluate the derivatives numerically. There are several ways to evaluate the derivatives numerically. Simple difference schemes [25] are disadvantageous in that the results can be highly susceptible to noise, particularly the second derivative; to obtain usable results, it is necessary to make the integration bracket sufficiently wide to average out the noise. Splines can also be used to differentiate data, but these do not always provide accurate derivatives from noisy data. A simple and more robust alternative to simple differencing is to use a three-point differencing scheme which for data that is unevenly spaced in time takes the form of

$$\left.\frac{dx_i}{dt}\right|_t = \frac{\alpha^2 x_i(t + \Delta t_a) - (\alpha^2 - 1)x_i(t) - x_i(t - \alpha\Delta t_a)}{\alpha(\alpha + 1)\Delta t_a} \tag{7.15}$$

$$\left.\frac{d^2 x_i}{dt^2}\right|_t = \frac{\alpha x_i(t + \Delta t_a) - x_i(t - \alpha\Delta t_a) - (\alpha + 1)x_i(t)}{\alpha(\alpha + 1)\Delta t_a^2/2}, \tag{7.16}$$

where $x_i(t)$ is a central data point, $x_i(t - \Delta t_p)$ is the point previous to the central, $x_i(t - \Delta t_a)$ is the point after the central point, and $\alpha = \Delta t_p / \Delta t_a$ is the ratio of the previous and after time steps. Note that for evenly spaced points $\Delta t_p = \Delta t_a$ and $\alpha = 1$.

In practice, for purposes of detecting and assessing dynamic regime changes, the integral in Eq. (7.12) can be evaluated in one of two ways. First, in a "block averaging" approach, it is computed over a window of data corresponding to n points in time encompassing at least one period of the system, and the average value of the FI is then assigned to all times in the window. Second, the integral value can be assigned to a central point in the data window, the window moved one or more points in time, and the procedure repeated so as to compute a "moving average." The objective of both procedures is to construct readable diagrams for FI versus time. The second approach seems to reveal more details of the system dynamics than the "blocky" diagrams resulting from the first procedure. For cases where the time of occurrence of the dynamic regime shift is known a priori, an average FI can be calculated for each dynamic regime independently. These averaging procedures have the effect of minimizing the noise inherent in the FI calculation, particularly when using noisy data.

The FI expression (7.12) is rigorously applicable only to systems for which the "shift invariance" assumption (7.2) holds. Typically, the shift-invariance assumption holds for systems in periodic steady states. Application of Eq. (7.12) to nonperiodic or complex multiperiodic systems is an approximation and can result in fluctuating or seemingly noisy FI where there is no change in dynamic order and no regime shift. This can occur when there is a mismatch between the integration bracket of the FI and the periodicity of the system, even when the data have no noise. However in practice, we have found that this does not obscure the observation of regime shifts under most circumstances, because most systems do tend to approach periodic or quasi-periodic steady states with time.

As already discussed, there are various challenges in estimating the FI from Eq. (7.12) and time-series data. These arise from noise, complex periodicities, nonperiodicity, and mismatches between the integration bracket and the system periodicity. For this reason it is not always possible to obtain readable block- or moving averages for the FI, where the regime shift signal is clearly distinguishable from otherwise irrelevant fluctuations. In such cases, if the location in time of the regime shift is known or at least suspected, an average FI for each of the two regimes can be computed before and after the shift. This process evens out a great deal of noise and fluctuation. If a significant change in the FI is found across the point in time where the regime change occurred, then the change in average dynamic order would lend credence to the existence of a "true" dynamic regime shift. Interpreted with the aid of the thesis already discussed (Section 7.7), this can be powerful and invaluable information. It is not, however, an assured detection of a dynamic regime shift. In other cases, one can also compute an average FI for a large block of data over a span of time for different systems and compare the average degree of dynamic order and FI between systems. Work is in progress on the development of more robust FI calculation procedures that are insensitive to noise and independent of shift invariance.

7.9. Applications to Model Systems

The application of FI to model systems is of significant interest. It could, for example, aid in providing model-assisted prediction of dynamic regime changes in very complex systems. This is important because otherwise, the interpretation of results from complex models involving many variables can be as daunting as the interpretation of data obtained from real systems. The application of FI to model systems, particularly simple model systems, allows us to confirm our expectations under strictly controlled conditions. These are the following: no noise, well specified system periodicities, and dynamic regime shifts caused purposefully by the modeler. We consider two such models: A simple, *two species ecosystem model* and a more complex, *multispecies ecosystem model* involving a pseudo-human compartment and a pseudo-economy with agriculture and industry.

7.9.1. Two-Species Model System

Several different types of equations have been developed to model species interactions [26] and build food webs [27, 28]. Lotka–Volterra (L–V) differential equations have been used by ecologists to describe and understand the dynamics of predator–prey systems [29]; their origin in mathematics makes these equations (and derivations of them) broadly applicable to many other types of dynamic systems as well [30, 31]. These equations describe the interaction between two or more species, dictating whether one species will cause the extinction of the other or whether the two species will coexist. The equations are typically used to model systems as they evolve through *time*; however, they have also been applied to model spatial distributions of species (e.g., [32]). Here we have used a two-species prey–predator L–V system of equations

$$\frac{dx_1}{dt} = g_1 x_1 \left(1 - \frac{x_1}{k}\right) - \frac{l_{12} x_1 x_2}{1 + \beta x_1} \tag{7.17}$$

$$\frac{dx_2}{dt} = \frac{g_{21} x_1 x_2}{1 + \beta x_1} - m_2 x_2 \tag{7.18}$$

to simulate a system closed to mass but open to energy. The system includes two interacting compartments with masses x_1 (the prey) and x_2 (the predator). Parameter g_1 is the prey growth rate, l_{12} is the prey loss rate due to predatory feeding, g_{21} is the predator feeding rate, m_2 is the predator mortality rate, k represents a population density limit to prey growth in the absence of a predator, and β is a predator satiation term.

Ecologically, predator abundance is dependent upon prey abundance, since predators must acquire energy in excess of their own needs to reproduce. However, very high predator populations can deplete prey populations, causing a prey population crash which then results in a predator population crash. Phase-space plots such as Figures 7.3 to 7.5 are an easy way to visualize these dynamics, graphing

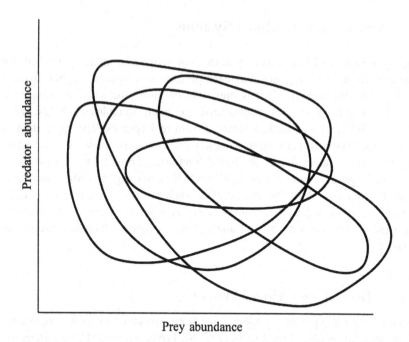

FIGURE 7.3. Phase-space plot illustration for a hypothetical prey–predator ecosystem, where the orbit varies in no discernable pattern (i.e., at random) and where the two species have little or no influence on each other's population.

the abundance of the predator versus that of the prey (time is implicit in these plots).

However, if two species do not interact, and if they are not affected by any shared external forcing function (which may cause their populations to cycle similarly), then they are completely independent of each other. A resulting phase-space plot for these two species would be completely random (see Figure 7.3). Here there are no constraints upon the populations, and any combination of their abundances is possible. In this case, predicting the abundances of both species would be impossible, and the resulting FI would be zero for the system.

Another scenario of zero information is as follows. As already discussed more abstractly in Section 7.6, if two species cycle regularly, two dynamic behavior extremes are possible (Figure 7.4). Species population abundances may increase or decrease smoothly over time, resulting in an equal probability of observing any particular state(s) of the system (i.e., combination of predator and prey abundance). In other words, the system would travel at a constant speed around the closed curve. FI would be near zero according to Eq. (7.12). In effect, this is because we have not accounted for the fact that the system is constrained to a closed curve of *allowable* predator and prey abundance combinations.

By comparison, if each species maintained a constant abundance with the exception of a small portion of time when the species rapidly increased and decreased

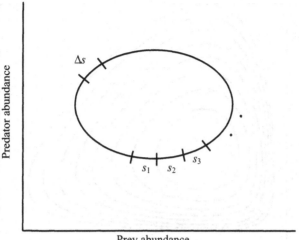

FIGURE 7.4. Phase-space plot for cyclic prey–predator ecosystem. The orbit for this system is a closed curve. Note that states are here identified as equal segments (Δs_i) of the system path (see Section 7.3.1). The system could travel around the curve at a constant speed where the likelihood of observing any particular state is the same for all states, resulting in low FI. Alternatively, the system could spend a large proportion of time (Δt) at any one state(s), resulting in high FI [17]. Used with permission from Elsevier.

back to its original abundance level, the FI for the system would be *very high*. There would be a high probability of observing the state in the most common species abundance combination. In this case, the system spends most of its time at one location on the circle, until some shift in predator or prey behavior causes it to traverse the circle very quickly and come to rest again at the original state. Note that whichever way the system moves around this closed curve, the system is still in the same dynamic regime.

Starting with the baseline or default parameter values $g_1 = m_2 = 1, l_{12} = g_{21} = 0.01, k = 625$, and $\beta = 0.005$, we perturb this system by altering parameter k, the carrying capacity for the prey. This forces the system to change dynamics from one limit cycle to another (see Figures 7.5a and b), i.e., changing from one closed curve to another. We, in fact, force the system to shift regimes, where predator and prey abundances cycle through different closed curves. FI in this scenario is calculated from Eqs. (7.12) to (7.14) and the model expressions of Eqs. (7.17) and (7.18). Note that as long as the system remains in the same circle/dynamic regime, FI remains constant.

However, once we perturb the system, it begins to visit new areas of the phase space and therefore FI changes (often by decreasing sharply); the predictability of the state of the system has declined. Once the system settles into the new regime, FI becomes constant. As to whether the new FI for the system is higher or lower than that for the original regime depends on the behavior of the system in each regime.

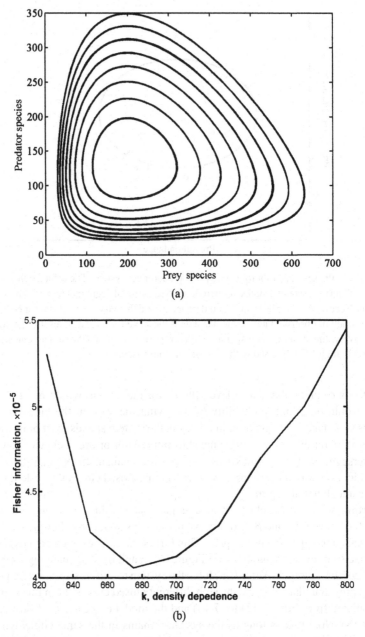

FIGURE 7.5. (a) A phase-space plot of a cyclic prey–predator ecosystem where a parameter (*k*) change has caused the system to shift to different dynamic regimes, with a subsequent change in FI [17]. (b) A FI plot of a cyclic prey–predator ecosystem where a parameter (*k*) change has caused the system to shift to different dynamic regimes, with a subsequent change in FI [17]. Used with permissions from Elsevier.

7.9.2. Multispecies Model: Species and Trophic Levels

While two-species models are adequate to understand how FI tracks dynamic system behavior over time, the information in this case does not add a new perspective since the behavior of this system is *easy to graph* in two dimensions. This type of system would also be *easy to manage* for sustainability (there are only two variables to manipulate). However, most systems in reality are orders of magnitude more complex, and for these systems FI can prove to be an invaluable tool.

Ecologists have many decades of experience with building computer simulations of food webs, either as abstract systems (where flows represent mass, nutrients, energy, or exergy) or as models of real systems such as estuaries or deserts [16, 27, 33]. Simple abstract food webs are easily simulated conveniently, the effect of the structure of the web itself can be assessed as it directs the flow of mass, energy, and nutrients through different trophic levels[2]. The number of trophic levels, and the degree of connectivity between the compartments, can vary in webs of the same number of compartments, and the behavior of these systems can be quite complex, even for systems with few trophic levels, compartments, and low connectivity [28].

We designed a 12-compartment model, with four trophic levels and two underlying resource pools, to be as structurally simple as possible while still resembling a potential real-world ecosystem (Figure 7.6; see [34] for the general form of the model compartment equations). The growth and mortality rates of the four plant species, three herbivore species, two carnivore species, and omnivore ("human") species are all determined by Lotka–Volterra equations, which use growth and mortality rates inherent to each species, plus mortality based on predation by other species.

For the omnivore species, mortality rates are inherent to the population, since no other species preys on it. Note that mortality due to predation transfers mass to the predator compartment, whereas mortality *not* due to predation transfers mass back into the accessible resource pool.

The inaccessible resource pool (IRP) could be a landfill, pollution spill, or other concentration where nutrients and mass are difficult to retrieve and use in the biological system. This compartment has a very slow return of mass to the resource pool, where it is then available to the rest of the system (such as phosphorus slowly reclaimed from the ocean via ocean food webs, after it runs off into the ocean from excessive fertilizer applications on terrestrial farms). Two of the wild plants are also able to recycle mass from this IRP, but again quite slowly (Figure 7.6).

For this model, FI is calculated using the speed and acceleration of the system in 12-dimensional space from the L–V differential equations [15, 34]. However, since the index we use here is based on the probability of observing the system in a particular state, the state must encompass a defined time period. Here, we average FI over 48 time steps, which is four times the periodicity of the forcing function we added to the model. We can of course calculate FI in 12 time steps (one period of the forcing

---

[2] A trophic level is a group of organisms that consumes resources on the same level; e.g., animals that eat plants only are all considered "herbivores."

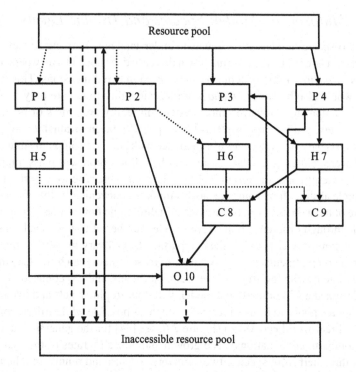

FIGURE 7.6. Diagram of a 12-compartment ecosystem model with a resource pool and IRP representing biologically inaccessible mass, as well as four plant species, three herbivore species, two carnivores and an omnivore human species [34]. Used with kind permission of Springer Science and Business Media. Dotted lines indicate mass flows influenced by the omnivore, while the dashed lines indicate wastefulness, as mass transfers from the resource pool (production wastefulness) and omnivore species (consumption wastefulness) to the IRP.

function). However, FI, as a *local* measure, is very sensitive to changes in the system and therefore some averaging is desirable to pick out major regime shifts[3].

We first found a set of parameters for which the model can run indefinitely without losing any species. We then perturbed several parameters to simulate real-world "desirable" and "undesirable" scenarios, and calculated FI using all 12 state variables to track the system's behavior. We found that FI was useful to illustrate the behavior of the system through time, and especially to *identify the time* at which perturbations occurred, and which led to a regime shift in the system.

For example, in one scenario *the wastefulness of production increased* (modeled as a direct transfer of mass from the resource pool to the IRP, with a resulting decrease in human mortality). Concurrently, *the wastefulness of consumption*

---

[3] Calculating FI over a time period of less than 12 steps introduces the effects of the forcing function into the behavior of FI, which causes the index to be responsive to this artificial forcing function in addition to the major system behaviors.

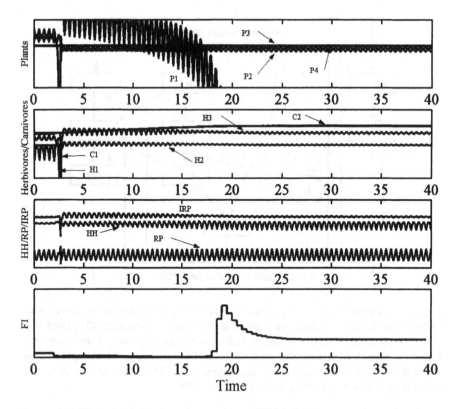

FIGURE 7.7. Time plots of compartment mass and FI for the 12-compartment ecosystem model with a resource pool and IRP. Note that the increase in production wastefulness with a decrease in consumption wastefulness is detected by FI (bottom graph), which spikes during the regime transition and then settles at a higher value [34]. Used with kind permission of Springer Science and Business Media.

decreased (modeled as the flow from the omnivore compartment O10 to the IRP). The FI measure detected the system change, as a spike in the information value, which then settled to a new (higher) equilibrium value (Figure 7.7). This scenario resulted in *the loss of three species* (P1, H5, and C8).

7.9.3. Ecosystems with Pseudo-Economies: Agriculture and Industry

In the previous model, human mass is dictated by births, growth, and death, the same as all other species in the model. However, in the real world, humans now appropriate up to 40% of net primary productivity on the planet [3, 35, 36] and construct roads, factories, machines, and other structures that require a large amount of energy and mass to build and maintain. The diversion of greater amounts of energy and mass of critical nutrients to urban, agricultural, and industrial activities, and the subsequent generation of pollution and environmental damage, has left the vast majority of other species on the planet vulnerable to extinction, and

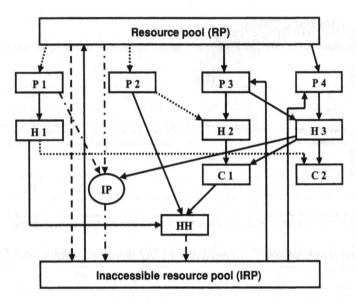

FIGURE 7.8. Diagram of a 12-compartment ecosystem model with attributes similar to those of the system in Figure 7.6, plus an embedded abstract industrial sector (IP) [43]. Used with permission from Elsevier. Dotted lines indicate transfers managed by humans, the dashed lines indicate production (a transfer of mass from the RP to the IRP) and consumption (from HH to IRP) wastefulness, respectively, and the dotted-dashed lines indicate wastefulness of the IP sector (mass flow from the RP and P1 to the IRP).

indeed *a mass extinction may be underway* [37–39]. The effect of the loss of a particular species in an ecosystem is difficult to predict. However, the cumulative loss of species often results in a noticeable decline in the quantity, quality, and consistency of goods and services (e.g., water purification, pollination, and raw materials for industrial processes) provided by ecosystems [40, 41].

Network models inspired by food webs have been used to model the flow of energy and mass through industrial systems, to increase the sustainability of industrial sectors with respect to social and environmental goals [42]. We developed a multispecies model, differing slightly from previous models, to simulate the effects of an industrial process in a food web [43]. This web consisted of two resource pools (one "inaccessible"), four plants, three herbivores, two carnivores, and a human population (Figure 7.8). In this system, a simple agricultural sector determines the flow of mass to two of the plants (to simulate crops), instead of representing an inherent growth rate; humans eat one crop directly (P2), and feed the other crop (P1) to a domesticated herbivore (H1), which has a growth rate also dictated by humans instead of an inherent rate. The humans also build fences to decrease hunting (mass transfer) of H1 by a wild carnivore (C2), while humans also hunt and eat wild carnivore (C1).

The two wild plants (P3 and P4) are the only species that can recycle mass from the IRP back into the system. With the exception of the industrial process

(IP) compartment, the amount of mass in each compartment is determined by birth, growth, and death rates of each species (or for the resource pools, death rates of all species). For the IP compartment, mass flows directly through the compartment from two raw materials (P1 and the resource pool) and into the IRP. The IP does not have a birth, growth, or death rate. However, as the flow of mass through the IP increases, the amount of mass flowing into the IRP increases, and the mortality rate of the humans decreases. This abstractly simulates the direct benefit of industrial processes to humans, while simultaneously generating an increased diversion of mass to an inaccessible compartment, similar to nonrecycled industrial and consumer waste dumped into a landfill.

We ran a variety of simulations in which the IP draws an increased proportion of mass from different trophic levels from a stable, baseline set of parameters [43]. If the IP draws most of its resources from a wild herbivore (H3) or a plant (P1), as many as four or five species go extinct. However, if the IP switches to extracting mass from the resource pool, extinction of most species can be avoided. Generally, extracting resources directly from the resource pool resulted in the fewest species lost, although the increased flow of mass tended to destabilize the system (Figure 7.9) without a compensating mass transfer from the resource pool to the IRP (this provides more mass to the wild plants, which are the only species that can utilize mass in the IRP).

In Figure 7.9, the utility of FI to track changes in multidimensional dynamic systems over time is readily apparent. At the initial parameter change at time step 2000, FI spikes as mass flows begin to shift. The system appears to stabilize for a brief period, and as mass increases in one of the herbivores (H1) the system appears to stabilize, and the information increases. However, after a threshold is reached, the system destabilizes again, most notably in the increasing oscillations in P3 and H2. FI again decreases with the lowered predictability of the system's trajectory.

7.10. Applications to Real Systems

Regime shifts have been identified in the global climate system [9] as well as in many ecosystems, including coral reefs, freshwater lakes, marine food webs, savannas, deserts, and forests, to name a few [7, 8, 13, 44, 45]. Indeed, ecosystems and the climate system are closely connected, and regime shifts in one system can trigger shifts in other systems at different scales [12, 46–48]. However, identifying a true regime shift in an ecosystem (as opposed to simply a large change in the system within a single regime) requires a combination of modeling and statistical, and empirical or experimental investigation [49, 50]. An index based on FI provides a much-needed tool for the identification of regime shifts in real-world systems. We used several ecosystems for which a true regime shift or shifts have been verified, in order to test and refine our FI index. This anticipates its use on data sets that tend to be noisy and lack a firm periodic cycle. For the following three examples, we use the FI index as computed using Eqs. (7.12) to (7.14).

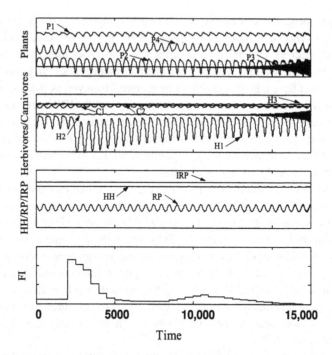

FIGURE 7.9. Time plots for compartment mass and FI for a scenario in which the mass feed into the generalized IP consists of 90% from the Resource Pool (RP), 5% from a plant (P1), and 5% from a wild herbivore (H3). The system begins at the baseline steady state, and then the mass throughput of the IP is doubled, between times 2000 and 2500. The system destabilizes, as evidenced by the growing oscillations, resulting in a much lower FI tending to zero [43]. Used with permission from Elsevier.

7.10.1. North Pacific Ocean

Hare and Mantua [51] identified what appeared to be two regime shifts in the marine ecosystem of the North Pacific Ocean, one in 1977 and the second in 1989, possibly driven by interdecadal climate fluctuations. The data set available for this system includes annual measurements for over 100 biological and environmental variables. However, the data spanned only about 30 years and data points were missing for some variables in some years. Given the short time span of data available, and the level of noise in the variables, the authors first used principal components analysis to isolate the variables with the greatest amount of uncorrelated variance, then normalized the remaining data, averaging both across variables within a year and then across years within each regime [51].

We applied FI on the normalized but otherwise raw data for 65 of the 100 variables for the system (35 of the variables had too few data points to allow us to fill in gaps through interpolation; [25]). When we calculated the information using all 65 variables, the regime shifts were apparent. However, the first regime shift (around 1977) was not well differentiated from the regime immediately following it (regime 2; Figure 7.10a). When we narrowed the calculation to the subset of

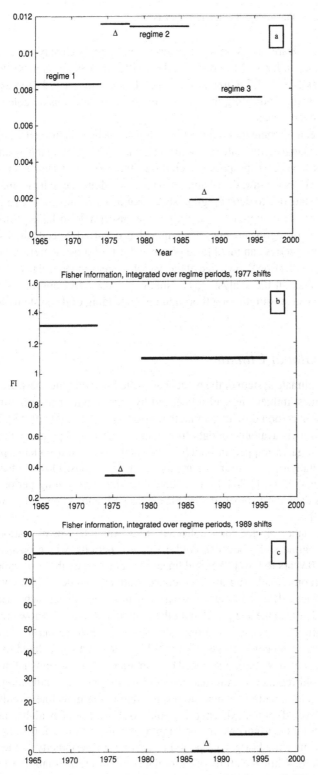

FIGURE 7.10. FI, averaged over each regime and regime shift, for (a) all 65 variables, (b) the 12 variables most closely identified with the 1977 shift, and (c) the 5 variables most closely associated with the 1989 shift. Regime shifts are indicated by Δ [25]. Used with permission from Elsevier.

variables that Hare and Mantua [51] identified as most closely associated with each shift, the ability of FI to pick out the 1977 and 1989 shifts in particular was enhanced, as expected (Figures 7.10b and c). In these subsets, FI decreases during the regime shifts, indicating that the system experiences considerable change in variability over the shifts.

In the oceanographic literature, regime shifts have been identified with points of significant change in the statistics of the data, mainly changes in the mean. Various approaches have been proposed, including Student's t-test and other statistical tests [51–55]. As we saw, the FI approach also uses dynamic order estimated from changes in statistics to detect regime shifts, but it is not limited to changes in the mean. All of these approaches, however, are susceptible to false positives—the identification of regime changes when none has occurred. Specifically, stationary red noise time series can yield false positives [56]. FI must, therefore, be used in conjunction with good knowledge of the overall system and the relative importance of the variables in the analysis [57]. Neither the use of FI nor any of the other approaches can substitute for a thorough understanding of the system dynamics.

7.10.2. Global Climate

The global climate system of the planet has shifted several times between cold and warm regimes, influencing and influenced by connectivity between atmospheric gases (mainly carbon dioxide and methane) and ocean currents [9, 58]. Evidence for these shifts is available in data from cores or probes of polar ice (samples of atmospheric gases trapped in bubbles), lakebeds (preserved pollen representing major vegetation), and ocean sediments (intensity of dust blown offshore from major deserts) [59–61]. While these cores provide a very long period of fairly continuous data, the number of variables measured tends to be quite few, usually less than 10.

We used data from ice cores collected at the Antarctic Vostok station to identify three regime shifts between cold and warm periods, which occurred in the last 160,000 years [62–67]. We used linear interpolation on three variables (atmospheric carbon dioxide and methane concentrations, and deuterium levels of the ice) to produce a data point every 50 years, and then normalized each variable [25]. We then calculated the average FI over the time period for each known regime and regime shift. The information dropped during each of three major regime shifts between warm and cold periods (Figure 7.11). The low FI calculated for the long cold period may have been influenced by a smaller shift between what may be two different cold regimes within that one cold period, perhaps indicating different ocean circulation patterns when the north Atlantic thermohaline circulation was either weak (cold) or nonexistent (very cold; [68]). A lack of available data for the most recent (and current) warm period may explain the low FI; not enough time has passed since the last glacial period to capture the periodicity of the variables was measured. While the utility of FI for data sets with few variables is less apparent than with many variables, it can nevertheless serve as a beneficial aggregate

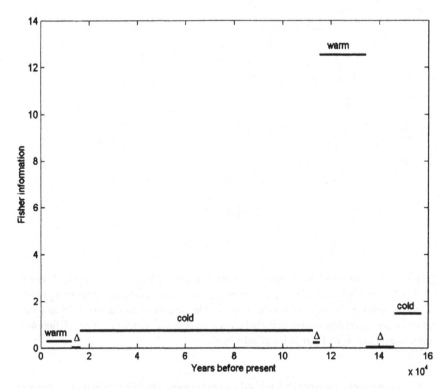

FIGURE 7.11. FI averaged over cold and warm global climate regimes from three ice core variables (atmospheric carbon dioxide and methane concentrations, and deuterium levels of the ice). Regime shifts are indicated by Δ [25]. (Notice very short curve segments below the Δ's.) Used with permission from Elsevier.

index for identifying potential impending shifts, and can perhaps suggest qualitative differences in dynamic behavior between regimes of the same system.

7.10.3. Sociopolitical Data

Historical data spanning over five decades from hundreds of countries have identified several events that can indicate the risk of a catastrophic collapse of a nation's governing body, also known as "state failure" (i.e., civil war, mass exodus; [10, 69–72]). Governments defined as stable and resilient provide effective security and social services to their citizens over their entire administered territory, and withstand a variety of stressors that would trigger state failures in weaker governments [11,73]. The level of infant mortality, government-sanctioned discrimination and trade openness, as well as state failures in neighboring countries, can broadly characterize a government's stability and its risk of regime shift [71, 72]. Some evidence suggests that democracies and autocracies are each stable regimes, characterized by the degree of representation afforded to minorities (both in terms of ethnicity and sociopolitical and economic traits) and the amount of power held by an executive authority [11].

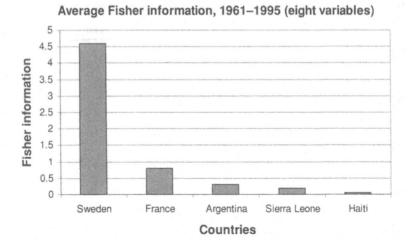

FIGURE 7.12. Average FI representing dynamic order calculated for five countries between 1961 and 1995, using eight normalized variables from the Polity database: democracy index, autocracy index, years in office for the head of state, percent of population in urban areas, infant mortality, relative degree of parliamentary responsibility, percent of population with military experience, and legislative effectiveness.

Using the eight variables identified as being most closely predictive of regime shifts or state failure, we calculated an average FI value for several countries using data from 1961 through 1995 (Figure 7.12). These annual data track variables that, in perfectly stable countries, should not change from year to year; FI for such countries would be very high. The greater the variability in these data for a country, the less stable it is likely to be, and this higher variability will lead to lower FI. Sweden experienced no state failures during this period and therefore represents a nation characterized by high stability (low variability). The lower average FI for France may be due to the collapse of France's government in 1958, shortly before the period in which these data were collected; many of the variables may have been measuring recovery after this collapse at the beginning of the data collection period. Argentina experienced military coups in 1966 and 1970, and after a period of brief stability, another coup in 1975–1976, which led to severe restrictions on civil liberties until democratic elections were resumed in 1981. Of the five countries included here, Sierra Leone and Haiti experienced the greatest level of instability over this time period. Sierra Leone suffered a series of coups between 1967 and 1968, and between 1991 and 1995 the "official" government controlled little more than the capital; a rebel force controlled most of the countryside. Between 1986 (when military rule ended) and 1991, a series of provisional governments ruled Haiti; a military coup in 1991 ended the string of democratically elected governments and ruled until 1994, when an international force restored the democratically elected leaders. While this FI analysis is fairly preliminary, we include a sample of it here to indicate the utility of the index to a broad variety of dynamic systems.

7.11. Summary

In this chapter, we introduced FI for the purpose of investigating the dynamic behavior of ecological, environmental, and other complex systems. The form of the FI that was used is based on the probability of observing a particular state of the system, which, for a cyclic system, is proportional to the amount of time the system spends in the state. Computationally, this varies as the integral of the ratio of the acceleration and speed tangential to the system path in phase space, where the integration is done over at least one cycle of the system. In a discretized phase space of the system dynamics, different states of the system are not in general observed with the same frequency. FI varies between zero and infinity depending on the degree of dynamic order. For example, in a perfectly ordered system, the system is seen to exist in only one state; the system is deterministic and the FI is infinite. At the other extreme, a perfectly disordered system exists in many equally probable states, so that no information can be obtained from observing states of the system, and the FI is zero. Therefore, we propose the use of FI as a quantitative measure of a system's dynamic order in the presence of phase-space oscillations and perturbations.

This ability to classify regimes on a continuum from "ordered" to "disordered" suggests the use of FI as a measure of overall system sustainability, where sustainability implies both stability (low variability) and desirability. This is summed up in the following three theses: (1) A change in the value of FI can signal a change in regime for a dynamic system. (2) FI associated with a particular regime of the system when averaged over sufficient system cycles will remain constant as long as the system does not leave the regime. (3) A trend with a steady decrease in the FI with time signifies a loss of dynamic order.

As demonstrated in this chapter, time-series data for FI calculations can come from model results or empirical sources. In particular, we demonstrated the change in FI using a two-species model and multispecies ecosystem model. The 12-compartment ecosystem model, with four trophic levels and two underlying resource pools, was constructed to be as structurally simple as possible, while still resembling a potential real-world ecosystem. We found that FI was useful to illustrate the behavior of the system through time, and *especially for identifying the time at which perturbations occurred that led to a regime shift in the system*. In an example of typical results from this model, FI detected the system change in one scenario, which lost three species as indicated by an initial spike in the information value, and then settled to a new (higher) equilibrium value.

In order to extend the ecosystem model to environmental management, we also linked the above multispecies ecosystem model to a human-industrial system such that the IP draws an increased proportion of mass from different trophic levels of a stable, baseline set of parameters. Human extraction from the environment resulted in the loss of over one-third of the model components. The utility of FI to track changes in multidimensional dynamic systems over time was readily apparent. Impact on the environment was less severe when the human extraction was directly from the *renewable* resource pool, although this too led to a destabilized system if there was not enough resource available to other species.

FI was also found useful when applied to data from real-world systems that have undergone regime shifts. The index always changed in value during transitions, usually dropping (and indicating higher variability in the data during these periods). Particularly for data sets with many variables, such as the North Pacific ecosystem, FI aggregated the multidimensional behavior of the system into one interpretable variable. However, ecological and social interpretations of the significance of different FI values for different regimes, both within and between similar systems or nations, require a great deal of knowledge about the systems.

Future work in this area will focus on (1) improving the techniques for estimating the FI, particularly from field or experimental data having significant noise or error, (2) understanding how to interpret FI results in the case of systems having complex dynamic behavior such as multiple attractors, (3) developing a theory that is not dependent on the assumption of shift invariance, and (4) applying the methodology to ecological, socioeconomic, and other natural systems.

Acknowledgments. The authors acknowledge the guidance of Professor B. Roy Frieden from the University of Arizona throughout this work. C.W. Pawlowski served as a postdoctoral research associate with the Oak Ridge Institute for Science and Education. B.D. Fath was a postdoctoral research associate with the US Environmental Protection Agency's Postdoctoral Research Program.

8
Fisher Information in Ecological Systems

B. Roy Frieden and Robert A. Gatenby

Fisher information is being increasingly used as a tool of research into ecological systems. For example the information was shown in Chapter 7 to provide a useful diagnostic of the health of an ecology. In other applications to ecology, extreme physical information (EPI) has been used to derive the population-rate (or Lotka–Volterra) equations of ecological systems, both directly [1] and indirectly (Chapter 5) via the quantum Schrodinger wave equation (SWE). We next build on these results, to derive (i) an uncertainty principle (8.3) of biology, (ii) a simple decision rule (8.18) for predicting whether a given ecology is susceptible to a sudden drop in population (Section 8.1), (iii) the probability law (8.57) or (8.59) on the worldwide occurrence of the masses of living creatures from mice to elephants and beyond (Section 8.2), and (iv) the famous quarter-power laws for the attributes of biological and other systems. The latter approach uses EPI to *derive the simultaneous quarter-power behavior* of all attributes obeyed by the law, such as metabolism rate, brain size, grazing range, etc. (Section 8.3). This maximal breadth of scope is allowed by its basis in information, which of course applies to all types of quantitative data (Section 1.4.3, Chapter 1).

8.1. Predicting Susceptibility to Population Cataclysm

8.1.1. Summary

We derive, and test by example, a simple preliminary test for determining whether a given ecology of animals is in jeopardy of a sudden drop in its population level, say by 50%. The approach uses the Cramer–Rao inequality (1.36) in combination with generalized L–V equations [2, 3]. The test verdict centers on whether an inequality growing out of the C–R inequality is obeyed. Somewhat surprisingly, there are only two major inputs to the test:

(a) The current population level of the test animals
(b) How long they have existed as a species

Detailed knowledge, such as of the population history of the animals, is *not* needed for purposes of this preliminary test. By the way, the test does not predict imminent extinction, but only the *possibility* of a substantial population loss. If a more definitive prediction is desired, the test should be supplemented by others that utilize the data.

8.1.2. *Introduction*

Consider an ecological system consisting of N classes, e.g., species. A case $N = 4$ of interest is an aquarium containing guppies, zebra fish, angel fish, and catfish, enumerated as classes $n = 1, 2, 3, 4$. This ecosystem is ideally assembled at a common time $t = 0$ of *inception* of these classes of fish. During the passage of an unknown amount of time t, the subpopulations m_n, $n = 1$–4 of the fishes change and evolve. Time t thereby has the biological significance of being an *evolutionary time*, i.e., the total time over which the four populations of the particular system have been evolving since inception. *Does knowledge of the populations at a given time have predictive value as to their potential for sudden population decline within a generation?* (Such declines are taken to be natural fluctuations due to existing ecological conditions. Declines due to obliteration of the entire ecosystem—say by dropping the aquarium—are outside the scope of the calculation.)

8.1.3. *A Biological Uncertainty Principle*

As usual, the analysis grows out of the taking of a datum. Suppose that this is the random observation of a creature of class type n (say, an angel fish, type $n = 3$) from the ecosystem. From this, the observer may estimate the age t of the ecological system. Denote the mean-square error in any such estimate, over all possible observations n, as $\langle e_t^2 \rangle$. Then by the C–R inequality (1.36)

$$\langle e_t^2 \rangle I(t) \geq 1, \tag{8.1}$$

where $I(t)$ is the Fisher information as a function of time. It is defined as (1.35), which for these particular parameters becomes

$$I(t) \equiv \left\langle \left[\frac{\partial}{\partial t} \ln p(n|t) \right]^2 \right\rangle = \sum_n p_n \left[\frac{\partial p_n / \partial t}{p_n} \right]^2 \tag{8.2a}$$

$$= \sum_n p_n (w_n - \langle w \rangle)^2 = \langle (w - \langle w \rangle)^2 \rangle \tag{8.2b}$$

$$\approx \frac{1}{(\Delta t)^2} \sum_n \frac{(\Delta p_n)^2}{p_n} \tag{8.2c}$$

with

$$p_n \equiv p_n(t) \equiv \frac{m_n}{M}, \quad 1 = \sum_n p_n \tag{8.2d}$$

$$0 \le m_n \le M, \quad M \equiv \sum_n m_n. \tag{8.2e}$$

Expectations $\langle \rangle$ and sums are over all possible species $n = 1, \ldots, N$. Equation (8.2b) is by straightforward use of the well-known L–V equations [2, 3]

$$\frac{\partial p_n}{\partial t} = p_n(w_n - \langle w \rangle), \quad n = 1, \ldots, N, \quad \langle w \rangle \equiv \sum_n p_n w_n. \tag{8.2f}$$

Approximate equality (8.2c) follows directly from Eq. (8.2a) in the presence of a small but finite intergenerational time Δt, such that

$$\frac{\Delta p_n}{p_n} \approx \frac{dp_n}{p_n} \to 0. \tag{8.2g}$$

In (8.2d) m_n is the number of animals of type n in the population, and by (8.2e) M is the total population in the system. By the form of the second Eq. (8.2d), p_n is the *relative* level of population of species n and, by the law of law numbers, this approximates the probability of randomly finding an animal of type n in the population.

A quantity w_n is a "fitness value," defined as the net increase $\Delta m_n/(m_n \Delta t)$ in occurrence of the nth population over a generation time Δt. Any fitness w_n is most generally function $w_n = w_n(p_1, \ldots, p_N; t)$ of all probabilities and the time.

The use of equality (8.2b) in Eq. (8.1) gives the *uncertainty principle of population biology* [1, 4]

$$\langle e_t^2 \rangle \langle (w - \langle w \rangle)^2 \rangle \ge 1. \tag{8.3}$$

Hence, this introductory problem of estimating t leads to an uncertainty principle. This will turn out to provide motivation for the approach that follows.

Note that (8.3) indicates a complementarity between the error in the age t of the system and its phenotypic variability $\langle (w - \langle w \rangle)^2 \rangle$. It is analogous in mathematical form to the Heisenberg principle of physics. That is, as one left-hand factor goes up, the other tends to go down in order to obey the inequality. (More rigorously, both factors cannot be arbitrarily small.)

8.1.4. Ramification of the Uncertainty Principle

The particular age t to which we apply (8.3) is that of a sampled specimen. The specimen can be alive or a *fossil*. If it is alive, then it is observed in the present, so that t is the current age of the system. If it is a fossil, then t is ideally the age of the system at the time the observed animal died.

In practice, of course this value of t can only be *estimated*. This is often from the difference between the observed geological depth of the specimen (which is at the surface if it is alive) and that of the furthest depth at which fossil members of the family are found. Hence the estimated value of t suffers from uncertainty due to imprecise knowledge of the observed depth values. Let the resulting relative error in the estimated t be denoted as quantity r,

$$\frac{\sqrt{<e_t^2>}}{t} \equiv r = \text{const.} \tag{8.4}$$

Quantity r is assumed to be at least roughly known. However, accurate knowledge of r is *not* critical to any of the predictions that are made below.

Combining Eqs. (8.1) and (8.4) gives

$$t^2 I \geq \frac{1}{r^2}, \quad \text{or } t \geq t_0, \quad t_0 = \frac{1}{r\sqrt{I}}. \tag{8.5}$$

The time t_0 is then, by definition, the threshold age necessary to achieve an information level I. It follows that *if a value of I is prescribed, this fixes a threshold age t_0 of the system.* That is, the system must be *older than t_0 to achieve* this I value.

The first inequality (8.5) indicates a complementarity between the age t of the system and the information level I. Note that t *is the age or duration of the system*, not backward time from the present.

8.1.5. Necessary Condition for Cataclysm

Let the term "population cataclysm," or just "cataclysm", define any population that has dropped by a sizeable fraction α, with $0 \leq \alpha < 1$, in a given generation $(t, t + \Delta t)$. The equality in (8.5) indicates that *a value of I that is characteristic of the cataclysm* defines a corresponding threshold age t_0. This threshold age then defines the indicated decision rule $t \geq t_0$ on vulnerability to cataclysm. If the inequality decision rule is obeyed then vulnerability is predicted; if it is not, then vulnerability is not predicted.

Hence the approach hinges upon fixing a value of I that is characteristic of a cataclysm. That is, the ecological model that will be used is the direct assumption that a cataclysm occurs. A strength of this approach is that it will not require an *assumed* form for the L–V fitness dependences $w_n = w_n(p_1, p_2, \ldots; t)$ upon probabilities of the fitness values, such as the usual linear dependence [3]. These fitnesses instead will follow from the assumption that the cataclysm occurs. Note also that *assuming* the cataclysm means that we are seeking a *necessary*, but *not sufficient*, condition on t_0 for occurrence of the cataclysm. The condition will be the decision rule. Sufficiency is outside the scope of this study. In summary, the assumed occurrence of the cataclysm by itself will fix the fitness values and then, by (8.5), the required threshold time t_0.

Hence, *suppose that a cataclysm of level α occurs.* We consider values of α that are substantial, say 20% or more, but not overly close to 1, i.e., do not define a complete extinction [5]. We are interested in any such steplike drop, by a relative amount α, i.e., discrete population changes $\Delta m_n = -\alpha m_n$ for all n. This is equivalent to fitness values $w_n = -\alpha/\Delta t$. What value of I is consistent with the cataclysm? We are interested in the minimum possible nonzero value of I, since by (8.5) this results in a *largest*, and hence, *conservative value* for the threshold time t_0. (That is, with t fixed as, e.g., the present time, the inequality $t \geq t_0$ tends *not* to be satisfied, meaning that the prediction of cataclysm tends *not* to be made.)

If *all* population components n suffer the cataclysm α, then $\Delta m_n = -\alpha m_n$, $n = 1, \ldots, N$, so that by (8.2d) all

$$\Delta p_n \equiv \frac{m_n + \Delta m_n}{M + \sum \Delta m_n} - \frac{m_n}{M} = \frac{m_n - \alpha m_n}{M - \alpha \sum m_n} - \frac{m_n}{M} \qquad (8.6)$$

$$= \frac{m_n(1 - \alpha)}{M(1 - \alpha)} - \frac{m_n}{M} = 0.$$

Then by (8.2c) the information $I = 0$, so that by (8.5) $t_0 \to \infty$. Unfortunately this would be useless as a threshold, since then inequality $t \geq t_0$ of (8.5) would always be violated by finite t, so that a cataclysm could never be predicted. Evidently the preceding requirement $\Delta m_n = -\alpha m_n$, $n = 1, \ldots, N$ was too strict.

8.1.6. Scenarios of Cataclysm from Fossil Record

As we saw, the preceding scenario of cataclysm was overly strict, or conservative, to be of use. That is, it never satisfies the decision inequality $t \geq t_0$ of (8.5). By comparison, a useful scenario would be one where t_0 is large but, yet, finite. Obviously, with t finite the inequality $t \geq t_0$ would still tend to be violated for t small, but now there also will be times t that are large enough to satisfy the inequality and, hence, *allow* cataclysm. Hence, the decision rule is now conservative but, usefully so, i.e., it does allow positive decisions. In these cases of t_0 large, by inequality (8.5) I must no longer be zero but must still be small.

Therefore, we seek a cataclysm scenario for which I is small but *finite*. Consider, then, the same loss condition $\Delta m_n = -\alpha m_n$ as before, but now where a finite, fixed number m of the creatures *do not* suffer the loss, with $1 \leq m \leq M$. This will result in a small but nonzero information, as we wanted, and hence a finite threshold. Is this nearly binary scenario realistic? That is, are there past cataclysms where a certain fraction of species suffered strong loss while the rest suffered nearly no loss? This model seems, in fact, to approximate many of the so-called "mass extinctions" of past epochs. For example, at the Cretaceous-Tertiary extinction it is believed that dinosaurs and marsupials suffered strong losses (large α), while amphibians and many aquatic reptiles were relatively unaffected ($\alpha \to 0$). Also, "land plants suffered only moderate extinction. This sort of selectivity characterizes all major extinctions" [5]. As to whether the approximation is good enough must ultimately rest upon its predictions, as follow from its resulting decision rule. Will it actually *predict* new vulnerabilities to cataclysm? Examples below seem to indicate that it will.

In summary, to attain instead a *finite* minimum value $I_{\min}$ we assume a scenario where *not* all population members suffer the relative loss α. In particular, let $(M - m)$ members suffer the loss and m *not* suffer it.

8.1.7. Usefully Conservative Scenario

It is convenient to classify all population types n (say, our guppies, zebra fish, etc.) according to whether they suffer the loss α or do not. Denote as $n \in \Omega_\alpha$ those population types n that suffer the loss, and $n \in \bar{\Omega}_\alpha$ those that do not. Thus,

$$\Delta m_n = -\alpha m_n \quad \text{for } n \in \Omega_\alpha, \tag{8.7a}$$

and

$$\Delta m_n = 0 \quad \text{for } n \in \bar{\Omega}_\alpha. \tag{8.7b}$$

For example, the creatures of type $\bar{\Omega}_\alpha$ might be those with fur or feathers, enabling them to escape unscathed ($\alpha = 0$) from the environmental challenge of an asteroid "winter". Also, with m total members of the unscathed type $\bar{\Omega}_\alpha$ at the time t just preceding the environmental challenge,

$$\sum_{n \in \bar{\Omega}_\alpha} m_n = m. \tag{8.7c}$$

These changes imply a total change in population of

$$\sum_n \Delta m_n \equiv \Delta M = -\alpha \sum_{n \in \Omega_\alpha} m_n = -\alpha \left(\sum_{n=1}^N m_n - \sum_{n \in \bar{\Omega}_\alpha} m_n \right) = -\alpha(M - m). \tag{8.7d}$$

Equations (8.2e) and (8.7c) were used. Equations (8.7a to d) define the *general scenario* for cataclysm that we will use.

Note that the number m of *unaffected* individuals that are present prior to the extinction interval is not fixed by the model. However, the event could only be considered a cataclysm if the large majority of the population suffered the attenuation α given in (8.7a). Thus $m \ll M$. Also, the value of I subject to (8.7a, b) is zero (as we saw) unless at least one creature $m = 1$ exists in the unaffected class $\bar{\Omega}_\alpha$. Thus m is bracketed as

$$1 \le m \ll M. \tag{8.7e}$$

The value of m is later fixed by the above requirement that I be minimized, which we saw leads to a conservative estimate of the threshold age t_0.

8.1.8. Resulting Changes in Population Occurrence Rates

By the first equality (8.6) the corresponding changes in probabilities obey

$$\Delta p_n = \frac{m_n + \Delta m_n}{M + \Delta M} - \frac{m_n}{M}. \tag{8.8}$$

The particular changes (8.7a to d) when used in (8.8) give (cf. (8.6))

$$\Delta p_n = \frac{m_n(1 - \alpha)}{M(1 - \alpha) + \alpha m} - \frac{m_n}{M} = -\frac{\alpha m m_n}{M[M(1 - \alpha) + \alpha m]} \quad \text{for all } n \in \Omega_\alpha \tag{8.9a}$$

and

$$\Delta p_n = \frac{m_n}{M(1-\alpha)+\alpha m} - \frac{m_n}{M} = -\frac{\alpha m_n(M-m)}{M[M(1-\alpha)+\alpha m]} \quad \text{for all } n \in \bar{\Omega}_\alpha.$$

(8.9b)

The information expression (8.2c) requires the *relative* changes in the probabilities, which by the preceding and (8.2d) obey

$$\frac{\Delta p_n}{p_n} = -\frac{\alpha m}{M(1-\alpha)+\alpha m} \quad \text{for all } n \in \Omega_\alpha$$

(8.10a)

and

$$\frac{\Delta p_n}{p_n} = -\frac{\alpha(M-m)}{M(1-\alpha)+\alpha m} \quad \text{for all } n \in \bar{\Omega}_\alpha.$$

(8.10b)

For general α, requirement (8.2g) is then obeyed for *all* n if, by the forms of Eqs. (8.10a, b),

$$M(1-\alpha) \gg 1.$$

(8.11)

This states that for the given population level M, the value of attenuation α cannot be overly close to 1, that is not overly close to complete attenuation (extinction). This describes a biologically meaningful scenario, since the fossil records of thousands of families of marine animals [5] indicate final extinctions only after a sequence of multiple steplike drops by *finite* fractions α.

8.1.9. Getting the Information

Using Eqs. (8.10a, b) in (8.2c) gives

$$I \approx \frac{1}{(\Delta t)^2} \left\{ \sum_{n\in\Omega_\alpha} \frac{\alpha^2 m^2 m_n}{M[M(1-\alpha)+\alpha m]^2} + \sum_{n\in\bar{\Omega}_\alpha} \frac{\alpha^2 m_n(M-m)^2}{M[M(1-\alpha)+\alpha m]^2} \right\}$$

(8.12)

By (8.2e) the sums are, respectively,

$$\frac{\alpha^2 m^2}{[M(1-\alpha)+\alpha m]^2}\left(\frac{M-m}{M}\right) \quad \text{and} \quad \frac{\alpha^2(M-m)^2}{[M(1-\alpha)+\alpha m]^2}\left(\frac{m}{M}\right).$$

(8.13)

Adding these, as in (8.12), gives

$$I \approx \frac{1}{(\Delta t)^2}\frac{\alpha^2 m(M-m)}{[M(1-\alpha)+\alpha m]^2}.$$

(8.14)

This is subject to inequality (8.7e), so that

$$I \approx \frac{1}{(\Delta t)^2}\frac{\alpha^2 m M}{[M(1-\alpha)+\alpha m]^2}.$$

(8.15)

The value of m has not yet been fixed except that it obeys inequality (8.7e). Recall that we are seeking a minimum (but finite) value for I. Equation (8.15) shows that I attains a minimum where m is a minimum. By (8.7e) m is a minimum at the lower boundary for possible values of m, that is, where $m = 1$. (Note: This is not a *local* or calculus extremum of (8.15); the latter instead maximizes (8.15)!) Thus,

$$I \equiv I_{\min} \approx \frac{1}{(\Delta t)^2} \frac{\alpha^2 M}{[M(1 - \alpha) + \alpha]^2} \tag{8.16a}$$

$$= \frac{1}{(\Delta t)^2} \frac{1}{M} \left(\frac{\alpha}{1 - \alpha} \right)^2 \tag{8.16b}$$

in view of (8.11) and because $\alpha \leq 1$.

8.1.10. Square Root Decision Rule

Using (8.16b) in (8.5) gives

$$t \geq t_0, \quad \text{where } t_0 = \frac{\Delta t}{r} \left(\frac{1 - \alpha}{\alpha} \right) \sqrt{M} \approx \Delta t \sqrt{M}. \tag{8.17}$$

The latter follows since typical values of α and r are such that $r^{-1}(1 - \alpha)/\alpha \sim 1$. Thus, t_0 depends upon population level M as its square-root, a slow dependence.

8.1.11. Final Decision Rule, Conditions of Use

The inequality (8.17) may be put in a simpler and more biologically meaningful form. Divide through both sides by Δt. The left-hand side, $t/\Delta t$, is approximately the number n_g of generations since inception of the population. Then the inequality becomes simply

$$n_g \gtrsim \sqrt{M}. \tag{8.18}$$

This is the decision rule we sought. In words, it states that *vulnerability to cataclysm exists if the mean number n_g of generations of the population since inception is greater than or of the order of the square root or of the existing level M of population.*

In most applications, the inequality (8.18) tends to be *decisively* satisfied or violated, by many orders of magnitude, so that the assumption $r^{-1}(1 - \alpha)/\alpha \sim 1$ is not critical to the prediction. Hence, this factor is ignored in the applications below.

The rule (8.18) indicates that long-evolved (high n_g) species that are currently small in number (small M) are susceptible to cataclysm. Or conversely, newly-evolved species that are modestly high to high in number are not susceptible to cataclysm. These trends make sense, and are further discussed below.

Note that the test (8.18) depends upon the total population M of the ecology, rather than upon its individual populations m_n. Specifically, M is to define the total population of all population types n that significantly interact in the L–V equations (8.2f). For example, when there is a single dominant species that effectively competes only with its own members for survival, as in the case below of an ideally breeding colony of rabbits, M will then be the population of that species. (The $m = 1$ individual that, by the cataclysm model of Section 8.1.7, is *not* subject to the loss α is then taken to be a member of any one of the other species of the ecology.) Of course any such assumption of significant noninteraction is only an approximation, and must affect the prediction to an extent. The strength of the affect has not yet been quantified.

Note that condition (8.17) (or (8.18)) has a limited scope of meaning. It is merely a *necessary condition* of the model approach, and therefore indicates merely a *potential* for the occurrence of the cataclysm α. That is, *if* the cataclysm occurred, the condition must have been satisfied. Conversely, if it is not obeyed then there is no (or rather, exceedingly low) potential for the cataclysm.Thus, the test (8.18) asks: Is an observed population level M low compared with its ecological age? If so, it is *vulnerable* to cataclysm. This conclusion is biologically intuitive and, in fact, underlies many modern attempts at species preservation. These assume that if the population level M of a relatively old (large n_g) species is observed to be very low, then the species is regarded as in danger of extinction.

This trend also makes sense on purely statistical grounds. The "strength" [6] of data, in this case population data, goes as their size M. Hence, the sample means m_n/M of low populations M are "weaker," i.e., more probable to suffer high fluctuations, than the sample means of high populations. Thus, simply on the basis of chance, a low-M scenario is more apt to suffer the postulated drop α than is a high-M scenario. As a matter of fact, this effect is utilized in the famous *chi-square* test, whose aim is often to predict significance (nonrandomness) of a hypothesis at, say, a 95% confidence level in a binary data set. For our population problem, the question the chi-square test addresses is the following:

Suppose, in a given generation consisting of M individuals, a number M_+ survive while M_- die off, $M_- > M_+$, with $M_+ + M_- \equiv M$. What population difference value $M_- - M_+ \equiv \Delta M$ is large enough to imply a hypothesis of "imminent cataclysm" with 95% confidence or more? The famous answer that the chi-square test gives [6] is

$$\Delta M \geq 2\sqrt{M}.$$

As with our decision rule (8.18), this is a square root test. Thus, the square root dependence of (8.18) has an important precedent in classical statistics.

Note that the *relative* population loss satisfying the preceding chi-square test is $\Delta M/M \geq 2/\sqrt{M}$. This is a quite slow dependence upon M, with important ramifications. It shows that for large M, a small *relative* population change satisfies the hypothesis (e.g., a 20% change for $M = 100$) while for small M, a large relative loss is needed (a 100% or total loss for $M = 4$). This chi-square test, aside from confirming the $\sqrt{M}$ dependence of our decision rule test (8.18), also provides a test that can be used as a supplement to (8.18). The two tests act independently, in that (8.18) just depends upon the absolute level M of the *current* population, whereas the chi-square test depends upon the observed *change* in the current population after an additional generation. However, the theoretical assumptions of the two

TABLE 8.1. Threshold values t_0 at successive generation times t of ideally breeding rabbits.

t	Δt	$2\Delta t$	$3\Delta t$	$4\Delta t$	$5\Delta t$	$6\Delta t$	$7\Delta t$	$8\Delta t$	$9\Delta t$	$10\Delta t$	$11\Delta t$
t_0	$1.41\Delta t$	$1.41\Delta t$	$2\Delta t$	$2.45\Delta t$	$3.16\Delta t$	$4\Delta t$	$5.10\Delta t$	$6.48\Delta t$	$8.24\Delta t$	$10.5\Delta t$	$13.3\Delta t$

tests are similar because both (8.18) and the above chi-square test are predicated upon a *hypothetical* population loss.

Returning to the decision-rule test (8.18), the cataclysm decided upon is only a potential one. It can only occur if, *in addition*, a sufficiently serious ecological challenge is present, such as a sudden flood or drought. We do not consider such events here.

The following example shows a typical history of test decisions, using decision rule (8.17), during the growth of an idealized ecological system. The aim is to test whether the decisions make intuitive sense.

8.1.12. Ideally Breeding Rabbits

Consider a system consisting of a colony of ideally breeding rabbits (no deaths, negligible interaction with other species, each pair produces another pair at each generation spacing Δt). At $t = 0$ the system is turned "on" (one pair of rabbits is born and progresses toward breeding age $t = \Delta t$). The negligible interaction with other species allows us to use the test (8.17) with M simply the total population level of the rabbits at a given generation, by Section 8.1.11. We can now track the decisions made by the rule (8.17) at each generation of the colony as it grows, to see if they make sense.

Suppose that at each generation time $t = n\Delta t$, $n = 1, 2, 3, \ldots$, a randomly chosen rabbit is observed. What prediction does the test (8.17) make about possible cataclysm at each such generation? This growth scenario is famous for giving rise to a Fibonacci sequence [7] of population *pair* values $1, 1, 2, 3, 5, 8, 13, 21, \ldots$ at consecutive generations. This series approximates exponential growth with time. The corresponding population values are twice these values, $M = 2, 2, 4, 6, 10, 16, 26, 42, 68, 110, 178, 288, \ldots$. Then by (8.17), threshold time values $t_0 \approx \Delta t\sqrt{M} = \sqrt{2}\Delta t, \sqrt{2}\Delta t, 2\Delta t, \sqrt{6}\Delta t, \sqrt{10}\Delta t, 4\Delta t, \sqrt{26}\Delta t, \sqrt{42}\Delta t, \ldots$ result. Note that this increases exponentially as well, since the square root of an exponential is an exponential. In Table 8.1, these (t, t_0) pairs are shown, where the square roots are approximated to facilitate comparisons between corresponding values of t and t_0. What prediction is made at each generation?

As was mentioned, the threshold values t_0 increase approximately *exponentially* with the time. Of course, by comparison, the generation time values themselves increase *linearly*, as $t = \Delta t, 2\Delta t, 3\Delta t, 4\Delta t, 5\Delta t \ldots$. These are also shown in Table 8.1. Comparing corresponding items (t, t_0) in the table shows that early in the age of the system (second through ninth column) the time values t *exceed* their corresponding threshold values t_0. Hence, *at these times the prediction (8.17) is that a cataclysm is possible.* This seems reasonable since at these early times the population levels are not yet high enough to be relatively stable. However, an exponential sequence is bound to ultimately overtake and pass a linear one. Hence, *beyond* a certain time ($t = 10\Delta t$ in the table), the values of t are *less* than their corresponding values t_0. *At all these times and beyond the test (8.17) predicts no cataclysm.* These predictions are also reasonable. With such high population values at a relatively young age, the species seems highly successful, and is therefore in no immediate jeopardy of cataclysm. This trend will continue, and become stronger, as long as the rabbits continue breeding ideally according to Fibonacci sequence.

Note that using a wide range of different values of r, α shows similar results since, again, an exponential sequence must ultimately pass a linear one.

In summary of the preceding tests, the decision rule (8.17) or (8.18) works by the following simple idea: *A biological system that is a candidate for cataclysm is one whose population size M is overly small considering its age.*

Of course a real rabbit population will ultimately saturate and start dropping, due to depletion of resources and/or increased predation by coyotes and the like. With M falling, a generation number n_g will eventually arise for which inequality (8.18) is *satisfied*, i.e. a catalysm is finally allowed. Of course any in-depth study of the health of an ecology should depend upon the use of other tests as well.

8.1.13. Homo Sapiens

Homo sapiens has reached a stage of development where there is a notable lack of competing species. We principally interact among ourselves, with populations obeying the L–V Eq. (8.2f). Hence, as discussed in Section 8.1.11, we can again use for M simply the total population of homo sapiens.

Hence, suppose that one member is randomly observed at the current time. (Note that there is still a finite error r in the system age t since the date of *inception* of *homo sapiens* is uncertain.) There are about $M = 5 \times 10^9$ members of the species worldwide, and the average age to maturity is about $\Delta t = 20$ years. Equation (8.17) then gives a time threshold of

$$t_0 \approx 10^6 \text{years}. \tag{8.19}$$

By comparison, the age t of the system homo sapiens is about 10^5 years. (Note that a factor of order 2 or 3 will not affect the result.) This is considerably *smaller* than t_0 given by (8.19), so that inequality (8.17) is not satisfied. The prediction is, then, that humans have little danger of suffering a population cataclysm in the coming generation. The species is *manifestly* too numerous, relative to the time it has existed; that is, it is too successful to suffer a downward fluctuation of that amount. Recalling the growth history of ideally breeding rabbits in Table 8.1, humans are at the relatively early and robust stage of their development, where growth is practically unrestrained. Of course, as discussed above for real (not ideal) rabbits, a population crash is to ultimately be allowed by (8.18) once human population saturates and starts falling due to depleted resources.

8.2. Finding the Mass Occurrence Law For Living Creatures

8.2.1. Summary

On a worldwide basis, smaller creatures tend to occur more frequently than larger creatures. There are many more mice than elephants. This indicates that the frequency of occurrence p_X of the mass x of a living creature varies inversely in some manner with x. The aim of this section is to quantify the relation. Is it in fact precisely inverse? The approach to the problem is based upon one biological effect, the special relation connecting the observed values y of attributes a of organisms with their masses x. This is a power law of the form $y = Cx^a$ with C a constant (the law derived via EPI in Section 8.3.) Examples of a are metabolic rate ($a = 3/4$) and age to maturity ($a = 1/4$). Numerous other examples are given in Section 8.3.1.2, and also in Table 5.2, Chapter 5. In general, a power law can govern a *recognition* problem, whereby one organism attempts to identify another through observation y of an unknown attribute a. A correct identification of the attribute requires a valid estimate $\widehat{a} \equiv \widehat{a}(y)$ of it, as formed from the datum y.

The unknown occurrence law $p_X(x)$ is computed on the following basis. An optimally valid estimate $\widehat{a}(y)$ of the attribute is called "efficient." Man is, of course, capable of forming efficient estimates, and man's mass x places him well within the population that obeys the law $p_X(x)$. Thus, nature *does* permit efficient estimates to be made (at least by humans, and probably by other creatures as well). Any efficient law $p_X(x)$ must satisfy a well-known condition in the form of a certain first-order differential equation. Requiring that it satisfy the equation, fixes the law to be of the form x^{-b}, $b = \text{const} > 0$, with $1 \le x \le r$, r some large upper mass bound. Another formal requirement of efficiency is that the estimator function $\widehat{a}(y)$ be *unbiased* (correct on average). This turns out to be satisfied by the requirement $b = 1$, in the asymptotic limit as the upper mass value $r \to \infty$. The result is a predicted probability law (8.57) for biological mass that is asymptotically of the form $p_X(x) \propto 1/x$. Thus the qualitatively inverse relationship mentioned at the outset is actually rigorously inverse. This is also equivalent to a log-uniform law ($\ln x$ uniformly distributed).

8.2.2. Background

Suppose that the value y of the attribute a of a living creature of mass x is measured. As derived by EPI in Section 8.3, many such attributes obey a well-known power-law relation [8–21]

$$y = Cx^a, \quad C = \text{const.}, \quad C, x, y > 0. \tag{8.20}$$

The constant C includes effects that are kept constant or vary by small amounts during the measurements y, such as temperature T (which otherwise contributes a Boltzmann factor $\exp(-E/kT)$ to the right-hand side of (8.20) with E an energy and k the Boltzmann constant; see [8]). Also, (8.20) only holds true as an average effect, so that individual measurement events y can randomly depart from it. For example, a given metabolic rate y can hold for organisms with randomly *different* mass sizes x.

Subject to these caveats, the power-law relation (8.20) holds for a remarkably broad range of attribute types a, including life span, growth rate, heart rate, DNA substitution rate, tree height, cross-sectional area of aorta, and mass of cerebral gray matter. Moreover, as also derived in Section 8.3, *for such attributes the corresponding parameter values are integer multiples of 1/4*, as $a = n/4$, $n = 0, \pm1, \pm2, \ldots$ The particular multiple n generally identifies not one, but, a class of attributes. For example, $n = 3$ identifies the attributes of basal metabolic rate, the cross-sectional areas of aortas and tree trunks, and others. For all members of this attribute class, $a = n/4 = 3/4$ in (8.20).

In general, some attributes obey (8.20) over a wider range of masses than do others. That is, the ranges of mass values $x = (r_{\min}(a), r_{\max}(a))$ obeying laws (8.20) generally depend upon the types a of attribute that are observed. But, we seek to use (8.20) to find the probability distribution function (PDF) law $p_X(x)$ on mass per se. This is an absolute property of biological mass, and should not depend upon any particular attribute value a under observation. Hence, to avoid any dependence upon a in the estimate of $p_X(x)$ following from (8.20), we limit our application of (8.20) to the population of masses x that obey (8.20) *in common*, irrespective of the value of a. This is the intersection of all the ranges of masses x, over all a, that obey (8.20). Hence the largest such mass value is $x = r \equiv \min_a r_{\max}(a)$, the minimum over a of all the maximum mass values that obey (8.20).

Also, for convenience, we arbitrarily choose as the unit of mass the *smallest* mass x in this range. Then each mass value is a multiple of this. It results that the masses we consider lie on the interval $x \equiv (1, r)$, where $r = \min_a r_{\max}(a)$. Masses outside this range are not further considered, and whatever PDF on mass they obey will remain unknown. It might be that the law $p_X(x)$ we derive for the limited interval $x \equiv (1, r)$ is analytic and, so, can be extended to a broader range of masses, but this is conjecture at this time.

The Second law of statistical physics generally causes information to be lost during a measurement. Remarkably, power laws (8.20) permit the observer to measure biological attributes with *minimal loss* of the information. In fact, the quarter-power laws (8.20) will be *derived* on this basis (Section 8.3; also [22]). This implies that nature has evolved into a state where its attributes are as transparent to observation as is possible (Section 1.4.5).

The statistical implications of a power-law relation (8.20) are quite strong. First, a system that obeys a power law does not have a preferred scale, i.e., is "scale free" in a certain sense. If the replacement $x \to bx$ is made in (8.20), with $b = \text{const}$, the resulting $y \to (Cb^a)x^a$, which is still dependent upon x as the power law x^a.

Indeed there is evidence that biological systems act in a scale-free manner with respect to many attributes [23–26]. As we saw from (8.20), mass x is another such attribute.

8.2.3. Cramer–Rao Inequality and Efficient Estimation

We briefly review the elementary estimation theory given in Section 1.2.1, Chapter 1. Consider the problem of estimating a system parameter value a, given a datum y from the system. The estimate $\hat{a}$ is some chosen function $\hat{a}(y)$ of the datum. The aim is of course to form an estimator function that *well-approximates the true parameter value* a. An estimator that achieves *minimum* mean-squared error is called *efficient* in classical estimation theory. An efficient estimate achieves equality in the C–R inequality (1.36) $e^2 I \geq 1$ of classical estimation theory. That is, the estimate achieves a minimum mean-squared error

$$e^2_{\min} = 1/I, \tag{8.21}$$

where in general the mean-square error obeys

$$e^2 \equiv \int dy p(y|a)[\hat{a}(y) - a]^2, \tag{8.22a}$$

and

$$I \equiv \int dy p(y|a) \left[\frac{\partial \ln p(y|a)}{\partial a} \right]^2 \tag{8.22b}$$

defines the Fisher information (1.35). The conditional probability law $p(y|a)$ is called the "likelihood function" for the system, in that it gives the spread in "likely" values of y that result from a fixed system parameter a. The relations (8.21) and (8.22a) assume the presence of *unbiased* estimator functions $\hat{a}(y)$, i.e., whose means equal the true parameter value a.

It should be remarked that *not* all data-forming systems allow for efficient estimation (see Section 8.2.11). Furthermore, even if a system permits an efficient estimate to be made, this does not mean that efficient estimation is actually carried through. *The choice is up to the user.* The user forms the estimator function, so whether it is the efficient choice or not is entirely at his/her disposal. Even if an efficient estimator exists, depending upon other prior knowledge the user might not choose to use it.

Likewise, getting away from the human observer, where whether to use the efficient estimate is a matter of purposeful choice, a predator observing the attribute a of a prey utilizes an *automatic* recognition system. For example, this could be its eye–brain for a visual attribute a. Here, even if an efficient estimate exists, its eye–brain might not be configured to form it. Thus, although a predator–prey system might allow for efficient estimation, it might not actually be utilized.

Nevertheless, as will be discussed, efficient estimation is actually carried through by some creatures, including humans. This factor will dominate in the derivation to follow.

8.2.4. Efficiency Condition

A necessary condition for the existence of an efficient estimator is that the likelihood law $p(y|a)$ obey the condition

$$\frac{\partial \ln p(y|a)}{\partial a} = k(a)[\widehat{a}(y) - a], \tag{8.23}$$

which is (1.39), Chapter 1. Here $k(a)$ is any function of a (but not $\widehat{a}$). (Likewise, by its notation, the conditional probability $p(y|a)$ is not a function of $\widehat{a}$.) What (8.23) says is that, in order for an efficient estimator $\widehat{a}(y)$ to exist, when the logarithm of $p(y|a)$ is differentiated the resulting expression in y and a must be expressible in the form of the right-hand side of (8.23). Here $\widehat{a}(y)$ is some appropriately identified function of y (alone) that defines the estimate.

As a simple example, consider the problem of estimating the mean a_0 of a Poisson process, defined by a likelihood function $p(y|a_0) = \exp(-a_0)a_0^y/y!$ One readily gets $\partial \ln p(y|a_0)/\partial a_0 = -1 + y/a_0 \equiv (1/a_0)(y - a_0)$. This is indeed in the form (8.23), if we identify $k(a_0) \equiv 1/a_0$ and $\widehat{a}_0(y) \equiv y$. In this simple case the efficient estimator exists, and it is the data value itself. For multiple independent Poisson data y_n, $n = 1, \ldots, N$ the same approach again shows that an efficient estimator exists, and it is now $\widehat{a}_0(y) = N^{-1} \sum_n y_n$, the arithmetic mean of the data y_n. This is in fact the most commonly used estimator, which is comforting to know (as declared the man who found that all his life he had been speaking prose).

Hence we proceed to evaluate the derivative left-hand side of (8.23) for the likelihood function $p(y|a)$ implied by the data relation (8.20). When this is done, its right-hand side will be examined to judge *whether* it can be placed in the form of the right-hand side of (8.23) for any one a. If it can, then an efficient estimator exists for that a, provided it is also shown to be unbiased (Section 8.2.9). Whether the efficient estimator is actually *used* is another matter.

8.2.5. Objectives

In our ecological problem, the parameter to be estimated is a (see preceding), and the datum y is a biological attribute, such as the radiated heat from a prey animal. The observer can be a predator animal. Its eye–brain system forms an estimate $\widehat{a}$ of a out of the observation y. We want to ascertain the conditions under which the chosen estimator function $\widehat{a}(y)$ *can be* an efficient one.

In this section through Section 8.2.10, we show the following:

(a) The special, power-law form (8.20) of a data law allows for an efficient estimate of a. That is, the *potential* for making an efficient estimate exists. Man, at the very least, is one creature that actually achieves the efficiency.
(b) This potential for efficiency *requires* the presence of a particular PDF on the mass x. This is the law $p_X(x) \propto 1/x^b$, $b = $ const.
(c) A property of *unbiasedness* in the estimate implies that the power b is close to 1, i.e., the PDF is $1/x^b$, where asymptotically $b \rightarrow 1$ (but does not equal it).

8.2.6. How Can the Efficiency Condition be Satisfied?

The entire calculation rests on Eq. (8.20). Subject to the caveats listed below (8.20), this states that there is one event $(x, x + dx)$ for each event $(y, y + dy)$. Then, by elementary probability theory [6], the two events $(x, x + dx)$ and $(y, y + dy)$ have the same probability,

$$p(y|a)dy = p_X(x)dx, \quad dx, dy > 0. \tag{8.24}$$

This allows us to solve for

$$p(y|a) = \left[\frac{p_X(x)}{|dy/dx|}\right]_{x=(y/C)^{1/a}} \tag{8.25}$$

after inverting (8.20) for x. Also, differentiating Eq (8.20) gives

$$\frac{dy}{dx} = aCx^{a-1} = aC\left(\frac{y}{C}\right)^{\frac{a-1}{a}}. \tag{8.26}$$

Using this in (8.25) and taking the logarithm gives

$$\ln p(y|a) = -\ln|a| - \frac{1}{a}\ln C + \left(\frac{1-a}{a}\right)\ln y + \ln p_X\left(\left(\frac{y}{C}\right)^{1/a}\right). \tag{8.27}$$

Differentiating, and using the identities

$$\frac{\partial}{\partial a}\left[\left(\frac{y}{C}\right)^{1/a}\right] = \frac{\partial}{\partial a}\exp\left[\left(\frac{1}{a}\right)\ln\left(\frac{y}{C}\right)\right] \tag{8.28}$$

$$= -\left(\frac{1}{a^2}\right)\left(\frac{y}{C}\right)^{1/a}\ln\left(\frac{y}{C}\right)$$

gives

$$\frac{\partial \ln p(y|a)}{\partial a} = -\frac{1}{a} + \frac{1}{a^2}\ln C - \frac{1}{a^2}\ln y \tag{8.29}$$

$$-\frac{1}{a^2}\left(\frac{y}{C}\right)^{1/a}\ln\left(\frac{y}{C}\right)\left[\frac{d}{dx}\ln p_X(x)\right]_{x=(y/C)^{1/a}}.$$

(Note that $\partial \ln|a|/\partial a = 1/a$ in both cases $a < 0$, $a > 0$.) Hence, the question is whether the right-hand side of (8.29) can be placed in the form of the right-hand side of (8.23) for some choice of $k(a)$ and $\hat{a}(y)$.

Considering all the factors $1/a$, denote

$$\frac{1}{a^2} \equiv k(a). \tag{8.30}$$

Then (8.29) becomes

$$\frac{\partial \ln p(y|a)}{\partial a} = k(a)\left\{-a + \ln\left(\frac{C}{y}\right)\left[1 + \left(\frac{y}{C}\right)^{1/a}\left[\frac{d}{dx}\ln p_X(x)\right]_{x=(y/C)^{1/a}}\right]\right\}. \tag{8.31}$$

Compare this with Eq (8.23). *It is of the same form* if there is an estimate obeying

$$\hat{a}(y) = \ln\left(\frac{C}{y}\right)\left[1 + \left(\frac{y}{C}\right)^{1/a}\left[\frac{d}{dx}\ln p_X(x)\right]_{x=(y/C)^{1/a}}\right]. \qquad (8.32)$$

8.2.7. Power-Law Solution

Of course any legitimate estimator $\hat{a}(y)$ can only be a function of the data, here y. For example, it cannot also be a function of the thing to be estimated, a. But, to the contrary, the far-right term of the proposed estimator (8.32) contains a. Therefore this far-right term must be made to somehow cancel out. This occurs *if and only if*

$$\left[\frac{d}{dx}\ln p_X(x)\right]_{x=(y/C)^{1/a}} = -\left(\frac{y}{C}\right)^{-1/a}b(y), \quad b(y) \geq 0 \qquad (8.33)$$

for some function $b(y)$. (Note: We chose a minus sign in $-b(y)$ for later convenience.) Using (8.20), this is the same as the condition on x

$$\frac{d}{dx}\ln p_X(x) = -x^{-1}b(Cx^a). \qquad (8.34)$$

(Note that $b(Cx^a)$ means function $b(y)$ evaluated at $y = Cx^a$.) Since a generally varies from problem to problem, condition (8.34) does not uniquely fix the distribution $p_X(x)$. The prediction would be that the PDF on mass should somehow depend upon the attribute under observation. This does not make sense, since $p_X(x)$ by hypothesis is the PDF on a range of masses, *irrespective* of the attribute under observation. That is, the data law (8.20) holds for each attribute a over, independently, a wide range of mass values x. The only way to avoid the inconsistency is if the function $b(Cx^a) = b = $ const. That is if, as used in (8.33),

$$b(y) = b = \text{const.} \qquad (8.35)$$

Now Eq. (8.34) *loses* its dependence upon a, becoming

$$\frac{d}{dx}\ln p_X(x) = -x^{-1}b. \qquad (8.36)$$

This has the direct solution

$$p_X(x) = Ex^{-b}, \quad \text{constants } E, b \geq 0. \qquad (8.37)$$

This is a simple power law in x. The negative of the power in (8.37) makes sense, in indicating that large masses occur less often than small masses. However, this result hinges upon the additional presence of efficient estimation. In the absence of such ideal estimation a power law would not have to hold.

We conclude that *if* nature is to allow for efficient estimates of one or more biological attributes a from observations y, the distribution of masses, over some finite range, should obey a power law. Again, the efficiency exists since humans accomplish it. However, it is of further interest that merely demanding *a potential* for efficiency likewise demands that the PDF on mass be of the form (8.37).

8.2.8. Normalized Law

At this point in the analysis, there could conceivably be a different power b for each of a sequence of mass ranges. As a matter of fact, the law (8.37) empirically holds, e.g., for pelagic populations across the range of masses from unicellular plankton to the largest animals across trophic levels (levels in the food chain). For these creatures empirically $b = 1$; or, for populations at equilibrium within a single trophic level, analysis (rather than empirical data) indicates that $b = 3/4$ [8].

However, the PDF $p_X(x)$ by definition describes the relative abundance of biomass as surveyed simultaneously over *all* trophic levels. Hence we expect b to turn out to be closer to 1 (the value for across trophic levels) than to 3/4 (the value for a single trophic level). This will turn out to be the case, again, under the assumption that an efficient estimate *can* exist.

The constants E, b remain unknown. E is found as follows. As discussed below (8.20), mass is defined over an interval

$$1 \leq x \leq r, \quad r \equiv \min_a r_{\max}(a). \tag{8.38}$$

Requiring the PDF (8.37) to obey normalization over the interval (8.38) fixes E, giving

$$p_X(x) = \left(\frac{1-b}{r^{1-b} - 1} \right) x^{-b}. \tag{8.39}$$
$$\scriptstyle 1 \leq x \leq r$$

8.2.9. Unbiasedness Condition

As discussed at the end of Section 8.2.1, for our estimator function (8.39) to be considered efficient it must be unbiased as well, i.e., obey

$$\langle \hat{a}(y) \rangle = a. \tag{8.40}$$

We show in this section and Section 8.2.10 that the solution (8.32), (8.39) only obeys this property in the limiting case $b \to 1$. Using conditions (8.20) and (8.36) in (8.32) gives an estimator function

$$\hat{a}(y) = \ln \left(\frac{C}{y} \right) \left\{ 1 - \left(\frac{y}{C} \right)^{1/a} \left[b \left(\frac{y}{C} \right)^{-1/a} \right] \right\} \tag{8.41}$$
$$= (1 - b) \ln \left(\frac{C}{y} \right).$$

This shows that the particular value $b = 1$ gives $\hat{a}(y) = 0$, which then violates the unbiasedness requirement (8.40) for an arbitrary parameter a. However, it will turn out that any value of b that does not equal 1 but is *sufficiently close to* 1 (depending upon the value of r) will suffice.

Taking the average of (8.41) gives

$$\langle \widehat{a}(y) \rangle = (1 - b)(\ln C - \langle \ln y \rangle) \tag{8.42}$$
$$= (1 - b)(\ln C - \ln C - a\langle \ln x \rangle) = -a(1 - b)\langle \ln x \rangle,$$

after use of (8.20). The PDF (8.39) is found to have a moment

$$\langle \ln x \rangle = \left(\frac{\ln r}{1 - r^{b-1}} \right) + \frac{1}{b - 1}, \quad b \neq 1. \tag{8.43}$$

Using this in (8.42) gives

$$\langle \widehat{a}(y) \rangle = a(1 + \delta), \quad \delta \equiv \left(\frac{b - 1}{1 - r^{b-1}} \right) \ln r. \tag{8.44}$$

For this result to agree with the unbiasedness requirement (8.40), the relative error δ must be very small. By sight, δ is small if b is close to value 1, and r is sufficiently large. Specifically, algebraically requiring $\delta \ll 1$ in the second Equation (8.44) gives a condition

$$|\epsilon| \ln r \gg 1, \quad \text{where } b \equiv 1 + \epsilon \text{ and } \epsilon \neq 0. \tag{8.45}$$

An example of a solution is $\epsilon = 0.1$, $\ln r = 100$. Here $b = 1.1$, $r = e^{100}$, and unbiasedness is satisfied to an accuracy of $\delta = 0.05\%$. Note that, for this value of b close to 1, the upper bound r on mass must be very large. In fact, from the second Eq. (8.45), letting $b \to 1$ requires that $\epsilon \to 0$ and hence, by the *first* Eq. (8.45), $r \to \infty$. This point will be returned to in the next section, where a *requirement* $b \to 1$ will result.

8.2.10. Asymptotic Power $b = 1 + \epsilon$, with ϵ Small

By Eqs. (8.23) and (8.30) our efficient estimator $\widehat{a}(y)$ obeyed

$$\frac{\partial \ln p(y|a)}{\partial a} = \frac{1}{a^2} [\widehat{a}(y) - a]. \tag{8.46a}$$

Squaring both sides and taking the expectation over all y gives directly

$$\left\langle \left[\frac{\partial \ln p(y|a)}{\partial a} \right]^2 \right\rangle = \frac{1}{a^4} \langle [\widehat{a}(y) - a]^2 \rangle. \tag{8.46b}$$

By definitions (8.22a, b) this is equivalent to

$$I = e^2 / a^4. \tag{8.47}$$

But, by our efficient estimate (8.21)

$$e^2 = 1/I. \tag{8.48}$$

(Recall that this requires the unbiasedness property (8.40), and hence inequality (8.45), to be obeyed. We return to this point below.) Eliminating e^2 between Eqs. (8.47) and (8.48) gives $I = 1/(Ia^4)$ or

$$I = 1/a^2. \tag{8.49}$$

We next compute I in terms of the free parameter b, and compare this with result (8.49), so as to obtain a condition on the value of b.

By Eqs. (8.20), (8.22a, b), and (8.24) the information can be evaluated in the x-domain as

$$I = \int dx p_X(x) \left[\frac{\partial \ln p(y|a)}{\partial a} \right]^2_{y=Cx^a} \qquad (8.50)$$

$$\equiv \left\langle \left[\frac{\partial \ln p(y|a)}{\partial a} \right]^2_{y=Cx^a} \right\rangle,$$

where the expectation is now over values of x. The integration limits are 1 to r, $r = \text{const}$, as discussed below (8.20). But from Eqs. (8.20), (8.30) and (8.31),

$$\left[\frac{\partial \ln p(y|a)}{\partial a} \right]_{y=Cx^a} = \frac{1}{a^2} \left\{ -a + \ln(x^{-a}) \left[1 + (x^a)^{1/a} \frac{d}{dx} \ln p_X(x) \right] \right\}. \quad (8.51)$$

By (8.37) $d/(dx) \ln p_X(x) = -b/x$. Using this in (8.51) gives

$$\left[\frac{\partial \ln p(y|a)}{\partial a} \right]_{y=Cx^a} = \frac{1}{a^2} \{ -a - (a \ln x) [1 + x(-b/x)] \} \qquad (8.52)$$

$$= -\frac{1}{a} [1 + (\ln x)(1 - b)].$$

This allows us to evaluate (8.50) as simply

$$I = \frac{1}{a^2} \langle [1 + (\ln x)(1 - b)]^2 \rangle. \qquad (8.53)$$

According to the plan we compare the result (8.53) with (8.49), requiring them to be equal through choice of b. By sight, one solution is

$$b \rightarrow 1. \qquad (8.54)$$

But, is this the only solution b? By (8.38), the mass $x \geq 1$, so that $\ln x \geq 0$. Consider, next, contrariwise, a candidate solution $b \neq 1$. Then, for *all* x on the range $(1, r)$ either

$$(\ln x)(1 - b) > 0 \quad \text{if } b < 1 \qquad (8.55)$$

or

$$(\ln x)(1 - b) < 0 \quad \text{if } b > 1.$$

Or equivalently,

$$[1 + (\ln x)(1 - b)]^2 > 1 \quad \text{if } b < 1 \qquad (8.56)$$

or

$$[1 + (\ln x)(1 - b)]^2 < 1 \quad \text{if } b > 1.$$

Then, in either of these cases b, the average of each over x is likewise not equal to 1. Then by (8.53), $I \neq 1/a^2$. This violates our requirement. It follows that the candidate solution $b \neq 1$ is rejected, leaving only the solution (8.54).

Does $b \to 1$ also satisfy the unbiasedness condition (8.45) of the preceding section? By that condition, with $b \to 1$ unbiasedness can only occur in the limit as the mass upper bound value $r \to \infty$. On the other hand, there is not a well-defined upper bound to the size of the largest conceivable organism.

We can conclude, then, as follows:

The data relation (8.20) admits of an efficient estimator, if the probability law $p_X(x)$ on the mass is a simple power law. The power b is close to 1 if the upper bound to the mass of an organism is exceedingly large. That is, $b = 1 + \epsilon, \epsilon \to 0$ (but not equaling zero). In this limit, the mass law (8.39) becomes, by the use of l'Hôpital's rule to obtain the normalization constant,

$$\underset{1 \leq x \leq r}{p_X(x)} = \frac{1}{(\ln r)x}. \tag{8.57}$$

In practice, the theoretical value of r is *unknown*, so that only approximations to this law can be known. As an example, if r is as large as the upper bound to mass for which the *metabolic* rate law (8.20) holds, $r \sim 10^{27}$, then (8.57) gives the approximate law

$$\underset{1 \leq x \leq 10^{27}}{p_X(x)} \approx \left(\frac{1}{27 \ln 10}\right) x^{-1} = \frac{1}{(62.2)x}. \tag{8.58}$$

Transformation from the mass variable x to a *log-mass* variable, as at the approach (8.24), (8.25), indicates that with mass obeying a $1/x$ law, log-mass obeys a *uniform* law, that is

$$\underset{z \equiv \ln x}{p_{\ln X}(z)} = \frac{1}{\ln r} = \text{const} \quad \text{for } 0 \leq z \leq \ln r \tag{8.59}$$

and zero for all other values of $\ln x$. A variable $z = \ln x$ obeying (8.59) is often called "log constant" or "log uniform" (its log is uniform).

8.2.11. Discussion

We showed that if a system obeys the special power-law form (8.20), then its attribute a has *the potential* to be efficiently estimated. And this is if and only if the PDF on mass x of the system is close to a simple $1/x$ law over a limited range of masses. But, is the premise correct that *nature requires a capability for efficient estimation of biological attributes?* That is, consider a system consisting of an observing creature and an observed creature. Depending upon its structure, does the system allow the observer, if he/she wants, to form an efficient estimate of a? Requiring a capability, but not necessarily its realization, is a weak requirement. Therefore it seems reasonable. First we consider whether there are precedents for this requirement.

Many phenomena of the physical sciences obey the requirement. That is, they often give PDFs $p(y|a)$ that obey the efficiency conditions (8.23) and (8.40). If the observer is a human, for example, he/she may *choose* to use the known efficient estimator in these cases. Examples are as follows.

All PDFs $p(y|a)$ belonging to the "exponential family" admit of efficient estimation [6]. This includes the Poisson, normal, exponential, binomial, Rayleigh, and many other laws (see discussion below Equation (8.23)). Exponential-family PDFs are hallmarks of the hard sciences, e.g., in the phenomena of electromagnetic radiation and quantum mechanics. As an example, photons commonly obey Poisson statistics, and this was identified in Section 8.2.4 as permitting efficient estimation. The resulting optimal estimator is the sample mean, and is commonly used in optics [6].

Also, the SWE often gives solutions $p(y|a)$ that permit efficient estimates, e.g., in the case of the simple harmonic oscillator ground state where $p(y|a)$ is a normal law. On the other hand, the SWE also gives solutions that *do not* permit efficient estimates to be formed. This ultimately depends upon the particular potential function $V(y)$ that is present. As a negative example, the scenario of a free particle in a box of length L is a case of $V(y, a) = 0$. The resulting SWE output is

$$p(y|a) \equiv |\psi(y, a)|^2 = K \sin^2 [\pi(y - a)/L], \quad K, L = \text{const.} \quad (8.60)$$

Taking its logarithm and differentiating does *not* result in the form of the righthand side condition (8.23). Thus, one can say that the SWE has the potential for providing efficient estimates, depending upon the particular scenario.

Finally, if but *one* case of efficient estimation of a biological attribute a exists, it proves the premise of its possibility. Certainly there is no law of physics that *prohibits* such a case. In fact, *human beings themselves carry through efficient estimation* in many scenarios. An example is that of additive-Gaussian noise, where the efficient estimator is the oft-used arithmetic mean [6] of the data. Gaussian noise occurs quite commonly owing to the central limit theorem.

In summary, the physical sciences and the central limit theorem allow a capability for efficient estimation. These offer a strong precedent for the same efficiency property to hold in estimating *biological* attributes. Efficient estimation is actually carried through in biology, at the very least by humans. And, as we saw, for this capability to exist uniquely requires a PDF on mass x that goes asymptotically as $1/x$.

8.2.12. *Experimental Evidence for a $1/x$ Law*

We derived a *theoretical* PDF law (8.57) or (8.59) for the occurrences of masses x of living organisms. How well does this prediction agree with an empirically known universal mass law? Unfortunately, establishing such a law, by surveying biomass on a worldwide basis, is yet to be done. An important complication is that more than half of all existing biomass is believed to be in the form of bacteria, and most of these live underground, to unknown depths [27]. These are bound to elude most present-day surveys. Therefore we only briefly address the subject of what *is* empirically known on this subject.

Limited surveys on mass occurrence have of course been made, e.g., on fish in confined bodies of water such as lakes, or on reptiles or mammals that exist on islands or other confined regions. A problem is that these surveys do not, then, include the smallest living masses (irrespective of species) within these regions, for example the insects or bacteria. For purposes of confirming the law (8.57), these smallest masses are crucial, as they strongly contribute to the great height of $1/x$ at small mass x. Hence, with these masses absent from the survey, the empirical result is that the PDFs drop—rather than rise—once mass is less than some characteristic size. Toward the other end of the curve, once mass is *greater* than some characteristic size, such PDFs must drop, since larger animals tend to be rarer. The net result is a "single humped", rather than uniform, distribution, and skewed to the right in many cases. See, e.g., [28, 29]. Perhaps unsurprisingly, such distributions often resemble a log-*normal* law, rather than the log-*uniform* law (8.59) we have derived. Nevertheless, an interesting characteristic of log-normal laws is that they do fall off for large x as $1/x$. Hence even these nonideal surveys agree with our result (8.57) at the larger mammal sizes *that they include*. This provides, then, a partial confirmation of our calculation.

In fact, the predicted log-uniform (i.e., $1/x$) laws have also been observed, even at a limited range of smaller masses. For example, in a survey [30] of North American mammals, log-uniformity (8.59) was observed for surveys even down to small scales. Log-uniformity was also found [31] in like surveys of South American mammals, and in other surveys [32–34]. A simple energy-dependent model was devised [35] for explaining such dependence. It will be interesting to see if future mass surveys of broader scope that, somehow, can include underground bacterial mass as well will increasingly approach the laws (8.57) or (8.59).

8.3. Derivation of Power Laws of Nonliving and Living Systems

8.3.1. Summary

It was assumed in Section 8.2.2 *et seq.*, and in Chapter 5, that many complex systems obey power laws $y = Yx^a$ (also sometimes called "allometric laws"). This holds for both nonliving and living systems. Here we set out to *prove* this, using EPI. The main reference is [22].

In a power law $y = Yx^a$, quantity $y \geq 0$ is the measured value of some system attribute a and quantity $Y \geq 0$ is a constant. Finally, x is a stochastic variable whose nature depends upon the application. For many *living* systems, x is the mass of a randomly selected creature in the population and, remarkably, the exponent a is limited to values $n/4$, $n = 0, \pm 1, \pm 2 \ldots$. These so-called "quarter-power" laws $y = Yx^{n/4}$ hold for many attributes, such as pulse rate ($n = -1$) and others shown in Table 5.2, Chapter 5. Allometry has, in the past, been theoretically justified on a case-by-case basis, for one attribute at a time. An ultimate goal is to find *a common derivation* of allometry of all types, and for both living and nonliving systems.

The EPI principle is shown next to provide the missing derivation. It describes the flow of Fisher information $J(a) \rightarrow I(y)$ from an attribute of value a on the *cell level* to an exterior, macroscopic space, where it is observed as a value y.

The latter is allowed to depend upon both a and the mass x of the observed system, through an unknown channel function $y \equiv f(x, a)$. EPI finds $f(x, a)$ by extremizing the information loss $I - J$ through variation of $f(x, a)$. The output is a general allometric law $f(x, a) \equiv y = Yx^a$, $a = \text{const}$.

For living systems, Darwinian evolution is presumed to cause a *second* extremization of $I - J$, now with respect to the choice of a. Accordingly, the EPI output $f(x, a) = Yx^a$ is substituted into I and J, and the condition $\partial/\partial a(I - J) = 0$ is imposed. The solution is $a = n/4$, $n = 0, \pm 1, \pm 2, \dots$. These define the particular *quarter powers* of biological allometry.

In addition, EPI generally allows properties of the source to be inferred by its outputs. In this problem they predict that the biological systems in question are controlled by *but two* distinct intracellular information sources. These are conjectured to be cellular DNA and cellular transmembrane ion gradients.

In fact the scope of the quarter-power effect is wider than these biological applications—quite literally much wider. In Chapter 5, the quarter-power law $y = Yx^{n/4}$ is shown to describe as well the unitless universal constants y of our universe. Here the mass x is, accordingly, that of the universe. This has some interesting ramifications to cosmology, including agreement with the modern notion of a multiverse of universes of (now) masses x.

8.3.1.1. General Allometric Laws

Allometric power laws have a general form

$$\mathbf{y}_n = \mathbf{Y}_n x^{a_n}, \quad a_n = \text{const}, \quad n = 0, \pm 1, \pm 2, \dots, \pm N, \qquad (8.61a)$$
$$\mathbf{y}_n, \mathbf{Y}_n \geq 0, \quad 0 < x < \infty.$$

These laws describe, to a good approximation, certain living and nonliving systems. In general n defines an nth class of observed attributes $\mathbf{y}_n \equiv y_{n1}, y_{n2}, \dots, y_{nK_n}$ of a system. Also, $\mathbf{Y}_n \equiv Y_{n1}, Y_{n2}, \dots, Y_{nK_n}$ is a corresponding vector of *constants*, and K_n is the number of attributes in the class. Thus there is a total of $K \equiv \sum_n K_n$ attributes over all classes. Quantity x is an *independent variable* of a system that is sampled for one of these attributes. The powers a_n in (8.61a) are empirically defined values of the various attributes and are regarded as ideal identifiers of these. The a_n are generally dimensionless numbers such as 2/3, 0.7, etc. Current approaches for explaining general allometry are "self-organized criticality" [25], Lande's model [13], the scale-free network property [26], and others [25].

8.3.1.2. Biological Allometric Laws

Likewise there are many living systems that obey allometry [9–12, 14–20],

$$y_{nk} = Y_{nk} x^{a_n}, \quad a_n = n/4, \quad n = 0, \pm 1, \pm 2, \dots, \pm N, \qquad (8.61b)$$
$$k = 1, 2, \dots K_n, \quad y_{nk}, Y_{nk} \geq 0.$$

Here x is specifically the mass of the organism and the dimensionless powers a_n identify attributes of the organism. Remarkably, *each power is always some integer multiple of* 1/4. Why this should generally be so, both within individuals and across different species, is a great mystery of biology [9], and is addressed by this chapter. Living systems have "extraordinary" complexity, and in fact are reputed to be "the most complex and diverse physical system[s] in the universe" [9]. This suggests that EPI—which applies to complex systems—is applicable to derivation of these allometric laws.

Note that the same power $n/4$ describes all K_n members of an nth class of attributes. For example, the class $n = -1$ has currently $K_{-1} = 5$ known attributes (see Table 5.2). The dynamic range of mass values x in (8.61b) by definition includes mass values that extend from some (unknown) *very small* and finite value to some (unknown) *very large* and finite value. Indeed, for the attribute $n = 3$ of metabolic rate, the dynamic range of x over which (8.61b) is known to hold currently exceeds 27 orders of magnitude [9–11].

Allometric laws (8.61b) describe both *individual* and *collective* properties of animals. Some examples from Table 5.2 are as follows. The attribute class $n = -1$ mentioned above has $k_{-1} = 5$ members. The class $n = 3$ has $K_3 = 6$ members. Equation (8.61b) even holds for a class $n = 0$, i.e., where the attributes do not vary with mass. An example is hemoglobin concentration in the blood, which does not vary appreciably with body (or mass) size. Other quarter-power examples [9] are "metabolic rate, life span, growth rate, heart rate, DNA nucleotide substitution rate, lengths of aortas and genomes, tree height, mass of cerebral gray matter, density of mitochondria, and concentration of RNA." The list of $K = 28$ attributes in Table 5.2 only scratches the surface.

On terminology: We interchangeably use the terms "allometry," "allometric scaling laws," "scaling laws," or "power laws" to describe Eqs. (8.61a, b).

8.3.1.3. On Models for Biological Allometry

Although many biological attributes obey the quarter-power law (8.61a, b), many *do not* (trivial examples are attributes that are the square roots of those that do). Nevertheless, many models exist for explaining cases of biological allometry [9–11, 13–16], as conveniently summarized in [9].

However [9], these models are lacking in not providing a *unified approach* to calculating the attributes. Instead, they were "designed almost exclusively to understand only the scaling of mammalian metabolic rates, and do not address the extraordinarily diverse, interconnected, integrated body of scaling phenomena across different species and within individuals...Is all life organized by a few *fundamental principles?*"

A general approach would also have to predict circumstances where allometry will *not* occur. A step in this direction is to find a model that establishes necessity for *allometry of all types*, biological and nonliving. That is, it would show that *if a given attribute obeys the model, then it must obey allometry*.

We next form such a model. This dovetails with the use of EPI, which always *requires* a model in order to define the source information J (Section 1.5).

8.3.2. Prior Knowledge Assumed

The high degree of *complexity* in allometric systems encourages us to attempt deriving the laws (8.61a, b) by the use of EPI. Indeed EPI has been success-fully used in a wide range of amplitude-estimation problems [1, 4, 36–44] for complex systems. These successes trace to its basis in Fisher information (1.35), which is a measure of complexity [23, 24] and other important physical properties [45, 46].

All uses of the EPI principle require prior knowledge. Here we use type (B), i.e. one or more *invariances* (see Section 1.5, Chapter 1). The general aim is to define the information functional $J(a)$ (see examples in Table 1.1, Chapter 1). In this problem the aim is to form a $J(a)$ that somehow represents the *full range* of biological and physical attributes $a \equiv a_n$ that obey laws (8.61a, b). *What can such a broad range of effects have in common?* One property is that of originating on the microlevel of biological cells or physical unit cells. An-other is asymptotic behavior near the origin. The following summarizes these properties:

(i) For all systems, information $J(a)$ originates on the *discrete microlevel*. For example, in nonliving systems such as regular crystals or irregular polymer chains, the sources are the unit cells or individual molecules, respectively. Likewise, in a living system, biological cells are the ultimate sources of information about a biological attribute a.

The information $J(a)$ is assumed to propagate as a superposition of plane waves, from a subset of cells and cell groupings to the observer. These waves originate at a "unit cell" $\Delta a = 1$ of a-space. (See alternative (v) below). The discrete nature of the "cell sources" will be essential to the derivation. The model will also make some useful predictions on *biological sources* of the information (Section 8.3.13).

(ii) The allometric laws obey certain *asymptotic behaviors* near the origin, as expressed next.

Differentiating allometric law shows that (8.20)

$$\frac{dy}{dx} \to \infty \quad \text{as } x \to 0, \quad \text{for } a < 1, \tag{8.62a}$$

but

$$\frac{dy}{dx} \to 0 \quad \text{as } x \to 0, \quad \text{for } a > 1. \tag{8.62b}$$

In words, as organism size x approaches zero, the rates of increase of certain attributes a *increase* without bound, while others *decrease* without bound. Since the size can never equal zero (as mentioned above) the trends are mathematically well defined. They also are intuitively reasonable in many cases. Hence we make these general requirement of our solution as well. Properties (8.62a, b) are key to the derivation in Sections 8.3.10.1 and 8.3.11.

(iii) In *general* cases (8.61a) of allometry, the powers a_n are regarded as *a priori fixed numbers* of unknown size (the view taken by classical estimation theory [6, 47]). In particular, these do not generally extremize the information flow $I - J$ of EPI (see Section 8.3.11).

However, in specifically *biological* cases (8.61b) Darwinian evolution is presumed to force a progressive drift of organismal attributes a_n toward those values that confer maximal fitness on the organism. And maximal fitness is taken to be achieved, in turn, by those attribute values a_n that extremize $I - J$. This model property is used in Sections 8.3.9 and 8.3.11.

(iv) (Only) in biological cases (8.61b), the independent variable x is the mass of the organism. That is, laws (8.61b) are scaling laws covering a range of sizes, where the sizes are specified by mass values x. Why specifically "mass," is discussed in Section 8.3.13. In nonliving systems the nature of x depends upon the system.

(v) (Only) in biological cases (8.61b), *alternative* to the unit-cell assumption (i) of $\Delta a = 1$, more generally allow $\Delta a = L$, some unknown constant. L should be fixed by some reasonable biological requirement. For example, the identification of the a_n with pure *numbers* requires that one be fixed as a boundary condition. Then let $a_1 \equiv 1/4$. In Section 8.3.12 it is found that on this basis $L = 1$ as before.

Note that these model assumptions are not in themselves sufficient to imply the allometric laws. For example, laws (8.61a, b) with x *incorrectly* replaced by $\sin(x)$ would still satisfy requirements (8.62a, b) of (ii).

Finally, not all systems obey allometry Eqs. (8.61a, b). Therefore, such systems do not obey this model, by the necessity condition above. This is further discussed in Section 8.3.13.

8.3.3. Measurement Channel for Problem

The EPI principle will be applied to both living and nonliving systems. Thus, the measurement channel described next is, in general, that of either a living or a nonliving system. However, for definiteness, biological terminology is often used.

8.3.3.1. Measurement, System Function

The measurement scenario is as follows. A fixed attribute a is measured, with value y, for an *independently* and randomly selected system of mass x. In general, the measured value y of a relates to a and to the mass x through a function

$$y = Cf(x, a), \quad -\infty \leq a \leq +\infty, \tag{8.63}$$

for some constant C. This fixed, deterministic function $f(x, a)$ is called the "system" or "channel" function. The system is a biological creature or a nonliving system such as a polymer. The source variable x obeys some unknown and arbitrary probability law $p_X(x)$. Its details will not matter to the calculation (although we found it to approach a simple $1/x$ law in biology; see Section 8.2).

The overall aim is to find the constants C and the channel functions $f(x, a)$, in the presence of any fixed but arbitrary PDF $p_X(x)$. Hence, the functions $f(x, a)$ will be varied to achieve the extremum in $I - J$ that is required of the EPI. The system function will turn out to be the allometric law (8.61a, b). In biological cases the attribute value a will be further varied, to further extremize $I - J$. The solution will equal $n/4$, for values of $n = 0, \pm 1, \pm 2, \ldots$.

Note that in general the *form* of a system function f defines the physics of a particular channel. As a simplistic example, for some channels, not considered here, $f(x, a) = a + x$. This would be the familiar case of additive noise corrupting a signal value.

8.3.3.2. Some Caveats to EPI Derivation

It should be noted that most past uses of EPI have been through variation of system *PDFs or amplitude functions*, not variation of the deterministic system function $f(x, a)$ as here. The success of the approach in a wide range of amplitude-estimation problems [1, 4, 36–44] implies that systems in general *obey EPI through variation of their PDFs or amplitude functions*. However, it is not yet well-substantiated that systems as well *obey EPI through variation of their channel functions*. However, one positive indication is the EPI-based derivation in Section 2.3, Chapter 2, of the known Tobin q-theory trajectory $K(t)$ of optimum capital investment. The derivation below will likewise be positive in this regard. It will show that *if* a system obeys EPI and also the model of Section 8.3.2, then it obeys allometry.

In the next two sections, we proceed to form the information functionals $I(a)$ and $J(a)$, and then use them in the EPI extremum principle. Since the extremum is to be found through variation of $f(x, a)$, both $I(a)$ and $J(a)$ must be related to this channel function. This is done next.

8.3.4. Data Information I

As defined at Equation (1.35), Chapter 1, the Fisher information obeys

$$I = I(a) = \int dy\, p(y|a) \left[\frac{\partial}{\partial a} \ln p(y|a) \right]^2. \tag{8.64}$$

Since x is random and a is fixed, Eq. (8.63) actually represents the transformation of a random variable x to a random variable y. Therefore, elementary probability theory [6] may be used to connect the respective probability laws $p_X(x)$ and $p(y|a)$, as

$$p(y|a)\, dy = p_X(x)\, dx, \quad dy > 0, \quad dx > 0. \tag{8.65}$$

Differentiating (8.63),

$$\frac{dy}{dx} = Cf'(x, a), \quad f' \equiv \frac{\partial f}{\partial x}. \tag{8.66}$$

Combining Eqs. (8.65) and (8.66) gives

$$p(y|a) = \frac{p_X(x)}{C|f'(x, a)|}. \tag{8.67}$$

This is to be used in (8.64) to form I. First, taking a logarithm and differentiating gives

$$\frac{\partial}{\partial a} \ln p(y|a) = -\frac{\partial}{\partial a} \ln |f'(x, a)|. \tag{8.68}$$

Conveniently, both $p_X(x)$ and the constants C have dropped out. Using the results (8.65) and (8.68) in (8.64) gives

$$I(a) = \int dx \ p_X(x) \left[\frac{\partial}{\partial a} \ln |f'(x, a)| \right]^2. \tag{8.69}$$

That is, the averaging $<>$ is now explicitly over the random variable x. Also, $I(a)$ is now related to the unknown function $f(x, a)$, as was required.

8.3.5. Source Information $J(a)$

The source information for the problem is next found. As with $I(a)$ it must be related to $f(x, a)$.

8.3.5.1. Microlevel Contributions

Recalling the model assumption (i) of Section 8.3.2, $J(a)$ originates at the cell level. In general, some cells and cell groups contribute independently, and others dependently, to $J(a)$. Then, by the additivity property of the Fisher information [37], the total information $J(a)$ is simply *the signed sum* of *positive* and *negative* information contributions from the independent cells and cell groupings of the organism. A well-behaved function $J(a)$ can of course be represented over a limited a-interval by a *Fourier series* of such terms. What interval size should be used?

Here we use model assumption (i) (Section 8.3.2) of a unit interval. A unit interval of a-space seems reasonable from various viewpoints. First, it is fundamental to many physical effects, such as in solid state physics where the number of degrees of freedom *per unit energy interval* is of fundamental importance. Second, a unit interval is certainly the *simplest* possible choice of interval, and hence preferred on the basis of Occam's razor.

The alternative model assumption (v) (Section 8.3.2) of a *general* interval size $\Delta a = L$ is taken up in Section 8.3.12.

8.3.5.2. Fourier Analysis

In Section 8.3.2, item (i), the information $J(a)$ was modeled as propagating waves. This can be substantiated. Heat or entropy propagates via plane wave Fourier series [48, 49] and Fisher information $J(a, t)$ is, like entropy, a measure of disorder, monotonically decreasing with an increase in time t [36, 37, 40]. Therefore $J(a, t)$ is, as well, meaningfully represented by Fourier series.

We assume steady-state boundary conditions so that $J(a, t) = J(a)$. (The attributes supply source information at a constant rate in time.) The Fourier series

[48, 49] for a unit interval $0 \le a \le 1$ is

$$J(a) = \sum_m F_m \exp(2\pi ima), \quad F_m = \int_0^1 da' J(a') \exp(-2\pi ima') \quad (8.70)$$
$$\underset{0 \le a \le 1}{}$$

$$J(a) \ge 0, \quad i = \sqrt{-1}.$$

However, this series is inadequate for our purposes. First, Eqs. (8.61a, b) hold over an *infinite* range $-\infty \le a \le \infty$ of attribute values, not a unit interval. Second, we expect function $J(a)$ to be an even function,

$$J(a) = J(-a) \quad (8.71)$$

since there is no reason to expect a negative attribute value to provide more information than its corresponding positive value.

One way to accomplish the infinite range $-\infty \le a \le \infty$ is to form the Fourier series for $J(a)$ over *a sequence* of symmetrically placed, half-unit interval pairs $(-1/2 \le a \le 0)$ and $(0 \le a \le 1/2)$; $(-1 \le a \le -1/2)$ and $(1/2 \le a \le 1)$; etc., each pair increasingly further away from the $a = 0$ origin. These are denoted as

$$a = \pm\left(\frac{j}{2}, \frac{j+1}{2}\right), \quad j = 0, 1, 2, \ldots. \quad (8.72)$$

Each interval number j defines in this way a total *unit* interval for a, as required. Note that the *half*-unit intervals (8.72) are contiguous and fill in all of a-space. The $J(a)$ for each interval obeys [48]

$$J(a) = \sum_m B_{mj} \exp(4\pi ima), \quad J(a) \ge 0 \quad (8.73)$$
$$\underset{\pm(j/2,(j+1)/2)}{}$$

$$B_{mj} = 2 \int_{j/2}^{(j+1)/2} da' J(a') \exp(-4\pi ima'), \quad j = 0, 1, 2, \ldots.$$

Thus each value of j identifies an interval over which $J(a)$ is defined by a distinct set of Fourier coefficients B_{mj}, $m = 1, 2, \ldots$. Since these intervals (8.72) are contiguous and span a-space, the resulting $J(a)$ is defined over all a-space as required. The factors 4 in the exponents, which will prove decisive, arise because each a'-integration (8.73) (second line) is over an interval of length $1/2$ (rather than 1 as in (8.70)).

This simplifies further. Because each $J(a)$ is an information and therefore *real*, (8.73) becomes

$$J(a) = \sum_m B_{mj}^{(re)} \cos(4\pi ma) - \sum_m B_{mj}^{(im)} \sin(4\pi ma), \quad j = 0, 1, 2, \ldots,$$
$$\underset{\pm(j/2,(j+1)/2)}{}$$
$$(8.74)$$

where (re) and (im) denote real and imaginary parts.

Requirement (8.71) of symmetry can only be obeyed if generally $B_{mj}^{(im)} = 0$ for all m, so that

$$J(a) = \sum_m A_{mj} \cos(4\pi ma), \quad A_{mj} \equiv B_{mj}^{(re)}, \quad j = 0, 1, 2, \ldots. \quad (8.75)$$
$$\underset{\pm(j/2,(j+1)/2)}{}$$

Next, using $B_{mj}^{(im)} = 0$ and that $J(a')$ is real in the second Eq. (8.73) gives

$$B_{mj} = 2 \int_{j/2}^{(j+1)/2} da' \, J(a') \cos 4\pi m a' = B_{mj}^{(re)} \equiv A_{mj}, \quad j = 0, 1, 2 \ldots \quad (8.76)$$

By the positivity of the definition (8.64) of $I(a)$, and the proportionality $I(a) = \kappa J(a)$, $J(a)$ must likewise obey positivity [36, 37]. Therefore, the coefficients A_{mj} must be constrained to give positive or zero values $J(a)$ at all a.

8.3.6. Net EPI Problem

For generality of results, in the analysis that follows we will regard the cellular contributions A_{mj} in (8.75) as arbitrary, except for giving symmetry (8.71) and positivity (8.73) to $J(a)$.

Using the particular informations (8.69) and (8.75) in the general EPI principle $I - J = \text{extrem}$ gives a problem

$$I - J = \int dx p_X(x) \left[\frac{\partial}{\partial a} \ln |f'(x, a)| \right]^2 - \int dx p_X(x) \sum_m A_{mj} \cos (4\pi m a)$$

$$= \text{extrem}, \quad j = 0, 1, 2, \ldots . \quad (8.77)$$

Here a choice of a defines 1:1 a choice of interval j, via Eq. (8.72), and therefore a choice of coefficients A_{mj}, $m = 1, 2, \ldots$. For mathematical convenience, we appended a multiplier of 1 (a normalization integral $\int dx p_X(x)$) to the second sum J.

As discussed in Section 8.3.3.1, we seek the channel functions $f(x, a)$ and (in biological cases) the system parameters a that extremize (8.77), in the presence of any *fixed* source PDF $p_X(x)$. Accordingly, the extremum in the principle (8.77) is first attained through variation of functions $f(x, a)$ and then, in biological cases, through the additional variation of parameters a. The mass PDF $p_X(x)$ is *not* varied, and turns out to not affect the answer. Thus, *the channel is optimized in the presence of a given source*. (The source PDF for biological mass x is found separately in Section 8.2.)

8.3.7. Synopsis of the Approach

The basic approach consists of three overall steps, as carried through in Sections 8.3.8–8.3.11:

(1) The information flow $I - J$ is extremized through choice of system function $f(x, a)$, in the presence of any fixed PDF mass law $p_X(x)$. This gives a general power law for its derivative $\partial f(x, a)/\partial x \equiv f'(x, a)$,

$$f'(x, a) = h(x)^{a-1}, \quad a \text{ real}, \quad (8.78)$$

(Eq. (8.87)). Quantity $h(x)$ is some unknown base function of x.

(2) The base function $h(x)$ is found, by further extremizing $I - J$ with respect to it, giving $h(x) = b_1 x$ (Eq. (8.105)). Using this in (8.78) gives

$$f(x, a) = x^a \tag{8.79}$$

(Eq. (8.109)) after an integration. An irrelevant constant is ignored. By Eq. (8.63), this achieves derivation of the general allometric law (8.61a).

(3) Finally, for a system that is biological, $I - J$ is extremized with respect to the choice of a, which gives $a = n/4$ (Eq. (8.92)). Using this in (8.79) gives

$$f(x, a) = x^{n/4}. \tag{8.80}$$

This is the biological allometric law (8.61b). The approach (1) to (3) is now carried through.

8.3.8. *Primary Variation of the System Function Leads to a Family of Power Laws*

The aim is to find the channel function $f(x, a)$ in the presence of a fixed source function $p_X(x)$. Hence we first vary $f(x, a)$, by use of the calculus of variations, holding the function $p_X(x)$ constant. Conveniently, it will drop out during the variation. The Lagrangian for the problem is, by definition, the integrand of (8.77)

$$\mathcal{L} = p_X(x) \left[\frac{\partial}{\partial a} \ln g(x, a) \right]^2 - p_X(x) \sum_m A_{mj} \cos(4\pi m a), \tag{8.81}$$

$$j = 0, 1, 2, \ldots.$$

We introduced a new function g defined as

$$|f'(x, a)| \equiv g(x, a). \tag{8.82}$$

In this way the function $g(x, a)$ replaces $f(x, a)$ as the quantity to vary in (8.81). Keeping in mind that the PDF $p_X(x)$ on mass remains a *fixed* function during the variation, the Lagrangian (8.81) is readily differentiated as

$$\frac{\partial \mathcal{L}}{\partial(\partial g/\partial a)} = 2 p_X(x) \frac{\partial g/\partial a}{g^2} \tag{8.83}$$

and

$$\frac{\partial \mathcal{L}}{\partial g} = -2 p_X(x) \frac{(\partial g/\partial a)^2}{g^3}, \quad g \equiv g(x, a).$$

Using these in the Euler–Lagrange equation [48]

$$\frac{\partial}{\partial a} \left(\frac{\partial \mathcal{L}}{\partial(\partial g/\partial a)} \right) = \frac{\partial \mathcal{L}}{\partial g} \tag{8.84}$$

gives, after some trivial cancellation,

$$\frac{\partial}{\partial a}\left[\frac{\partial g/\partial a}{g^2}\right] = -\frac{(\partial g/\partial a)^2}{g^3}, \quad g \equiv g(x, a). \tag{8.85}$$

Thus, the unknown PDF $p_X(x)$ has dropped out, as we anticipated above. Doing the indicated differentiation gives after some algebra

$$g\frac{\partial^2 g}{\partial a^2} - \left(\frac{\partial g}{\partial a}\right)^2 = 0. \tag{8.86}$$

The general solution to this can be found by using $g \equiv \exp(k)$, $k \equiv k(x, a)$, in (8.86) and solving the resulting differential equation for k. The answer is $k = K(x)a + L(x)$, with $K(x)$, $L(x)$ arbitrary functions. Exponentiating back to g gives an answer

$$g(x, a) = h(x)^{a-1}, \tag{8.87}$$

where $h(x) \equiv \exp(K(x))$ is an arbitrary real function of x called the "base function," and we took $L(x) \equiv -K(x)$. The latter choice gives the term -1 in the exponent of (8.87), for later numbering of the attributes $a \equiv a_n$ (see also (v), Section 8.3.2). The solution (8.87) may be readily shown to satisfy differential equation (8.86), keeping in mind that its derivatives are with respect to a and not x.

Hence the solution to the problem has the general form of a *power law*. That is, on the basis of optimal information flow $J \to I$, nature generally acts to form power-law solutions for the rate of change $g(x, a)$ of the channel function.

The general solution (8.87) contains a general base function $h(x)$ of the mass. This function will be found in Section 8.3.10. Also, the values of the power $(a - 1)$ of $h(x)$ to be used for the biological laws are not yet fixed. These unknown powers will next be fixed, as the second optimization step.

8.3.9. *Variation of the Attribute Parameters Gives Powers* $a \equiv a_n = n/4$

Here, by premise (iii) of Section 8.3.2, we vary a, for use in the biological laws. (Note that this will not affect the general law (8.61a) derivation since a so obtained (Eq. (8.92)) will *not* be used in that derivation.) Since a is a discrete variable, ordinary calculus is used, differentiating $\partial/\partial a$ (Eq. (8.77)) and equating the result to zero. This gives, after use of (8.82),

$$\frac{\partial}{\partial a}\int dx p_X(x)\left[\frac{\partial g(x, a)/\partial a}{g(x, a)}\right]^2 \tag{8.88}$$

$$-\frac{\partial}{\partial a}\left[\int dx p_X(x)\sum_m A_{mj}\cos(4\pi ma)\right]$$

$$= 0, \quad j = 0, 1, 2\ldots.$$

The first derivative term in (8.88) is next shown to be zero. Its derivative $\partial/\partial a$ operation may be moved to within the integrand, giving

$$\frac{\partial}{\partial a}\left(\frac{g_a}{g}\right)^2, \quad g_a \equiv \frac{\partial g(x,a)}{\partial a}. \tag{8.89}$$

Carrying out the indicated derivative $\partial/\partial a$ gives

$$2\left(\frac{g_a}{g}\right)\frac{\partial}{\partial a}\left(\frac{g_a}{g}\right) = 2\left(\frac{g_a}{g}\right)\left(\frac{gg_{aa}-g_a^2}{g^2}\right) = 0. \tag{8.90}$$

We used Eq. (8.86) since the biological optimization requires the *simultaneous* satisfaction of both the Euler–Lagrange condition (8.84) and the special extremum condition (8.88).

We showed in the preceding paragraph that the left-hand term in (8.88) becomes zero after the indicated differentiation, that is, $\partial I/\partial a = 0$. This has two important consequences. First, as will be shown below, I then does not depend upon a for the power-law solution (8.87).

Second, only the second term of (8.88) now remains. This defines a problem

$$\frac{\partial}{\partial a}\left[p_X(x)\sum_m A_{mj}\cos(4\pi ma)\right] \tag{8.91}$$

$$= -p_X(x)\sum_m A_{mj}(4\pi m)\sin(4\pi ma) = 0,$$

$$j = 0, 1, 2, \ldots.$$

(Note that $\partial\cos(4\pi ma)/\partial a = -4\pi m\sin(4\pi ma)$ within any interval j.) For arbitrary coefficients A_{mj}, the required zero is obtained if and only if

$$a \equiv a_n = \frac{n}{4}, \quad n = 0, \pm1, \pm2, \pm3, \ldots, \tag{8.92}$$

since then the sine function in (8.91) becomes $\sin(mn\pi) = 0$ for all integers m, n. Note that the solution values (8.92) form *in sequence* for the different unit intervals j given by (8.72). As examples, the interval for $j = 0$ is $(-1/2, 0), (0, 1/2)$ and contains solution values (8.92) $a = 0, \pm1/4, \pm2/4$; the interval for $j = 1$ is $(-1, -1/2), (1/2, 1)$ and contains solutions $a = \pm2/4, \pm3/4, \pm4/4$, and so on, thereby forming *all* solutions (8.92).

Result (8.92) shows that the attribute value a must be a multiple of $1/4$, or *the powers a in the law (8.87) are multiples of $1/4$.* This is an important milestone in the biological derivation. We emphasize that it only could follow because of the *discrete* nature of the sum over m, which follows from the model assumption (i) (Section 8.3.2) that information originates on the level of the discrete cells.

8.3.10. Secondary Extremization Through Choice of $h(x)$

The solution (8.87) to the extremization problem (8.77) of $I - J = $ extrem was found to contain an arbitrary function $h(x)$. Clearly, the appropriate $h(x)$ is the one that further extremizes $I - J$. We seek this function in this section. First we establish a general property of $h(x)$.

8.3.10.1. Special Form of Function $h(x)$

Here we show that $h(x)$ can be expressed as a linear term in x plus a function that is at least quadratic in x. Function $h(x)$ can be generally expanded in Taylor series as

$$h(x) = b_0 + b_1 x + b_2 x^2 + b_3 x^3 + \cdots . \tag{8.93}$$

Differentiating (8.61a), and then using (8.93) in (8.82) and (8.87) gives, in sequence,

$$\frac{dy_{nk}}{dx} = C_{nk} \frac{df(x, a_n)}{dx} = C_{nk} g(x, a_n) = C_{nk} h(x)^{a_n - 1} \tag{8.94}$$
$$= C_{nk}(b_0 + b_1 x + b_2 x^2 + \cdots)^{a_n - 1}.$$

Then

$$\lim_{x \to 0} \frac{dy_{nk}}{dx} = C_{nk} b_0^{a_n - 1} \equiv \frac{C_{nk}}{b_0^{1-a_n}}. \tag{8.95}$$

We now use the model properties (8.62a, b). If $a_n < 1$, then limit (8.62a) holds. This can only be obeyed by (8.95) if

$$b_0 = 0. \tag{8.96}$$

Consequently, by (8.93) $h(x) = b_1 x + b_2 x^2 + b_3 x^3 + \cdots$ or

$$h(x) = b_1 x + [k(x)]^2, \quad k(x) \equiv x \sqrt{b_2 + b_3 x + \cdots} \tag{8.97}$$

for some function $k(x)$. By the square root operation in (8.97), the latter is in general either pure real or pure imaginary at each x; it is found next.

8.3.10.2. Resulting variational principle in Base Function $h(x)$

Using definition (8.82), and Eq. (8.87) in Eq. (8.69), gives an information level

$$I = \left\langle \left[\frac{\partial}{\partial a} \ln \left(h(x)^{a-1} \right) \right]^2 \right\rangle = \left\langle \left[\frac{\partial}{\partial a} (a - 1) \ln h(x) \right]^2 \right\rangle \tag{8.98}$$
$$= \langle \ln^2 h(x) \rangle$$

after obvious algebra. Quantity a has dropped out.

The information difference $I - J$ is to be extremized in a *total sense*. The base function $h(x)$ that defines I in (8.98) has been expressed in terms of a new function $k(x)$ (Eq. (8.97)). Hence $I - J$ must be further (secondarily) extremized *through variation of function $k(x)$*. Using EPI result (8.97) in (8.98), and combining this with (8.75) and (8.92), gives a new problem

$$I - J = \langle \ln^2[b_1 x + k^2(x)] \rangle - \sum_m A_{mj}(-1)^{mn} \equiv \text{extrem}, \quad j = 0, 1, 2 \ldots$$

$$(8.99)$$

in $k(x)$.

8.3.10.3. Secondary Variational Principle in Associated Function $k(x)$

Since the coefficients A_{mj} are independent of the function $k(x)$, the net Lagrangian in (8.99) for varying $k(x)$ is

$$\mathcal{L} = p_X(x) \ln^2[b_1 x + k^2(x)].$$
$$(8.100)$$

Function $p_X(x)$ arises out of the expectation operation $< >$ in (8.99), and is likewise independent of $k(x)$. The general Euler–Lagrange equation for problem (8.100) is [48]

$$\frac{d}{dx}\left(\frac{\partial \mathcal{L}}{\partial k'(x)}\right) = \frac{\partial \mathcal{L}}{\partial k(x)}, \quad k'(x) \equiv \frac{dk}{dx}.$$
$$(8.101)$$

Since $\mathcal{L}$ in (8.100) does not depend upon $k'(x)$, the left-hand side of (8.101) is zero. Also, differentiating (8.100) gives

$$\frac{\partial \mathcal{L}}{\partial k(x)} = \frac{2 p_X(x) \ln[b_1 x + k^2(x)]}{b_1 x + k^2(x)} 2k(x) \equiv 0.$$
$$(8.102)$$

Once again, $p_X(x)$ is merely a constant multiplier, dropping out of the problem. Equation (8.102) has two obvious formal solutions.

8.3.10.4. Result $k(x) = 0$, Giving Base Function $h(x)$ Proportional to x

The first formal solution is

$$b_1 x + k^2(x) \equiv h(x) = 1,$$
$$(8.103)$$

the middle identity by (8.97). The second solution is

$$k(x) = 0.$$
$$(8.104)$$

(Note that this holds regardless of whether $k(x)$ is pure real or pure imaginary.)

However, one solution is readily eliminated. The candidate (8.103) when used in (8.98) gives $I = \langle [\ln 1]^2 \rangle = 0$. This extremum is the *absolute minimum* value possible for Fisher information. However, a scenario of $I = 0$ is rejected since then the observed value y of the attribute would unrealistically provide no information about the attribute. Hence the solution (8.103) is rejected.

By comparison, the candidate (8.104) when used in (8.97) gives

$$h(x) = b_1 x,$$
$$(8.105)$$

and consequently

$$I \equiv I_{\text{extrem}} = \langle [\ln(b_1 x)]^2 \rangle,$$
$$(8.106)$$

by (8.98). Information (8.106) is generally nonzero, thereby representing a subsidiary minimum, which makes sense on the grounds that the observation must contain at least some information about the parameter. Hence the solution (8.104), (8.105) is accepted.

8.3.11. Final Allometric Laws

We are now in a position to form the final allometric laws (8.61a, b) for, respectively, general and living systems. Substituting the solution (8.105) into Eqs. (8.82) and (8.87) gives $|f'(x, a)| = (b_1 x)^{a-1}$ or

$$f'(x, a) = \pm (b_1 x)^{a-1}. \tag{8.107}$$

Indefinitely integrating gives

$$f(x, a) = \pm b_1^{a-1} \int dx x^{a-1} \equiv \pm a \int dx x^{a-1}, \tag{8.108}$$

for a suitably defined b_1. Doing the integration trivially gives $f(x, a) = \pm x^a + C$, $C = \text{const}$. Is this constant of integration finite? In all attribute parameter cases $a > 1$, as $x \to 0$ it is required by (8.62b) that the attribute value $y_{nk} \to 0$, and hence by Eq. (8.63) that $f(x, a) \to 0$. Therefore, $C = 0$, and we have

$$f(x, a) = x^a. \tag{8.109}$$

We ruled out the negative alternative by using the fact that the attribute values y are positive (Eqs. (8.61a, b)).

The *general* allometric law (8.61a) is to hold for a priori empirically defined powers a_n (see (iii), Section 8.3.2). Here the specific powers (8.92) that held for optimization of $I - J$ do *not* apply. The solution is more simply the combination of Eqs. (8.109) and (8.63). Reinserting subscripts gives

$$y_{nk} \equiv C_{nk} f(x, a_n) = Y_{nk} x^{a_n}, \tag{8.110}$$

so that

$$Y_{nk} \equiv C_{nk} , \quad n = 0, \pm 1, \pm 2, \ldots.$$

This confirms the general allometric law (8.61a) for empirically known a_n.

Next we turn to the *biological* allometric law, which is modeled ((iii), Section 8.3.2) to hold for the particular powers a_n given by (8.92) that enforce a *further extremization* in the problem (8.77). Using powers (8.92) in the power-law solution (8.109), and also using (8.63), gives

$$y_{nk} \equiv C_{nk} f(x, a_n) = Y_{nk} x^{n/4}, \tag{8.111}$$

so that

$$Y_{nk} \equiv C_{nk}, \quad n = 0, \pm 1, \pm 2, \ldots.$$

This is the law (8.61b). As contrasted with laws (8.110), the powers a_n are here purely multiples of $1/4$.

8.3.12. Alternative Model $\Delta a = L$

The preceding derivation assumed a priori a unit fundamental length $\Delta a = 1$ (model property (i), Section 8.3.2). A stronger derivation would allow $\Delta a = L$, with L general. With $\Delta a = L$, the half-unit interval pairs in Section 8.3.5.2 are replaced with pairs of length $L/2$. Also, the Fourier properties (8.72) to (8.76) now hold [48] under the replacements $j \rightarrow jL$, $(j + 1) \rightarrow (j + 1)L$, and $m \rightarrow m/L$. Consequently the requirement of zero for Eq. (8.91) now becomes one of zero for $\sin(4\pi ma/L)$. The solution is $a \equiv a_n = nL/4$. Hence, by (8.110) the biological power law is now $y_{nk} = Y_{nk}x^{nL/4}$ instead of (8.111). Also, now $a_1 = 1 \cdot L/4 = L/4$. But by model assumption (v) of Section 8.3.2, $a_1 \equiv 1/4$. It results that $L = 1$. Consequently the quarter-power law (8.111) results once again.

8.3.13. Discussion

Section 8.3 has had the limited aim of *establishing necessity* for allometry. Accordingly, it showed that if a system obeys the model of Section 8.3.2 and also obeys EPI through variation of its channel function $f(x, a)$, it must obey allometry. However, this does not necessarily imply the converse—that any system that obeys allometry must also obey EPI and the model. (Note that this in fact might be true, but is regarded as outside the scope of the chapter.) Also, of course *not all* systems obey allometry. Then, by the necessity proven in this chapter, such systems do not obey the model of Section 8.3.2 and/or EPI.

By the overall approach, the allometric laws (8.61a, b) follow as the effect of a flow of information $J \rightarrow I$ from an attribute source to an observer. We saw that the derivation for general laws (8.61a) slightly differs from that for biological laws (8.61b). Each general law (8.61a) accomplishes an extremization of the loss of information $I - J$ through variation of the system function $f(x, a_n)$ and its subfunctions $h(x)$ and $k(x)$. By comparison, each biological allometric law (8.61b) accomplishes the extremization with respect to these functions *and* its system parameters a_n. The extra optimization with respect to the a_n reflects the specialized nature of biological allometry. But, why should biological systems be so specialized?

The answer is that, as compared to nonliving systems, biological systems have resulted from natural selection, i.e., Darwinian *evolution*. Thus, evolution is postulated ((iii), Section 8.3.2) as selecting particular attribute parameters a_n that *optimize the information flow loss $I - J$*. The postulate is reasonable. Survival and proliferation within an adaptive landscape favors optimization of phenotypic traits which, in turn, confers maximal fitness on the individual. Here the phenotype traits are, in fact, the attribute parameters a_n. Therefore, the a_n will evolve into those values that favor maximal fitness. Meanwhile, maximal fitness has been shown [4, 37] to result from optimal information flow loss $I - J = extrem$. (The latter gives rise to the L–V equations of growth which, in turn, imply maximal fitness through "Fisher's theorem of genetic change.") Therefore, it is reasonable that the same parameter values a_n that evolve via natural selection will also satisfy $I - J = extrem$.

In a related derivation [43], under the premise that in situ cancer is likewise in an evolutionary extremized state—now of transmitting *minimal* information about its age and size—the EPI output result is the correct law of cancer growth, *again a power-law form* (8.61a). However, here x is the time and $a_n = 1.618\ldots$ is the Fibonacci golden mean. Also, as here, *the information is optimized with respect to the exponent a_n*. This is also further evidence that the premise (iii), Section 8.3.2, of evolutionary efficiency is correct.

It was assumed as prior knowledge ((iv), Section 8.3.2) that in biological cases (8.61b) the independent variable x is the *mass* of the organism. That is, laws (8.61b) are scaling laws covering a range of sizes, where the sizes are specified uniquely by mass values x. Aside from being a postulate of the derivation, this is reasonable on evolutionary grounds. By its nature, the process of evolution favors systems that are close to being *optimized* with respect to the energy (and information) they distribute [9–11] to phenotypic traits at its various scales. On this basis only a dependence upon absolute size or mass would remain.

The precise form of the biological function $J(a)$ is unknown. It is possible that it is periodic, repeating itself over each fundamental interval j. This implies that all $A_{mj} = A_m$, $m = 1, 2, \ldots$, irrespective of j. Interestingly, for such periodicity $J(a)$ breaks naturally into two classes. Back-substituting any one coefficient (8.92) into the Fourier representation (8.75) now gives

$$J(a_n) = J(n/4) = \sum_m A_m \cos(mn\pi) = \sum_m A_m(-1)^{mn}. \tag{8.112}$$

Since the A_m remain *arbitrary*, this still represents an arbitrary information quantity $J(a_n)$ for $n = 0$ or 1. However, for higher values of n, the form (8.112) repeats, giving

$$J(\pm a_3) = J(\pm a_5) = \cdots = J(\pm a_1), \tag{8.113}$$

and

$$J(\pm a_2) = J(\pm a_4) = \cdots = J(\pm a_0). \tag{8.114}$$

Hence the odd-numbered attributes $n = \pm 1, \pm 3, \pm 5, \ldots$ all share one fixed level of ground truth information J about their values a_n, and the even-numbered attributes $n = 0, \pm 2, \pm 4, \pm 6, \ldots$ share another. Consequently, the source information of the channel is specified by but *two* independent values, (say) $J(a_0)$ and $J(a_1)$. That is, the allometric relations result from two basic sources of information. As we found, the numerical values of the two information levels remain arbitrary, since the coefficients A_m are arbitrary.

It is worthwhile considering why *biologically* there should be but two classes of information. The postulate (i) of Section 8.3.2 that discrete cells are the sources of information enters in once again. This ultimately gave rise to the sum (8.75) representing the source information $J(a)$ for the attribute. The sum is over the biological cells, and here there are only two independent information sources, of types (8.113) and (8.114). On this basis each cell must provide *two* independent sources of attribute information. The existence of two such sources is, in fact, consistent with recent work [50], which concludes that *cellular DNA* and *cellular transmembrane ion gradients* are the sources (see also Chapter 3 in this regard).

Finally, returning to a point made at the outset of Chapter 1, this derivation of allometry did not succeed out of a phenomenon-based analysis. It did not analyze, say, the single attribute of metabolism rate out of considerations of energy and momentum flow. Rather, it worked using the general concept of *information* as the system specifier. The former, phenomenon-based approach can only be applied on a case-by-case basis. The latter, by comparison, works simultaneously for a broad class of systems and attributes. Thus, the derivation applies to a wide category of living and nonliving systems. This is a definite benefit of the use of EPI.

Acknowledgments. We thank Professor Daniel Stein of the Courant Institute, New York University, for checking the math and making valuable suggestions on the modeling. Professor Patrick McQuire of the Center for Astrobiology (CSIC/INTA), Madrid, provided valuable comments on biological aspects of the manuscript. Finally, valuable support was provided by H. Cabezas of the Sustainable Technologies Division, US Environmental Protection Agency.

9
Sociohistory: An Information Theory of Social Change

M.I. YOLLES AND B.R. FRIEDEN

The aim of this chapter is to provide a framework, or paradigm, for estimating the logical past or future possibilities of a socioculture. These possibilities are stochastic in nature and form a "sociohistory." A framework for the sociohistory must therefore be statistical in nature and, to help in identifying its cultural trends, also epistemological. The trends that are so identified can help an inquirer pose sensible new questions, of the "What if ... ?" form, about the past or future dynamics of the socioculture.

The epistemological content of these dynamics is constructed out of extreme physical information (EPI) theory. This approach utilizes, as well, a frame of reference provided by social viable systems (SVS) theory. In turn, SVS theory incorporates the sociocultural dynamics of Sorokin. The coupling of these three theories has the potential for explaining and possibly predicting long-term and large-scale, or short-term and small-scale, sociocultural events. Thus, sociohistory analyzes the history of a culture out of a combined theory of social interaction and quantitative information (as we noted, EPI).

9.1. Summary

The emphasis of the presentation that follows is twofold: Upon the philosophical positioning of the sociohistory approach and upon its relationship with that of physics through EPI. The latter relationships are developed as follows: In Sections 9.3.2 to 9.3.10 informations I and J, and the EPI principle, are introduced. In Sections 9.5.5 to 9.5.10 it is seen that informations I and J can be used to measure opposing (enantiomic) forces in a society, in the form of its levels of sensate activity and ideate values, respectively. On this basis the efficiency constant $\kappa \equiv I/J$ measures the balance that is achieved between the two opposing forces in the society. In Sections 9.6.1 to 9.6.3 the value of achieving such a balance is discussed, ideally resulting in a "Hegelian alliance" between ideals and practice. In Sections 9.7 and 9.8 one set of physical aspects of a society that are to be described by the EPI-based sociohistory are arrived at, namely, the relative occurrences of its various populations and resources. The analysis is quantified generally in

Sections 9.8.1 to 9.8.4, and applied postfacto and semiquantitatively in Section 9.9 to a particular conflict. This was between the Shah and his opponents in post-colonial Iran.

9.1.1. Philosophical Background

By the time the West entered the eighteenth century, it had formulated some relatively successful paradigms in applied physics that were able to describe, explain, and predict the movements of material bodies through physical space. The aim was to make "positive" statements that assert things about the world and its universe. Such statements may be empirically true or false, so that what is asserted may or may not be true. If it is true, then it adds to our knowledge; if it is false then it does not. The propositions embedded in positive propositions support the notion that if a statement can be proven wrong by empirical evidence, then it is a testable statement that can be related to data that has been or will be acquired. This assumes that the data is well collected and a "true" representation of the phenomena that it concerns.

By the end of the nineteenth century, many scientists thought that everything there was to be discovered had been discovered. No new positive statement of a fundamental nature would be made. However, it seems a maxim of science that the more absolute a truth appears, the surer we can be that ultimately it will be undermined. Of course such was repeatedly the case in the year 1905, the "anno mirabilis" of A. Einstein.

Positive statements have the useful goal of objectively observing objects of attention, measuring them, and creating value-free knowledge deemed to be independent of an observer. However, within physics this philosophical stance has given way to postpositivism. Whereas a positive perspective permits material objects to be viewed as having innate properties, in postpostivism the properties are dependent upon the vantage point from which they are observed. Hence, observer participation is added into the observation process [1]. From this perspective, EPI is a postpositive approach to science. However, EPI is not so observer-dependent that observers of the same experiment see *different* physical laws. Nature is, after all, repeatable on this level (were it not, could we exist?) Wheeler's dictum of the "participatory universe" in Chapter 1 is likewise so constrained in meaning.

The relative historical success of positive physics resulted in attempts to transfer some of its principles to social science. Thus for instance Auguste Comte (1798–1857) called for a "social physics" that could claim its place alongside celestial, terrestrial, mechanical, and chemical physics [2]. Such transfer usually resulted in failure, and not only because of the limitations of positivism. The idea that theory from an existent paradigm in physics could be directly transferred into an existent paradigm in the social science, with its own distinct proprietary axioms and propositions, proved to be invalid. It results in what is today known as "paradigm incommensurability" [3].

Despite this, approaches from physics are still being successfully adopted to social theory. For example, see Chapter 7. Another example is Ball [4], who discusses the application of material dynamics of physics to social systems, and explains that one approach is to consider individuals as particles (as we do in a later section). He explains reasonably that this is problematic to some researchers, in particular those who decry a behaviorism that ignores the human dimensions of people [5]. However, given an appropriate paradigm, it is possible to enable both arguments to maintain analytic validity. In physics various levels of analysis are usually possible, and all need to be combined in any coherent formulation of a phenomenon. The same is true of the social sciences; the levels of analysis are simply different. Thus, for instance, in the study of consumer- or crowd behavior, it is *not unusual* for people to be regarded as particles in a "social mass."

9.1.2. Boundary Considerations

Given a complex situation, an inquirer in a social situation may encounter at least two types of boundary problems. One concerns *information boundaries*. That is, inquirers are always uncertain to an extent about how a boundary will condition their views of a subject.

Another boundary of knowledge is defined through the subjective nature of the *worldview* of an individual. Worldviews of individuals define different sets of prior knowledge. However, a coherent theory can result when people come together to form a consensus, or common paradigm, out of these worldviews.

The social dimensions set by knowledge boundaries also affect the physical sciences. Distinct paradigms that are used to describe particular phenomena can give rise to interparadigmatic conflicts [3, 6] or incommensurabilities. This might happen, for instance, when the knowledge embedded in one paradigm contradicts that of another. This happened historically when the Christian religious paradigm conflicted with the paradigm of science in connection with evolutionary theory.

Other boundary factors also condition views of situations, including factors that may derive from psychological, emotional, or cognitive functions. Such issues also enter the domain of physics, but they are not direct parts of physical paradigms. These ideas of perspectivism drive us away from notions of positivism within which many physical approaches are set, and can lead us to constructivist views of science that go beyond postpostivism [7, 8]. Yolles and Dubois [9] have also discussed this, for instance, within the context of second and third cybernetics. While postpositive second cybernetics entertains subjective perspectives, constructive third cybernetics goes a step further. While in second cybernetics we distinguish between two frames of reference—that of the observer[1], and that of the observed—a third frame of reference can exist that is neither of these. Rather it is a construction that comes from them both, and creates a new and distinct emergent frame of reference.

---

[1] Within the context of critical theory we should really talk about viewers rather than observers. Viewers are creative or participant observers in the sense discussed in Chapter 1.

9.1.3. Sociohistory: Historical Aspects

We have noted that Comte's approach to forming a social physics is problematic, and is likely to result in paradigm incommensurability. One way of overcoming this is to allow a paradigm shift to take place, whereby a new paradigm[2] evolves out of appropriate principles from the existent paradigms as well as supplemental principles. A well-known example of this is the past formation of the quantum theory of particles out of the ashes of the preexisting classical mechanics and the wave theory of light.

In this chapter, this approach is used in the development of sociohistory. This is a constructivist paradigm that has its antecedents in three others: (i) The theoretical construction of Sorokin, who explains how sociocultural change derives from the interaction of dichotomous cultural forces. In turn, this theory has been expressed through (ii) SVS theory, which explores the geometry of social dynamics through a position that is consistent with phenomenology [10]. The phenomenology entails ideas about the mediating role of consciousness in exploring the dynamics of material objects, as developed by early twentieth century philosophers like Edmund Husserl[3], Martin Heidegger, and Maurice Merleau-Ponty. (iii) The paradigm of EPI, as discussed later.

One of Husserl's interests in phenomenology was the study of the structures of consciousness that enable it to refer to objects outside itself. This does not presuppose that anything exists, but rather amounts to setting aside the question of real existence. Such ideas highlight a tension between objectivism and subjectivism. This was also an interest of Whewell [11][4], who proposed that all knowledge has both a subjective (or ideal: a view held by the so-called Idealists) and an objective (or sensation: a view held by the Sensationalists) dimension, and they were therefore commensurable conceptualizations. He argued that this was so because one can reconceptualize subjectivism such that it is seen to arise not just from experience but, rather, as an *interaction between* the phenomenon experienced and the consciousness that experienced it. Anticipating the use of EPI later in this chapter, one such interaction is a measurement-induced *perturbation* of the observed phenomenon (as described in Section 1.4, Chapter 1).

The mind, therefore, is not merely a passive recipient of sense data but, rather, *an active participant* in our attempts to gain knowledge of what we perceive as the world around us. This is, again, an underlying premise of EPI as previously

---

[2] Since a paradigm is a group phenomenon with practical application, following Yolles [3] the evolutionary process is that a virtual paradigm emerges which may or may not develop into a full-fledged paradigm, depending upon its level of support.

[3] In the phenomenology of Husserl (Grolier online Encyclopedia, 2002), reference to reality also occurs through the experience of that reality, but Husserl divided this into the "noesis" (act of consciousness) and the "noema" (object of consciousness). Here the line between idealism (the view that the mind or spirit constitutes the fundamental reality) and phenomenology became blurred, although the suspension of belief in the reality of an object of consciousness is not the same thing as denying that it exists.

[4] *Stanford Encyclopaedia of Philosophy*, on William Whewell (http://plato.stanford.edu/entries/whewell/).

expressed by Wheeler (Chapter 1). In sociological phenomena, as well, different ways of viewing can give rise to individually invented meanings [12].

9.1.4. Kant's Notion of the Noumenon

A student of philosophy will recognize what we are heading toward. In developing a philosophical basis for the development of physics, Kant introduced the notion of the *noumenon*, i.e., the "thing in itself." This is contrasted with the *phenomenon*, which is the generally imperfect projection of the noumenon that we observe. By this philosophy, man can only imperfectly sense (view, hear, smell, ...) the "true" reality that exists outside of him.

Kant argued that physical systems are a function of the mind that can be expressed in objective Sensationalist terms. This differs from the perspectivism of SVS where the noumenon is subjective, but relative to a defined frame of reference. In SVS, the material domain is phenomenological in nature and is mediated through a noumenal domain that, unlike Kant's notion, is constructivist in nature [13]. The interaction between SVS's phenomenal and noumenal domains is conditioned by a third existential domain that constrains and facilitates the relationship between the two. The whole assembly constitutes an autonomous world with its own identifiable frame of reference that is cybernetic in nature, and can link with other relatable and interactive worlds, where they can be argued to exist.

9.1.5. Extreme Phenomenal Information

Wheeler (Chapter 1) in effect supercedes Kant in stating that there is no distinction between a human being and its "outside." That is, the distinction is artificial. Thus, man is part of the noumenon he observes as a phenomenon. Hence, humans have a participatory role to play in forming the laws connecting noumena and phenomena, i.e., the laws of nature.

These effects carry over particularly well into sociohistory, whose laws, likewise, are formed in participatory fashion by all its players, human and otherwise. Hence, the *mathematical* aspects of the sociohistory paradigm come from EPI, and indeed this in turn has had an influence on the development of the SVS theory. An instance of this is the representation of its perspectivism in terms of the idea of a *global* noumenon, quite distinct from Kant's universal noumenon. While EPI traditionally stands for extreme *physical* information, within the context of the sociohistory paradigm its constructivist nature [8] leads us to regard it alternatively as extreme *phenomenal* information. Hence it may be seen that SVS has also had a least some influence on EPI. Indeed, some of the questions raised in developing sociohistory have challenged EPI to new conceptualizations of quantum processes, and indeed, the relationship between quantum and classical processes (as in Section 5.2, Chapter 5). This highlights clearly the power of paradigmatic convergence.

The sociohistory paradigm developed here arises out of the *complexity* stable. It also recognizes that yet another boundary condition arises for the system under consideration. This is whether they are qualitative, and analyzed through mere discussion, or quantitative, and analyzed by a mathematical model. Any proposed analysis should always be exposed to a form of boundary critique [14–16].

9.1.6. Complex Systems and Chaos

As in the EPI approach to statistical mechanics (Chapter 4), sociohistory should be viewed within a *non-equilibrium* social geometry. Sociohistory is able to explore microcosms of social interaction that obey its own statistical mechanics. It is, thus, presumed that individuals and groups can affect social and political movements in a complex and often chaotic way.

A sufficiently complex system will be chaotic, i.e., unstable to even the smallest, or apparently least significant, perturbing event. This is also called the "butterfly effect" [17], since a butterfly stirring the air today in Beijing can transform storm systems next month in New York. This was aptly discovered by a meteorologist, Edward Lorenz, in about 1961, and was first reported on at the December 1972 meeting of the American Association for the Advancement of Science in Washington, DC Since then, chaos has been found to be widespread in nature, and many books on fractals and chaos have been published, including a popular one by Mandelbrot [18]. Indeed, sociohistory is sensitive to historical volatility, and would seem to be another effect dominated by chaos and complexity.

9.1.7. Yin–Yang Nature of Sociohistory

Sociohistory derives its initial conceptualization from Sorokin's theory of history, originally published as a four-volume set between 1937 and 1941, and concerned with the rise of different cultural supersystems in the West. His theory proposes that all social cultures can be defined in terms of what we call cultural "enantiomers"[5]. These are two dichotomous forces of Being called the *ideational* and the *sensate*. These polar cultural forces interact in the same way as other enantiomers, like male/female, light/dark, Being/not Being. They also have a transcendent function that, according to Jung[6] (who uses the Greek version of the word enantiodromia), comes from experiencing the conflict of *opposites*.

---

[5] The term enantiomer (also enantiomorph that in particular relates to form or structure) means a mirror image of something, an opposite reflection. The term derives from the Greek enantios or "opposite," and is used in a number of contexts, including architecture, molecular physics, political theory, and computer system design. We use it in the sense of complementary polar opposites. The related word enantiodromia is also a key Jungian concept used in his notions about consciousness (e.g., http://www.endless-knot.us/feature.html): it is the process by which something becomes its opposite, and the subsequent interaction of the two: applied especially to the adoption by an individual or by a community, etc., of a set of beliefs, etc., opposite to those held at an earlier stage. For Jung the word enantiodromia represents the superabundance of any force that inevitably produces its opposite. Consequently the word enantiodromia often implies a dynamic process, which is not necessarily implied by the word enantiomer. By using the simpler word enantiomer we shall not exclude the possibility of any dynamic action that may have been implied by the term enantiodromia.

[6] In a letter on May 3, 1939 that discusses Psychological Types.

Interestingly, Jung seemed to abandon the perhaps difficult word "enantiodromia," and instead adopted the Chinese term *yin–yang,* which he used to explore personality traits [19, 20]. Yin–yang is a Chinese metaphysical concept developed in the Han dynasty (207 BC–9 AD). It comes from the perception that the universe is run by a single principle, the Tao, in which two principles exist that oppose one another in their actions: yin and yang.

9.1.8. Dialectic Process

In essence we can think of yin and yang as dichotomous primal opposing enantiomers that in interaction can form a global whole that symbolizes Tao. All change in the whole that it produces can be explained by the workings of yin and yang as, through a *dialectic process,* they either produce or overcome one another. Since each of these enantiomer opposites produces the other, the production of yin from yang and yang from yin occurs cyclically and constantly, so that no one principle continually dominates the other.

However, this cyclic symbiosis can be interrupted and overcome. To consider the yin–yang conceptualization as fundamental to the description of a culture may seem arbitrary and ad hoc. However, this provides a powerful and basic approach that has been successful throughout the history of social and human science. An example is the notion by Hegel of dialectics. It also has a parallel in physics, where EPI originates as an interplay between informations I and J, as we shall discuss in due course.

Like the interactive nature of yin and yang, Sorokin identifies ideational and sensate enantiomer forces that together enable a continuum of "mixed" conditions to arise. This defines an idealistic condition that he calls "integral." The integral cultural mentality does not appear to have much of a role in Sorokin's theory, and is used principally to explain the rise of the Western industrial revolution. We shall, however, generalize on this concept by using *joint alliance theory* [21–23], and illustrating how this develops through SVS theory.

9.1.9. Hegelian Doctrine of the Dialectic

The concept of a joint alliance traces rather directly from the philosophy of G.W.F. Hegel (1770–1831). Like the earlier principle of yin–yang, he believed in the alliance or synthesis of enantiomic opposites, called the doctrine of the "dialectic." The Hegelian dialectic is in particular concerned with a synthesis of ideas, by which factors working in opposite directions are, over time, reconciled. Hegel argued that *history is a constant process of dialectic clash.* One idea or event forms a thesis or intellectual proposition, and an opposing idea or event will be its antithesis or opposing idea of action. The two clash, and inevitably result in a synthesis or integration.

Modern corporate alliance theory does not express the optimism of Hegel's inevitability of synthesis, as is illustrated by the statistic that at least half the attempted joint alliances between enterprises fail. However, in contradistinction to

Hegelian theory, the notions that underpin Tao and its yin–yang dialectic seem more capable of exploring the creation *or failures* of corporate alliances. The interest here is to develop a theory of sociohistory that provides a capacity to explore the formation and maintenance of viable, and therefore sustainable, systems, and the pathologies that led to their failures.

9.1.10. Principle of Immanent Change

Unlike Spengler and Toynbee, who are interested in the decline and fall of societies, Sorokin's [24] work concentrates more generally on historical transitions. It argues, and has been empirically demonstrated, that social and cultural history can be represented as a dynamic system, whose dynamics come not from the external needs of society but, rather, from within; specifically, from the interaction of its enantiomers. This is what he calls the "principle of immanent change."

As a yin–yang process, the enantiomer forces of culture are in continual interactive conflict. Where they find balance, one or the other emerges in a society with some degree of dominance to create a "cultural disposition"[7]. This will determine the direction that the society takes. Another form of expression for cultural disposition is "cultural mentality" [26, 27] or equivalently "cultural mindset" [3]. Both cultural disposition and cultural mentality suggest a social collective with shared norms, and the terms can be used interchangeably. Remarkably, while the idea of the cultural mindset can be applied in the large, to large-scale social groups like societies, it also has the capacity to be applied in the small, to small-scale cultural groups like organizations.

9.2. Social Cybernetics

Sociohistory should be seen as a theory with a long history that embraces a social cybernetic. Rosenblueth et al. [28] were interested in the teleological properties of systems, those that relate to their identity and degree of autonomy and coherence. In particular they were interested in biological, physiological, and social systems, and their control and feedback processes. These authors formed the Teleological Society and, after Wiener coined the term cybernetics, changed its name to the Cybernetic Society. Such important beginnings have led to some significant ideas.

One of these is by Schwarz [29]. This provides a modeling approach that can explain why chaotic events should *not* just be seen as temporary accidental fluctuations that occur in our complex social systems but are, rather, caused by the inadequacy of our worldview and our methods of managing complex situations.

---

[7] We use the word "disposition" here to mean a characteristic or tendency of the collective being. It is consistent with the psychological use of the term "mental disposition" by Wollheim [25]. Within the context of culture we take it as a collective mental condition that embraces beliefs, knowledge, memories, abilities, phobias, and obsessions, that has duration, history, and inertia.

He argues that explicative frameworks like religious or political ideologies are not pertinent tools to understand these developments. Furthermore, *mono*disciplines like economic science, sociology, psychology, anthropology, etc., are unable to apprehend *hybrid* systems. These are rather like the phenomenological approaches discussed in Chapter 1.

Rather, he argues, a linguistic framework is needed that comes from a suitably coherent model. It should be able to describe and interpret complex situations that are part of autonomous, *complex* systems. He argues that various theoretical developments have occurred to address such approaches, including general systems theory, nonlinear dynamics (chaos theory), complex adaptive systems research, cellular automata, recursive and hyperincursive systems, and artificial life. The frame of reference developed by Schwarz is intended to interpret complex systems with more or less autonomy or operational closure (like self-organization), and possessed of other related facets as self-regulation, self-production, and self-reference.

The approach adopted here is one that likewise comes out of the "complexity systems" stable. In particular, it concerns how cybernetic principles may be applied to social systems [30, 31]. According to Beer, complex systems embody a special feature: *recursion*. Recursion is described by Espejo [32] as enabling autonomous social organizations to handle the complexity they create by establishing autonomous subsystems. Specifically, "recursion" refers to the process by which the autonomous subsystems come into being. It is interesting this may equivalently[8] be quantified in terms of fractal patterns: where a self-similar system looks approximately like itself at different levels of inspection [18]. The organization then has the potential to fulfill the purposes of the whole system through those of its fractal parts, and this contributes to the development of cohesion.

9.2.1. SVS Theory

Recursion is a fundamental feature of both of the two formal theories that we shall introduce here. SVS theory is a graphic approach that is able to explain how systems survive and change. It is based on the theory of self-determining autonomous systems devised by Schwarz [33]. More recently this has been developed by Yolles for social systems in a variety of publications (e.g., [34]).

The expectation about the predictability of human societies already has some credence. For instance Ball [4] has explored the phenomenal nature of large groups of people whose behavior can be understood on the basis of very simple rules of interaction. In support of this, there has been a series of recent papers attempting to provide evidence of large-scale human behavior. For instance in financial markets, crashes and large corrections are often preceded by speculative bubbles with two main characteristics: a power-law acceleration of the market price decorated with log-periodic oscillations [35]. Also, a related model has been produced by Yu [36].

---

[8] By this we are recognizing that the concepts of recursion and fractal structures are closely related.

It has also been shown that crowd behavior has complex phenomenological properties of fractal patterns [37, 38].

These ideas would at first seem to support a "behaviorist" formulation of social processes in which individuals act essentially as *automata* responding to a few key stimuli in their external environment. Such behaviorism has of course been decried by many systems thinkers (e.g., [5]), so one should always seek a more complete explanation for the apparent effects. Firstly, it should be realized that *all* complex autonomous[9] viable organizations (whether animate or inanimate) have internal structures from which automative properties arise. These structures are ultimately responsible for their behaviors[10] and enable organizations to be viable and to have complex structures. It should therefore be recognized that while behaviorism sees animate objects on purely inanimate terms in relation to an *external* environment, this provides only a partial view of the situation. It does not account for an *internal* environment defined by its inherent structure and the potential of that structure for morphogenesis. What appears, therefore, to be an automative response to stimuli is often simply an indication of the *capacity* of an organization to respond to stimuli, given its current structure and behavior.

The capacity to *change composition* is also relevant to some organizations, and this capacity ultimately determines their structure. In social collectives, this capacity is often referred to as transformational or dramatic change, and it's normally accompanied by metamorphosis. In many cases the metamorphosis is self-determined through the metasystem (e.g., the privatization of public companies like British Telecom during the 1980s), but in many cases it is not (as in the case of corporate hostile takeovers). The capacity to change composition is fundamentally existential. In *biological* organizations, composition and, therefore, structure and potential for behavior, are largely determined by DNA. This is of course susceptible to evolutionary change. In contrast, in social collectives, composition is largely due to culture, i.e., memes rather than genes, and this similarly maintains a

---

[9] The nature of autonomy is not a function of the interaction between some organized object of attention and its environment, but is rather a function of its internal structure that determines its behavior as it experiences that interaction. Viable objects of attention have the capacity to survive under adverse environmental conditions given that they have a structure that has the potential to change.

[10] Yolles [13] argues that there is a tight connection between culture, structure, and behavior in coherent social collectives. Culture provides a capacity to create meaning for a given structure, and thus there is a distinction between apparent and affective structure. Structure facilitates behavior, and the implicit or explicit process controls that are embedded within a given structure determine what behaviors are permitted. In human societies, there is a further complication, in that individuals who populate the roles offered by formal or informal structures each have a (local) worldview, and that worldview is in a structural coupling with the (global) host paradigm of the social collective. This structural coupling provides a means by which knowledge can be migrated between the local worldviews and the global paradigm. This coupling also contributes to the definition of the paradigm, which is intimately connected to organizational culture. As role populations change (for instance when people retire and their jobs are repopulated), so this coupling can impact on the global paradigm. There are short-term and long-term impacts of this, the latter often being referred to as evolutionary change.

capacity for structural definition and change. It is ultimately culture that provides the capability for an autonomous social collective to self-determine its morphology and, hence, its inherent potential for criticality and change.

A social collective that has developed and maintains a durable structure is sometimes referred to as a social system, and its existential component is often called a metasystem [13, 39]. The latter is a *global* entity that acts for the *system as a whole* to control structure and processes. It is through the metasystem that the "collective mind" arises.

9.2.2. Collective Mind

The notion of the collective mind is also consistent with that of the *noumenon* of Kant, as previously defined. He took this to exist in the unknown realm of the mind.

The notion of the collective (or group) mind was originally proposed by Espinas in 1878 and more recently reasserted by Le Bon ([40], Chapter 1). This is not exactly the same as Jung's notion of the collective unconscious, which is inherited and is associated with psychological archetypes [13]. However, it leads to the argument that there is a difference between individual (or unitary) and collective (or plural) agents. The former has a personal unconscious while the latter has a collective psyche. From this, cultural structures emerge and give rise to normative perspectives and normative reasoning. In this sense, it may well be seen that plural agents have a constructive frame of reference that occurs through the function of collective *associative projection*. We have already said that this occurs when the collective mind is active in forming an image of phenomenal reality (rather than being simply a passive receptor). This is through its reasoning and perspective-generating capacity [13], and results in patterns of behavioral coherence.

The emergence of a collective mind involves the development of collective culture that establishes a capacity for the collective to develop a global noumenon.

9.2.3. Global Noumenon

The content of the noumenon is defined at an epistemological horizon, the word "horizon" connoting unknowable and indescribable. It is defined to be a *consummate*, or perfect, expression of a positivist material reality. It expresses universal and absolute truth. Since it is at a horizon it is unknowable and indescribable. With these limitations, the notion of the noumenon can be regarded as no more than a *visualization* of reality.

However, such absolute idealism acts to provide entry into the constructivist frame of reference, and allows us to propose the notion of global (or as a logical subset of this, *local*) *noumena*. Global noumena are constituted by mental (and therefore virtual) *ideates*. Used as a verb, *to ideate* is defined[11] as the capacity to think and conceptualize mental images. It is also sometimes used as a noun,

---

[11] American Heritage Dictionary, 2004.

and we shall define *an ideate* in the following way. It is a valued (and perhaps complex) system of thought expressible as logical or rational structures that may be formulated as, or associated with, sets of often relatable images. Ideates are formulated by social collectives, which intentionally construct them over time through the influence of social factors like culture, politics, ideology, ethics, social structures, and economics. The aim is to *overcome chaos* and to create conceptual order.

9.2.4. Relative Noumenon

The *relative* noumenon is composed of a set of more or less isolated *inconsummate* ideates. These are systems of thought or images with cognitively (and imperfectly) defined structures. In a global situation they are normatively agreed upon as reflecting *phenomenal reality* rather than noumenal reality. We adopt here the phenomenological proposition that our experiences of phenomenal reality are mediated by relative noumena. Phenomenal reality is ultimately consciousness-mediated. Most cultures adapt the noumenal axiom that phenomenal reality evolves continuously over time. Their people continuously seek patterns, connections, and causal relationships in the phenomenal events that they perceive. This permits them to infer the existence of unknown ideate structures at noumenal horizons. One purpose of scientific inquiry is to identify, voyage to, explore, and thus make known, these horizons.

9.3. Developing a Formal Theory of Sociohistory

9.3.1. The Ontological Basis for SVS Theory

Schwarz's viable systems theory has been developed in a variety of publications within the context of sociocultural systems as SVS by Yolles [13]. Some of its antecedents are explored in Yolles [3]. Yolles and Guo [41] explain that its approach for social phenomena is metaphorical, but that this in no way diminishes its usefulness, as also indicated by Brown [42]. Basically the general autonomous system that we are talking about has three interactive ontological dimensions, or domains, as shown in Figure 9.1. Each domain may host a unitary system or plurality of related type systems.

The *phenomenal domain* hosts phenomenal, or operational, systems that are seen as system behavior. The *noumenal domain* hosts noumenal or virtual (ideate) systems that are supported by host ideology, ethics, or forms of rationality. We noted before that the term ideate is usually used as a verb, when *to ideate* is defined[12] as the capacity to think and conceptualize mental images. It is also sometimes used as a noun, with an ideate being a generally complex system of thought expressible as logical or rational structures associated with relatable images.

---

[12] American Heritage Dictionary, 2004.

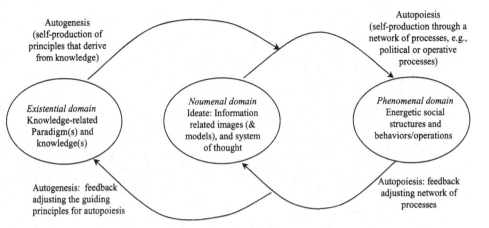

FIGURE 9.1. An autonomous holon.

Finally, the *existential domain* hosts metasystems. These ultimately have the task of controlling the nature of, and interaction between, virtual and phenomenal systems.

As illustrated, when two domains are coupled their realities are symbolically connected such that ontological migrations occur between them. Such migrations define the capacity for the reality of one domain to be manifest in another. This is, for example, through the creation of channels. These permit epistemological migrations that constitute the content of an ontological coupling, and occur through autopoiesis and autogenesis. The virtual and phenomenal domains are interconnected as a first-order ontological couple. In practice, the epistemological migrations that occur through autopoiesis can be thought of as an operative network of processes that may manifest ideate images or thoughts phenomenally. Thus, a conceptual idea can be manifest as part of a social structure through the appropriate application of a network of processes. These processes vary according to the nature of the epistemological migrations.

The three domains in their pattern of interaction are together referred to as an autonomous *holon*. This metaphorically describes social agents [43]. The holon is recursive in nature, with no beginning or end. Where the phenomenal domain applies to individual and social behaviors, the three domains can be assigned properties that are described in Table 9.1.

Autopoiesis—a self-sustaining network of processes—allows the content of the ideate that populates the noumenal domain to be manifest phenomenally. Here, images of a noumenal domain are migrated to the phenomenal domain by a network of processes that include political, operative[13], and communicative attributes.

---

[13] The term "operative" comes from Schwaninger [44] who proposes the idea of operative (as opposed to operations) management, and that Yolles [13, 45] has expressed in terms of autopoietic processes.

TABLE 9.1. Domain properties for the SVS model.

	Sociality (Nature of the organization)		
Cognitive properties	Kinematics (through social motion)	Direction (determining social trajectory)	Possibilities/potential (through variety development)
Cognitive interests	Technical	Practical	Critical deconstraining
Phenomenal (conscious) domain Activities energy	Work. This enables people to achieve goals and generate material well-being. It involves technical ability to undertake action in the environment, and the ability to make prediction and establish control. *Develops routines for communication. Causal explanations important. Use of empirical-analytic methods also important.*	Interaction. This requires that people as individuals and groups in a social system to gain and develop the possibilities of an understanding of each other's subjective views. It is consistent with a practical interest in mutual understanding that can address disagreements, which can be a threat to the social form of life. *Uses symbols, energy from leader(s)/facilitator(s), encourages appropriate behavior, seek descriptions of perceived situation and practical understanding.*	Degree, of emancipation. For organizational viability, the realizing of individual potential is most effective when people (i) liberate themselves from the constraints imposed by power structures and (ii) learn through precipitation in social and political processes to control their own destinies. *Needs rewards for behavior, disengagement from present constraining conditions uses critical approaches.*
Cognitive purposes	Cybermetical	Rational/appreciative	Ideological/moral
Noumenal or organizing (subconscious) domain Organizing information	Intention. Within the governance of social communities this occurs through the creation and pursuit of goals and aims that may change over time, and enables people through control and communications processes to redirect their futures. *Defines logical processes of communication and feedback, design of transition processes, arrangements for transition, facilitation of support.*	Formative organizing. Within governance enables missions, goals, and aims to be defined and approached through planning. It may involve logical, and/or relational abilities to organize thought and action and thus to define sets of possible systematic, systemic, and behavior possibilities. It can also involve the (appreciative) use of tacit standards by which experience can be ordered and valued, and may involve reflection. *Involves key power group support, stability processes are built in, and reflection and aesthetics are encouraged.*	Manner of thinking. Within governance of social communities an intellectual framework occurs through which policymakers observe and interpret reality. This has an aesthetical or politically correct ethical positioning. It provides an image of the future that enables action through politically correct strategic policy. It gives a politically correct view of stages of historical development, in respect of interaction with the external environment. *Social dissatisfaction where it exists should be seen in ideological terms, change can be mobilized through participation.*

	Socio	Base	Political
Cognitive influences creating cultural disposition Existential or cognitive (unconscious) domain Worldviews knowledge	Formation. Enables individuals/groups in a social collective to be influenced by knowledge that relates to its social environment. It affects social structures and processes that define the social forms that are related to collective intentions and behaviors. *An important basis for the creation of images of the future in the management of social processes. An understanding of the basis for cybernetic purposes is also important to enable technical aspects of the organization to materialize. Motivations, that should be understood, and drive purposes.*	Belief. Influences occur from knowledge that derives from the cognitive organization (the set of beliefs, attitudes, values) of other worldviews. It ultimately determines how those in social communities interact, and it influences their understanding of formative organizing. *Use of language and related concepts are needed that can give meaning to knowledge, and provide a basis for metaknowledge. Myths that are supported can misdirect the social collective. The propositions that underpin the paradigm of the social collective are defined here, those that give meaning to its existence. This influences social purposes.*	Freedom. Influences occur from knowledge that affect social collective polity determined, in part, by how participants think about the constraints on group and individual freedoms; and in connection with this, to organize and behave. It ultimately has impact on unitary and plural ideology and morality, and the degree of organizational emancipation. *Creates a culture's normative boundaries through its beliefs, values, symbols, stories, and public rituals that bind people together and direct them in common action. These determine the creation of ideological/ethical and power constraints. They connect to the structure of social collective and the way that power is distributed and used.*

Autogenesis enables principles to be generated that guide the development of the system and its behavior. The autonomous holon can also be described in terms of processes of change. Some conditions for change are identified in italics in Table 9.1. The three domains are formulated within the context of a constructivist philosophy, and in particular the noumenal domain is relative and constructivist [13].

Remarkably, *the flow of logic in Figure 9.1 follows that of the EPI principle* as well. In fact that process was likewise noted to be autopoietic (Section 1.4.6, Chapter 1). This correspondence is our first inkling that the EPI measurement process, originally conceived of as describing physics, might also be used to describe social systems. Hence, the inquisitive observer introduced in Chapter 1 can be taking data about a physical effect *or about a social system*. In fact, social interactions are themselves basically "measurements," in the sense of samplings of a social system and, therefore, naturally fit within the framework provided by EPI.

It is necessary at this junction to consider a little further the noumenal domain, the conceptualization of which is relatively new. While both the notions of noumenal realm of Kant and the noumenal domain considered here are related to the mind, they are distinct in other respects. In Kant's positivist stance the *noumenal realm* is a universal, i.e., both consummate (or perfect) and consisting of absolute truth. The noumenal domain, however, is relative to the autonomous agent with which it is associated. It is also constructivist in nature in that it is a global (collective), with individual localities, rather than being a universal entity. Ideate formulations that reside in a given global noumenon are therefore likely to vary relative to that of other global agents. While the noumenal ideate of a given global agent may be partly known, it is never certain that ideate content is the absolute truth, as supposed in Kant's noumenal realm. This is even though much of the content may be maintained at an unknown noumenal horizon, as posited by Kant. In other words, the autonomous holon may not know its own ideate content. This comes in part from the notions in Yolles [45] where the autonomous holon is explored as an intelligent collective, and where the phenomenal and noumenal domains are respectively assigned the psychological metaphors of the conscious and subconscious.

The formulation in Figure 9.1 can be used recursively (e.g., Figure 9.5) in, e.g., an enquiry process about the nature of the sociocultural change effects in a society. An illustration of this plurality is the constructivist approach to scientific enquiry shown in Figure 9.2. The process is embedded in the noumenal domain because the measuring process is a creator of noumenally defined physical laws that arise because of our participation in the measuring process. This involves Frieden's [46, 47] idea of the *creative observer*—the inquirer whose worldview influences the way that information is acquired. We also introduce the notion of structural coupling that occurs for structure-determined/determining engagement in an interactive family of systems. According to Maturana and Varela ([48], p. 75) the engagement creates a history of recurrent interactions that leads to a structural congruence between the systems, and to a spatiotemporal coincidence between the changes that occur in the family of systems ([49], p. 321).

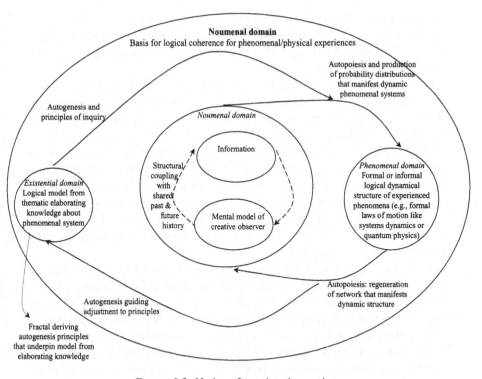

FIGURE 9.2. Notion of creative observation.

9.3.2. Extreme Phenomenal Information

The transfer of EPI from a Kantian noumenal framework to one that adopts the notion of an autonomous system suggests that we shift its name from extreme physical information to extreme *phenomenal* information. EPI now becomes, in this context, a theory [46, 47, 50] that can be used to construct the unknown probability laws of *social* science.

With a view toward minimizing page turning, we next give a capsule review of EPI. Full details of the approach are in Chapter 1.

9.3.3. System Informations I, J

EPI has two forms of information, I and J. These will be seen to be indicators of complexity. The informations are of the type invented by R.A. Fisher in the early 1920s and are commonly called "Fisher information." The premise of EPI is a self-evident one—that data comprise information "about" an observed effect. The latter takes on the formal role of a *noumenon* in the philosophy of I. Kant. As previously discussed, the noumenon is defined as the "thing-in-itself," that is, the intellectual conception of a thing, with complete precision. The information that is *intrinsic to* the noumenon, fully describing it, is denoted as J. This is to be compared with how the noumenon is *known*, i.e., through sense-based perception as a *phenomenon*. This is an inconsummate[14] or imperfectly *observed* local effect carrying a level of Fisher information level I. As will be discussed in Section 9.3.10, informations I and J may also be regarded as enantiomic opposites, in the sense defined by Sorokin.

From the preceding, J is the total amount of information that is needed to describe the noumenon with perfect accuracy. Hence J is called the "bound" information, in the sense of being "bound to, or descriptive of" the noumenon. It also may be regarded as a "potential" information, i.e., potentially the largest value that the acquired information I can attain. By the way, the possibility that I can achieve the value J was in effect forecast by Spinoza, who believed that humans have the potential to completely understand nature (or God, by his philosophy). Of course Plato and, later, Kant famously denied this possibility of complete knowledge. Having noted this, we should remind ourselves that we are not concerned with the universalism of Kant's noumenal realm, but rather the globalism of a constructivist noumenal domain including EPI.

To clarify this, note that knowledge of the value of J does not equate to absolute knowledge of the *physical form* that the noumenon takes. In the famous allegory of Plato, the shadows on the cave wall describe only projections of people, not the people themselves in their three-dimensional, flesh-and- blood (and beyond) forms. Thus, EPI can (when $I = J$) create a complete *mathematical* description of that physical form, *but the form itself remains unknown* (and probably unknowable). This agrees with the views of many people, who view knowledge of what is "really out there" as fundamentally unattainable.

9.3.4. Information Channel

The basis for the EPI principle is simply that whatever information is at hand arises out of a flow of information

$$J \to I \tag{9.1}$$

from the effect (sometimes called the "source effect") *to* the data collectors. The medium between the source effect and the data collectors is called an *information channel*. An example of a source effect is the law governing the distribution in space and time of photons radiating from the subject, or object, of an image-forming camera. This requires information J to define it perfectly. In this system the channel is the material camera, so that the data are the intensity values at the pixels in the image plane of the camera. These convey information level I to the observer. Because of the Second law of statistical physics, information is generally lost in transit from source to data collectors. Therefore the received or "acquired" information I is generally *some fraction* of J, as we shall see in a moment.

---

[14] The dictionary definition of the word consummate is unqualified, perfect, or complete. We therefore use the word *inconsummate* as its constructivist opposite, to mean that the perception by the creative observer of noumena is qualified so that phenomena are perceived partially or incompletely, and in some circumstances even "imperfectly."

More generally, the EPI view of an unknown[15] noumenon governing a system is that it is prescribed by its ability to convey information about a parameter value a that defines the system under observation. For example, the noumenal effect called "quantum mechanics" can be derived by analyzing its ability to convey information into any observation of the position of a particle about the its true position. This sensate version of the phenomenon is generally partial or "inconsummate," generally suffering from data noise, and if a number of such measurements $\mathbf{y}$ of a are made these will all randomly differ.

9.3.5. Information I

We next use some introductory material about Fisher information from Chapter 1.

In general the measurements $\mathbf{y}$ contain an amount I of Fisher information about a defined as

$$I = \langle [d/da(log(p(\mathbf{y}|a)))]^2 \rangle \qquad (9.2a)$$

The notation d/da means a mathematical derivative with respect to a, and brackets $\langle \rangle$ indicate an expectation, i.e., multiplication by the law $p(\mathbf{y}|a)$ and integration over all $\mathbf{y}$ (not shown).

Probability law $p(\mathbf{y}|a)$, called the "likelihood function" in statistics, defines the probability of each possible vector $\mathbf{y}$ of measurements in the presence of the ideal parameter value a. Hence, in applications to physics, the law reflects "the noumenal nature" of the observed effect. Analogously, when the data and parameter are suitably defined below in terms of sociocultural effects, the probability law $p(\mathbf{y}|a)$ will take on the role of describing a statistical, sociocultural physics. Our aim will then be to estimate the probability law and, therefore, quantitative aspects of the socioculture that gave rise to the data.

Evaluations of I for various probability laws via Eq. (9.2a) disclose that the broader and (by a normalization requirement) lower it is as a function of the $\mathbf{y}$, the smaller is I. A broad, low-likelihood function indicates close to equal probability for all values of the $\mathbf{y}$, i.e., a maximally disordered system. Thus, a *small* value of I indicates a *high* level of disorder; similarly it can be shown that a large value of I indicates a small level of disorder.

As an example, if the probability law is normal with variance σ^2 then $I = 1/\sigma^2$. Thus, if I is small σ^2 is large, indicating high randomness or high disorder. These effects allow us to express the famous Second law of thermodynamics in terms of Fisher information (as opposed to the usual measure called "entropy"). See Eq. (1.75).

As another example, in the context of the sociocultural problem below where there are discrete probabilities $p_n(t)$, $n = 1, \ldots, N$ corresponding to population

---

[15] In Section 9.1.4 we indicate that, even taking a constructivist representation of Kant's noumenon, there still exists an ideate horizon that is not accessible to the mental process.

components, the general information Eq. (9.2a) becomes

$$I = I(t) = \sum_{n=1}^{N} \frac{[p'_n(t)]^2}{p_n(t)}, \quad p'_n(t) = \frac{dp_n(t)}{dt}. \tag{9.2b}$$

The latter is the slope of the probability law at a given time. Thus, Eq. (9.2b) shows that the information at each time analytically increases with the slopes (either upward or downward) of the probabilities at that time. Thus, I is a *local* measure of disorder (also see Chapter 4 in this regard).

The Second law of thermodynamics states that disorder must inevitably increase. Disorder can be measured in many different ways. The usual measure is entropy. Entropy increases when disorder increases, so the Second law is usually expressed by the statement that the rate of change of entropy with time is positive (i.e., it increases).

9.3.6. Fisher I as a Measure of the Arrow of Time

However, the state of disorder, as we discussed, may also be expressed in terms of Fisher information I. For probability distribution functions in space-time that obey a Fokker–Planck (diffusion) equation, the rate of change of I is negative,

$$\frac{\Delta I}{\Delta t} \leq 0 \tag{9.3}$$

[51, 52]. That is, with an increase in time $\Delta t \geq 0$ the change in information must be negative,

$$\Delta I \leq 0. \tag{9.4}$$

Or, Fisher information *monotonically decreases* with time. This is called the *Fisher I-theorem*, in analogy to the conventional Boltzmann H-theorem that utilizes entropy instead.

On the level of the observables or data $\mathbf{y}$, this means that ever more randomness monotonically creeps into them. The system defined by the $\mathbf{y}$ becomes ever more disordered.

The information transition (9.1) represents a change in information obeying

$$\Delta I = I - J. \tag{9.5}$$

9.3.7. EPI Zero Condition

Combining Eqs. (9.4) and (9.5) gives the result $I - J \leq 0$, or equivalently,

$$I = \kappa J, \quad \text{or } I - \kappa J = 0, \quad 0 \leq \kappa \leq 1. \tag{9.6}$$

The middle equation is called the EPI *zero condition*. It is one of the two equations comprising the EPI principle. It shows that the constant κ is a measure of the efficiency with which the information is transferred from the effect to the observer. The efficiency parameter is always between 0 (0% efficiency) and 1 (100%

efficiency). Its value depends upon the quality of the detectors and the particular effect that is under observation.

For example, when observing *quantum* effects, if the detectors are perfect then $\kappa = 1$. No information is lost. By comparison, classical effects such as gravitation or electromagnetism turn out to arise out of imperfect observation due to detectors that lose half the information, with $\kappa = 1/2$. The data are too coarsely spaced in space-time to sense the much finer, quantum gravitational fluctuations. In our sociocultural application, the value of κ will vary from one sociocultural system to another.

Equation (9.6) states that the acquired information I has the potential to equal the value of J at most. This is permitted by the philosophy of Spinoza, as previously mentioned. In applications where $I = J$, all the information necessary to describe the noumenon is now available in the observations. A result is that a noumenal, and not merely phenomenal, description of the system can be known. That is, the probability law $p(\mathbf{y}|a)$ that is the output of EPI now mathematically describes the noumenon as well as the phenomenon (keeping in mind, as above, that a mathematical description is not equivalent to absolute knowledge of the noumenon.) This happens, for example, for measurements that are on the quantum level as indicated above.

9.3.8. I is General, J is Specific

EPI regards data information as *generic*. Regardless of phenomenon, I always obeys the fixed form of Eq. (9.2a or b). By contrast, EPI regards noumenal information J as *specific*, and particular, to the observed effect. Thus, information J varies in form from problem to problem. In fact, as a useful rule, it is always found by the use of an appropriate invariance principle. This is an invariance that is appropriate in characterizing the *particular* measured effect. This invariance principle must of course be known, as a form of prior knowledge (sometimes called "physical insight"). See Sections 1.5.1–1.5.4. We will use a very simple invariance principle in our problem below, that of unitarity (invariance of length).

The "physical information" K is defined to be the change ΔI in the information that is incurred during its transit (9.1) through the channel from the source to the data collector(s),

$$K = I - J. \tag{9.7}$$

Therefore, by Eq. (9.4), K is always zero or negative, indicating that it is generally an information *loss*. As we saw, this means that the state of disorder of the system increases. As shall be explained in Section 9.3.10, where we introduce the "knowledge game," carrying through a measurement necessarily incites the activities of a "demon," whose aim is to minimize the received information.

9.3.9. EPI Extremum Principle

The information involved in the transition (9.1) is necessarily carried by a physical entity, e.g., a photon in the case of observation with a microscope. The carrier of information illuminates the source, which inevitably perturbs it. Correspondingly, its information level J is likewise perturbed. But then the process (9.1) indicates that the received information I is perturbed as well. A basic premise of EPI is that the two perturbations are equal, so that the difference of the perturbations is zero. This implies that the perturbation of the difference $I - J = K$ is zero as well. This is another way of saying that K is at some extreme value,

$$K = I - J = \text{extremum.} \tag{9.8}$$

This is a principle of extreme physical (or, in these applications, phenomenal) information. The extreme value is attained through variation of the likelihood law $p(\mathbf{y}|a)$, subject to the relation (9.6) connecting I and J. In classical problems, such as this one, the extremum is a *minimum* in particular.

As an example taken up below, the ideal parameter a is the unknown time t at which a system of populations $n = 1, \ldots, N$ is sampled for a population member of random type n. The datum n can be of a person or a resource (say, a cubic meter of water) of type n. Hence the generic datum $\mathbf{y}$ is here n, and the likelihood function $p(\mathbf{y}|a)$ is a probability law $P(n|t)$ on n if t, which we denote as $p_n(t)$. This represents the "growth law" for population component n. The totality of such growth laws $p_n(t), n = 1, \ldots, N$ describes the overall system of peoples and resources, which can be a sociocultural one. *Our aim will be to compute these laws.*

It is important to note that since the time t is *general*, we are not limited in this approach to seeking equilibrium states of these probability laws. Equilibrium states are defined at the particular limiting time $t \to \infty$. Instead, the EPI solutions will be expressed as functions $p_n(t)$ of a general time value. Thus, they represent in general non-equilibrium solutions. Such functions of the time are also termed "dynamical" solutions, as in problems of Newtonian mechanics. EPI is eminently suited to finding such non-equilibrium solutions, having already done so in problems of statistical mechanics [46, 53–55], econophysics [56], and cancer growth [57]. See also Chapters 2–4 in this regard.

9.3.10. Knowledge Game

The EPI principle (9.8) has a useful interpretation. The solution $p_n(t)$ that accomplishes the extremum can represent the payoff of a mathematical game called the knowledge game. The game is played between the "observer" recipient of the information I and the "constructor" of the source effect of level J (which can only be nature). This constructor is personified as an "information demon," in analogy to the famous "Maxwell demon" of thermodynamics (Chapter 3). The basis for the game is supplied by the working hypothesis of EPI, that *the aim of observation is to learn*, i.e., to gain knowledge. In this knowledge game then, the aim of both players is to maintain a maximum level of information to sharpen knowledge. However, the amount I of information that the observer receives is purely at the expense of the demon's information level J (called a "zero-sum" game). As a result the observer tries to *gain* a maximum amount of information (I) while the demon tries to *pay out* a minimum amount (J). Hence information measures I and J act in opposition, that is, they are *enantiomers*. This forms a bridge between EPI theory and aspects of Sorokin's theory.

It should be mentioned that the association of EPI with the particular descriptive word "enantiomer" is *not necessary* to the theory that will follow. The informations *I* and *J* are well defined anyhow, as used throughout this book. Rather, the enantiomer label is a helpful device for learning and "picturing" this particular, sociocultural application of EPI. As a matter of fact, the knowledge game is another such device, although it does have practical uses (see *game corollary*).

9.4. Sociocultural Dynamics

Through his principle of immanent change, Sorokin ([58], vol. 4, p. 590) perceives that social groups with coherent cultures function as autonomous bodies. Sorokin ([58], vol. 4, pp. 600–601) also states that change in a socioculture occurs by virtue of its own internal forces and properties. It cannot help changing even if all external conditions are constant. Its decisions, we are told, are necessarily imputed to it, and this occurs *without* the benefit of conscious decision. One of the specific forms of this immanent generation of consequences is an incessant change of the system itself, due to its existence and activity.

Sociocultures therefore have dynamic systems that are constantly in a state of *flux*, generated proprietarily from within, and which would today be referred as *autonomous*. Thus cultures exist only as they are now because of their histories, inertia, and futures. The same can be said of *economics* (Chap. 2).

9.4.1. Cultural Driving Forces

One of the other key aspects of Sorokin's work is the identification that in a given human society there appears a degree of logico-reasoning underpinning many of its artifacts, laws, and institutions. These, together with structures and the outcomes of its behavior, may be termed the phenomenal manifestations of a given culture. Quoting Winston, Sorokin ([24], vol. 1, p. 5) states that culture should *not* be seen as "a simple mathematical addition of individual parts" but, rather, a system of interactive components.

He identifies ([24], vol. 1, p. 55) what we might call *collective cultural mentalities*. These derive from mind, value, and meaning. While they are based on individual participants in a culture, the predominant statistical consequence is a dominating cultural condition that defines the collective (or subcollective, e.g., a class or a sociocultural faction). However, it is normally the case that a socioculture maintains a plurality of cultures. Where there are common elements that can be identified within that socioculture, these can be referred to as the "dominant" culture. This is often manifest by a ruling class.

Sorokin defines *cultural mentalities* as the elements of thought and meaning that lie at the base of any logically integrated system of culture, belonging to the realm of inner experience. These occur either in a coordinated form of unintegrated images, ideas, violations, feelings, and emotions, or in an organized form of systems of thought woven out of these elements of that inner experience. These cultural mentalities are dispositions that "characterize" phenomenal manifestations, and provide constant internal forces of social dynamics. These dynamics are sensate and ideational mentalities that oppose and may balance each other, i.e., are enantiomers.

So, while a society can be perceived to be in a constant flow of change, shaped, directed, or characterized by different cultural dispositions, it can also be seen in terms of social and cultural dynamics that are constituted by differing polar cultural forms.

9.4.2. Sensate and Ideational Aspects

The sociocultural dynamics that develop are a consequence of the shifting relationship between the cultural sensate and ideational macroscopic variables. Sorokin ([24], vol. 3, pp. 511–512) notes that the relationship between these two macroscopic variables characterizes the dominant culture and the conduct of the people that live in it. The relationship between the dominant culture and the behavior of its bearers is not always close, but it does exist.

Following Sorokin ([24], vol. 1, p. 70), the dynamics of human society derive from the characteristics of the polar variables, and these may be seen to include ontological and epistemological attributes. The *ontological attributes* consist of the cultural perceptions (by those who constitute it) of the nature of reality. Ontologically, belief within sensate disposition allows realities to be deemed to exist only if they can be sensorially perceived. Such belief does not seek or believe in a supersensory reality, and it is agnostic toward the world beyond any current sensory capacity of perception. Its needs and aims are mainly physical, that is, that which primarily satisfies the sense organs.

By comparison, the *epistemological attributes* include the nature of the needs and ends to be satisfied, the degree of strength in pursuit of those needs, and the methods of satisfaction. The means of satisfaction occur not through adaptation or modification of human beings, but through the exploitation of the external world. It is thus practically orientated, with emphasis on human external needs. With reality as perceived from the senses, it also views reality through what can be measured and observed, rather than reasoned. Sorokin identifies the degree of strength in pursuit of these needs as "maximum."

Ontological, ideational disposition sees reality as nonsensate and nonmaterial. By comparison, epistemological needs and ends are mainly spiritual, and internal rather than external. The method of fulfillment or realization is self-imposed minimization or elimination of most physical needs, to promote the greater development of the human being as a Being. *Spiritual needs* are thus at the forefront of this disposition's aims rather than human physical needs. As with sensate disposition, the degree of the strength in pursuit of these needs is also a "maximum."

The two mentalities are enantiomer contradictions that are antagonistic to each other and hold different priorities, aims, and needs for human society. They also come to reflect the didactic of the cultural evolutionary process that encourages social complexification[16].

Since each has a maximum degree of desire to pursue their aims, worldview holders that maintain these polar mentalities do not compromise with each other. Rather, they engage in conflict. Sorokin notes that in a society where they coexist they create "latent antagonism" that can flare "up into open war" ([58], vol. 1, p. 75). Such "open war" would presumably be one of the non-equilibrium solutions that the EPI principle would produce.

9.5.   The Paradigm of Sociohistory

9.5.1.   The Propositional Base Through SVS

It is essential at this point to consider a way of explaining how joint alliances are able to form through the interaction of a pair of enantiomers. An enantiomer within the Sorokin context is a cultural orientation that may be thought of as an individual, and collective, disembodied existential, and cognitive construct that operates as a social force. As such it influences patterns of thought and behavior. We will therefore consider it as an autonomous, dispersed[17] social agent that has the potential to operate as a viable system. It is disembodied because it is not normally possible to associate it with a single named structured social organization that constitutes that construct, even though there may be individual organizations with a given orientation that constitute it. This is because it is constituted as what we shall call a *dispersed collective agent*, having the capability of spontaneously establishing local social (phenomenal) agents with that particular cultural orientation. Some of these agents may rise to bid for social power and the control of the social collective.

9.5.2.   Dispersed Agents

A dispersed agent is composed of a plurality of individuals, who may be interconnected by communication that is either indirect (e.g., passively sequential through books) or direct (e.g., dynamically interactive). It has an existential domain where beliefs (including beliefs about behavioral norms) and values exist. Behavioral norms are usually more or less adhered to by members of a cultural orientation due to a shared history. From this we can conceive of an implicit social structure that limits the individual's potential for behavior. It may be expressed, for instance, as a moral code that may or may not be enforced by law. Orientational beliefs can

---

[16] As a point of information, Fisher *I* measures the degree of complexity of a system. Hence the Fisher *I* of the overall sociocultural system would rise at this point of increased "social complexification."

[17] The dispersed agent in the context of Sorokin's theory is deemed to exist either (a) in an ideational world because it is an essence that can be manifested in its ideate or (b) in a sensate world if it can be identified phenomenally and measured. Other dispersed agents may also exist in other contexts.

also limit the ideate content of the noumenal domain, this being populated by ideate images or systems of thought that may be maintained by constructed information.

An enantiomer cultural disposition may be thought of as an autonomous, virtual[18] social agent that has the potential to operate as a viable system. It is disembodied because it is not normally possible to associate it with a single named structured social organization that constitutes that construct, even though there may be individual organizations with a given disposition that constitute it. This is because it is constituted as a *dispersed collective*, having the capability of spontaneously establishing local social organizations of that particular cultural disposition, some of which may rise to bid for social power and the control of the social collective.

The *dispersed agent* has the potential to formulate an autonomous holon, as illustrated in Figure 9.1. It is autopoietic because it is able to self-produce phenomenally its own components (like patterns of communications or behavior). This is according to its local proprietary orientational principles (autogenesis), and through a distributed network of processes.

9.5.3. Ideational vs Sensate Dispersed Agents

The distinction between ideational and sensate dispersed agents is that in the former the cultural disposition has values that are grounded in *noumenal* domains, i.e., virtual ideational aspects like ideas and principles; while in the latter the values are grounded in phenomenal domains, i.e., observables or the sensate. Thus, sensate and ideational concepts derive from different ontological domains that constitute a fraction of the existential domain. This difference is illustrated in Figure 9.3, using the constructions of SVS.

Further, sensate disposition is concerned with survival, is connected with external relationships, and tends to be concerned with the pathologies of *Doing* (e.g., how can we improve the survivability of a particular organization). This is in contrast to ideational disposition that is connected to the generation of ideas independent of immediate needs, to internal condition, and tends to be concerned with the pathologies of *Being* (e.g., how can we improve the likelihood of achieving enlightenment or nirvana[19]).

The manifestation of such a social collective develops a metasystem that arises from the dispersed collective, and a behavioral system that spontaneously arises. These have a basis in what we can refer to as a dispersed system. Information from this world is constructively acquired by them and is used to support and maintain ideate structures.

---

[18] The dispersed agent, once it is conceptualized, is deemed to exist either (a) in an ideational world because it is an essence that can be manifested in its ideate, and (b) in a sensate world if it can be identified phenomenally and measured. It should be realized that in this chapter we are attempting to create a social physics that can make the measurements for cultural enantiomers that a sensate disposition would require.

[19] Any place of complete mental bliss and delight and peace.

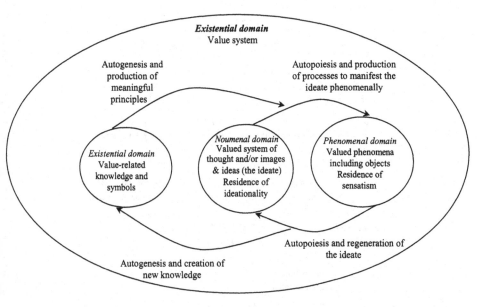

FIGURE 9.3. Basis that distinguishes ideational and sensate values.

9.5.4. The Dynamics of Viable Holons

To explain how such agents emerge systemically, we really have to explore the notion of the dispersed system a little more fully. A dispersed system provides a global cultural potential for the capacity to manifest local social collectives. The potential has a proprietary cultural orientation within which, under the right conditions, a set of local cultural singular identities[20] can be manifested. The singular identities adopt the ambient proprietary cultural orientation of the dispersed system, perhaps with some variations, the reason for which will become clear in a moment. They may also locally evolve independently and, therefore, maintain further differentiation from other such localities.

Processes of change and complexification, as occur with the formation of agent factions, can be represented in different theoretical (but often relatable) ways. Two relatable approaches to dynamic evolutionary change can be commented on here. A popular one is related to the work of Kauffman [59], who proposes the idea of a "fitness landscape" in which "species" survive according to random processes. This approach appears to sit well with complexity theorists, and developments like the Bak–Sneppen model of evolutionary change [60] are of interest, used to test the punctuated equilibrium theory of Eldredge and Gould [61] and to demonstrate the processes of self-organizing criticality. To do this it defines two rules of species

---

[20] By the term singular identity we mean an identity that has a quality of being one of a kind, in which one entity is distinguishable from all others in a given context, or the quality or state of being of that singular identity: some character or quality of a thing by which it is distinguished from all, or from most, others.

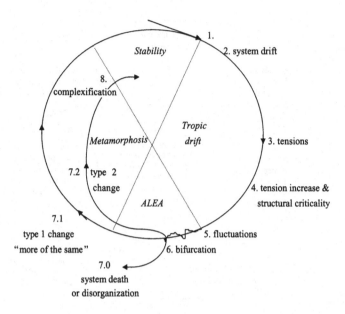

FIGURE 9.4a. The spiral of self-organization.

selection, and these have been used to simulate evolutionary processes[21] that appear to generate sensible outputs.

The other approach that we shall refer to, and indeed prefer, is connected with our ontological viable systems. Viable systems are those systems that are durable and sustainable, and survive internal and external environmental challenges through processes of change. Here, principles for spontaneous emergence can be described in terms of the model by Schwarz [3, 29, 33, 62, 63]. In the social context, a viable system can be portrayed as a local agent (or a social subagent or faction) within a global environment. The Schwarz model is shown as a spiral of self-organization in Figure 9.4a, and its elaboration across the three connected ontological domains is shown in Figure 9.4b, modified from Schwarz [33]. The phases of Figure 9.4a are explained in Table 9.2.

In the context of the cultural enantiomer that is the interest of this chapter, when we are referring to structural criticality we are referring to the structured patterns of knowledge or belief that become critical and thereby reach a bound in their stability. In practical terms, when such a structure becomes critical it will likely mean that the recognized value of the patterns of knowledge (or belief) being held by agents will be in doubt.

In terms of complexity theory, one explanation for the rise of new local agents is to argue that new local patterns of knowledge (or belief) form as an *attractor*, and it is from this that a local singular agent emerges. This new local agent will have knowledge that is connected to, but differentiated from, that of other singular

---

[21] These rules have been programmed into a computer program that can be downloaded from the Web site http://www.jmu.edu/geology/evolutionarysystems/programs/baksneppen.shtml, accessed January 2005.

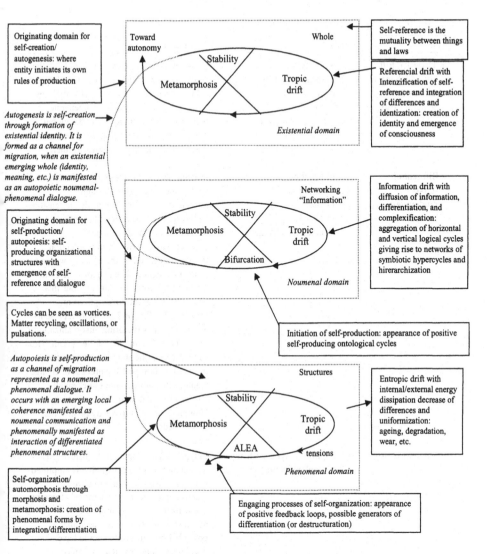

FIGURE 9.4b. Generic scheme for evolutionary change in self-organizing systems for complex systems, adapted from Schwarz [33] and developed in Schwarz [64].

agents that may emerge due to the process of *migration*[22]. The different agents would arise from the cultural enantiomer stock of knowledge, and this explains the idea of the rise of social factions.

---

[22] The term "migration" as coined by Yolles [13] means that if we can distinguish between a global and a local context, then when global characteristics are manifested locally their semantic morphology changes and takes on a local character. As an illustration, we can apply the idea to communication processes across a social collective/community. During the communication process, a global message is transmitted by a source and received by a local sink. However, the knowledge embedded in the message is understood in local (not global) terms, due to the differentiated patterns of knowledge that exist in that locality.

TABLE 9.2. The phases of self-organization.

Phase	Steps	Explanation
1. Disorder drift (of which tropic drift is the general case)	1. Stability 2. Spontaneous disordering drift 3. Tropic drift 4. Increase in tensions	Disorder leads to the more probable, to the actualization of potentialities. It is often the coherent actualization of the potentialities of the parts of the system that generate tensions and eventually break the global homeostatic or even autopoietic networks that hold all the social agents together.
2 Bifurcation (crisis, randomness, hazard)	5. Fluctuations 6. Bifurcation 7.0. Option 0: decay 7.1. Option 1: type 1 Watzlawick change	Fluctuations occur internally, or in the environment as noise. Through amplification of fluctuations due to tensions following entropic drift, a discontinuity occurs in the causal sequence of events/behavior. "Stochastic" selection occurs, influenced by the tensions within a problem situation. The tensions correlate to the amplification of the fluctuations that occur. At this point three options are possible: 7.0, 7.1, or 7.2. Decay represents a process of destructuring, disorganization, regression, or extinction of the system. This can be seen as the start of a catastrophe bifurcation. In type 1 the process of change begins with "more of the same" small changes that maintain it in current state. However, such changes may be in some way bounded.
3 Metamorphosis	7.2. Option 2: type 2 change 8. Complexification	In type 2 change, metamorphosis begins as a local morphogenic event that is amplified within a critical structure to have a macroscopic effect. In the critical structure a new form can arise initiated by the nonlinear condition. It is one of many possible bifurcations that could have developed. Complexification can occur during iteration of spiral. Autonomy may develop.
4 Stability	9. Dynamic stability	Occurs through self-regulation and/or existential self-reference.

9.5.5. Emergent States and EPI

Let us now explore the nature of these emergent agents, with particular reference to EPI. Since we are dealing with sensate and ideational dispersed agents, two classes of information consequently arise, denoted by I and J. These informations are housed in the noumenal domain (Figure 9.5). *Information I represents the*

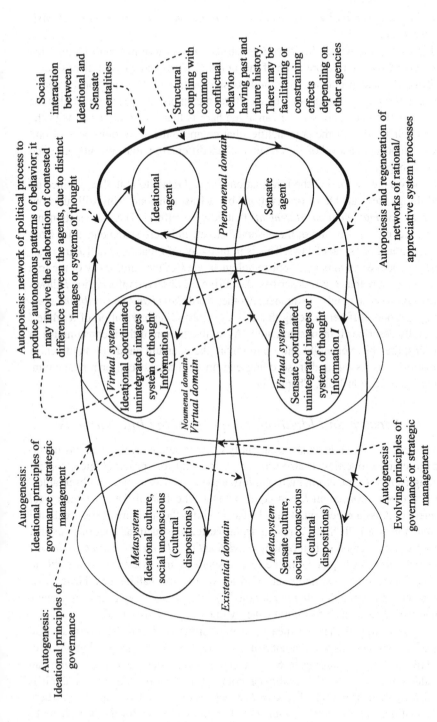

FIGURE 9.5. Interaction between distinct ideational and sensate mentalities and their social impacts.

acquired information by sensate culture, and information J the theoretical basis for the society.

Let us discuss the validity of this construction. The argument will require that we create a special interpretation of the nature of ideational cultural disposition.

It may be added that we have only considered the mechanisms for the development of a joint alliance, not why a cultural metamorphosis occurs such that the alliance is formed. The best we can argue here is that it is likely to do with the "social chemistry" of the personalities that compose the enantiomer agents, and this must occur at a level of SVS recursion that differs from that currently initiated in this chapter.

Now we are aware that the informations *I and J* relate to real effects, since all real things are "capable of observation" by at least someone. This still rules out as "real" the *mathematical statements* that describe effects. These are statements of the theoretical structure of the observed effect, and these statements trace from the ideational or "source" laws. This is in fact why informations *I* and *J* are relatable. *I* generates an *imperfect*, epistemological view of the complexity of the true source structure that *J fully* describes. Acquiring the observed information *I* allows the source to be learned or estimated. In other words *both I and J describe the same thing*, the only difference being that *J* is an idealized representation of it while *I* generally describes it inconsummately. As an illustration of the inconsummate in physics, the observer might choose to ignore the quantum nature of an effect, as previously discussed, and in so doing end up with a classical, or nonquantum, theory (see Section 1.5.3, Chapter 1).

9.5.6. Sensate and Ideational Aspects of the Informations

To review, a noumenon exists as an entity of information level *J*, and has intrinsic structural properties. It also provides a potential for epistemological content through a flow or transformation of information $J \rightarrow I$ to a space of observables. The received information *I*, consequently, is a reflection of the epistemological content of *J*. As a reflection, it may be used to infer the structure of the ideational noumenon, via EPI. Since *J* results from particular structural relationships that vary from noumenon to noumenon, *J is defined by EPI theory in a variable way* (see Section 1.5, Chapter 1). In the case of sociocultural dynamics we are able to define the variables as variable sets of ideate values. We shall define *I*, however, as the acquired information that is patterned by all sensatists, regardless of society. Therefore, *I* has a fixed form in the theory (a point previously made).

We have argued that the measures of information *I* and *J* for sensate and ideational cultural disposition constitute part of a single, noumenal domain, necessarily so since according to SVS theory it is only here that information can reside. Sensatism and ideationality cannot exist one without the other, and this is shown by their proportionality $\kappa = I/J$, where κ is finite (see Eq. (9.6)). These informations are universal measures of structure and order (or the lack thereof). Ideationality only includes all theoretical or ideational aspects. It does not include sensate. Sensate arises out of it, and as an expression of it.

There is therefore a relationship between ideationality and sensatism. Ideational cultural disposition is everything that defines ideal *rules* of being and activity for human kind. As such, there are at least two ways of interpreting this assignment. As the social collective through its cultural disposition attempts to carry through these rules, it can only do so inconsummately. The attempted carry-through is called sensate activity. Equivalently, but seen from a different perspective, *sensate cultural disposition represents an inconsummate attempt to carry through ideational rules.*

These dispositions are, however, disembodied. They are manifested through agents of cultural disposition as illustrated in Figure 9.5. These manifestations are expressed as a phenomenal social conflict process. For example, within the context of EPI, *J* represents the total level of structural information in the given system, with complete accuracy relative to the holon. This structural information exists as ideals and laws. People act in response to culturally transmitted ideas that are embedded in symbolic representations of structural information.

Ideational and sensate cultural dispositions are therefore enantiomic distinctions, i.e., between what people might aspire to (ideationality) and what they actually achieve (sensatism). There is in a sense a "conflict" between the two, in that they do not completely agree. Also, perhaps more to the point, people *purposely* do not carry through the ideational most of the time. This purposefulness might indicate the conflict that they seek. This is also sometimes termed "pragmatic" activity, as in a "white lie" which we "shouldn't" do (ideationally) but do anyhow on the grounds of a "better good" for everyone concerned. A white lie is an example of a small difference between the ideational and the sensate. When the difference becomes large enough, according to some interpretations this would indicate a kind *of conflict called a "sin."*

9.5.7. Role of Efficiency Constant κ

The degree to which the sensate aspect of a society agrees with its ideational foundation varies from one society to another. By Eq. (9.6), the agreement is measured by the size of the efficiency constant κ. The size of κ is determined by the attitudes and practices of the members of the society toward its laws and ideals. Thus, the value of κ varies from one society to another. A "traditionalist" or "law abiding" society has a κ close to 1, while a society that ignores its ideational foundations has a κ close to 0. The former is "conservative" and the latter is "liberal" in the usual senses of the words (although there are exceptions).

The natures of the sensate and ideational dispositions can vary absolutely or relatively. Let us consider the absolute first. In general, the enantiomer dispositions change in their levels of complexity. The interest is to identify the nature of that complexity, because this has an impact on the way the society behaves. Since *I* and *J* are measures of these dispositions, we wish to evaluate their changes relative only to themselves (we refer to this as absolute change). A low value of *I* implies a simple sensate society, and a high value of *J* implies a complex ideational society. We shall refer to this combination of conditions as *primitive* in the sense of taking one of the simplest possible or extreme forms, including the fact that a near-zero sensate capability would imply an inability to cope well with complex change. It must be stressed that by using the word "primitive" in a social collective, we are adopting terms that do not imply value judgments about the nature of a society. The term is meant in the sense of taking one of the simplest possible or extreme forms. As we mentioned, a near-zero sensate capability I would imply *an inability to cope well with complex change*. Also, ideational primitiveness (high J) suggests a society that is so bound up by complex ritu al that it dominates people's lives, either by its conspicuous absence (in atheists or agnostics) or by its conspicuous presence (in priests or zealots). Also, with $I \ll J$ the high complexity of the ideational rules are not being practiced on the sensate level.

We offer a caveat here, which is to emphasize that these are only qualitatively expected trends that appear reasonable at this time. These arguments, those given below, and indeed the entire theory to be developed, are as yet largely untested. We do not regard any of the results or interpretations as decisive. The aim is to initiate the theory, with the hope that it is reasonable and self-consistent, and will stimulate further inquiry into the subject along these lines.

9.5.8. Coefficients of Information

It is possible to make these trends more transparent by rescaling I and J as follows. Let us define "coefficients of information" $\mathcal{I}$, and $\mathcal{J}$, that are scaled versions of I and J, as $\mathcal{I} = 2I/(I + J)$ and $\mathcal{J} = (J - I)/(I + J)$. By the use of Eq. (9.6), these may be expressed in terms of the information efficiency constant κ as $\mathcal{I} = 2\kappa/(\kappa + 1)$ and $\mathcal{J} = (1 - \kappa)/(1 + \kappa)$. Note that, as coefficients, or ratios, of informations $\mathcal{I}$ and $\mathcal{J}$ have no units. Also, since κ lies on the interval $(0,1)$, both $\mathcal{I}$ and $\mathcal{J}$ likewise are positive numbers that lie in the interval $(0,1)$. These values of $\mathcal{I}$ and $\mathcal{J}$ are also complementary to each other in that $\mathcal{I} + \mathcal{J} = 1$, this complementarity highlighting their enantiomic relationship.

It may be noted that when I and J take their primitive values, $\mathcal{I}$ and $\mathcal{J}$ become close to their extreme values, 0 and 1, respectively. Further, I and J have units (reciprocals of variances), so that their values are always relative to a choice of units. By comparison, the unitless coefficients $\mathcal{I}$ and $\mathcal{J}$ have *absolute* values, i.e., values that are independent of choice of units. This allows $\mathcal{I}$ and $\mathcal{J}$ to be compared across *different* sociocultural holons[23]. This includes absolute comparisons of small-scale variations, or even periodic oscillations, *away from* primitive conditions in I and J. Such a potential for oscillation occurs, for instance, when large-scale enantiomer conflict progresses, or when one enantiomer succeeds in dominating and, in so doing, marginalizing the other.

While the unitless information coefficients $\mathcal{I}, \mathcal{J}, \mathcal{K}$ are of use, to keep notation to a minimum in this chapter they will not be used further. While their construction does provide general transparency, it will perhaps be clearer to continue with the basic absolute information quantities I, J, and K that we had already introduced.

---

[23] This is similar to the way that the Shannon information capacities of different kinds of systems may be compared.

9.5.9. Role of κ in Defining States of Society

Let us now consider changes in κ. First consider where κ is close to unity, so that I is very close to J. Here, in essence every sensory experience fully agrees with an existing ideational law or principle. Indeed the agreement is so good that the experience is merely a reexpression of an existing law or principle[24]. Thus, there are no fundamentally "new" experiences or ideas. The society stagnates and, ultimately, runs out of ideas.

Second, consider the other extreme when $I \ll J$. This is an ideational-dominated society, where κ is close to zero. The received information I is now very low, describing a noisy, chaotic holon. Here sensory experience randomly and widely diverges from the social norms of the ideational aspect. This might be manifest in a breakdown of morality, a high crime rate, etc. The society cannot function in accordance with its own rules. Consequently it will be too impractical to exist.

Thus *a society in which κ is close to zero or to unity is dominated either by ideational or sensate disposition, respectively. With a sensate disposition, the society will ignore its own principles, possibly suffering a breakdown of morality, etc. At the other extreme, an ideationally dominated society will become more and more impractical. It likely fixates on ritual rather than be responsive to internal or environmental pressures, eventually running out of ideas.*

In either case the society will start to stagnate and become "structurally critical," increasingly unable to cope with problems and crises that it will face (e.g., famine, infiltration, external war, inflation). In this increasingly critical state even small perturbations of the system can affect it in a major way.

An example of the shifting relative values of I and J can be suggested through a conjecture. In early Greek society the slave and merchant classes took care of the culture's operational and external needs, while the ruling class pursued its ideational needs. The conjecture is that as more slaves were acquired, the ruling class had more time to build on its ideology. Thus, the ideational level J of information went up. But also, since the slaves added degrees of freedom to the system, this increased the sensate level of Fisher information. Thus, both J and I went up in value. This is quite in agreement with the EPI Eq. (9.6), which states that I maintains proportionality to J over time. Taking an autopoietic view, ambient political or operative processes were consequently dominated by ideational disposition and benefited those pursuing an ideational future.

In summary, ideational or sensate *dominated* cultural disposition will fail to meet the needs of its members. This will lead to a loss of confidence by society in the direction that cultural disposition takes it. The debate and conflict will reopen, other mentalities will reassert themselves, and the chaotic state will return. Or, this might not happen, an existing dominant cultural disposition simply reasserting

---

[24] It may be noted that when this occurs in physical applications of EPI, it expresses a situation of "quantum entanglement" of realities.

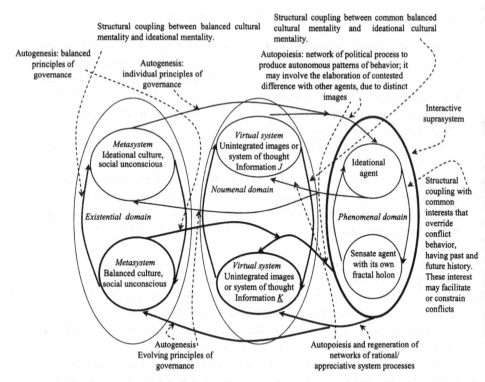

FIGURE 9.6. Relationship between ideational and/or sensate cultural dispositions and an emergent "alliance" or balanced culture.

itself. But in doing so, society will still remain structurally critical. Inevitably, it will change its disposition or the society will simply fail.

9.5.10. Emerging Balance Between Cultural Dispositions

The *political* dimension of dynamic sociocultural processes illustrated here is important. Political debate by members of a society constitutes an innate conflict on the value of these mentalities in its social and cultural development. The resultant is a pulling of society in many directions, which may become chaotic and unstable.

One of the outcomes of the innate conflict (and therefore the political processes that accompany them) can be a balanced cultural disposition, as the agents come together and an emergent joint alliance is formed (Figure 9.6). By this we are referring to Sorokin's *integral* notion, but broaden it so that it can develop a variable cultural disposition determined not only by the state of the enantiomers, but also by *the mix* that results between them. This notion of a mix is consistent with the development of phenomenal joint alliances in small-scale social collectives [15, 22, 23], and there is no apparent reason to argue that it cannot also be valid for large-scale societies.

In terms of EPI, a balance is always maintained between information *I* (measuring "survival ability") and information *J* (measuring "degree of structure" of the internal conditions of a socioculture). This continuous maintenance of balance in the face of change means that the theory is one of instantaneous, dynamic non-equilibrium. By comparison, equilibrium, if it is attained, is usually attained only at large time values.

9.6. Exploring the Dynamic of Cultural Disposition

The general solution to the EPI problem is a differential equation (Section 1.4, Chapter 1). This generally has multiple solutions. The size of I relative to that of J depends upon which solution state the system finds itself in. The so-called "ground state" solution is the smoothest, but higher-order or "excited state" solutions that are more erratic also exist. Each of these solutions incurs a different value of $I - J$. One solution in particular makes it an absolute minimum. This might be the ground state, but this would have to be verified. In any event, the size of $I - J$ can indicate the degree of a "Hegelian alliance."

Alternatively, regarding "catastrophes," these can be mathematically present in any solution, depending upon the problem. For example, in the application we cover below, one population ultimately "crashes," i.e., suffers a catastrophe (see also Chapter 8). Hence the approach allows for catastrophes, predicting them as particular trajectories.

9.6.1. Time Evolution of the Informations

Let us consider the population dynamics introduced in Eq. (9.2b). We should note that in general the word "population" has a generic meaning that may include volumes or densities of resources as well as populations of people. These are lumped together and arbitrarily called population types $n = 1, \ldots, N$. The *values* of the populations are denoted as the $p_n(t)$. These evolve over time t according to various non-equilibrium distributions. These distributions may change rapidly enough to more aptly be called "states." These states are characterized by attaining, at each instant of time, a total information I that is maximally close to information level J. The states are, by construction, stationary, but not, however, stable. Their total information level I, which changes with time and generally decreases according to the Second law Eq. (9.4), is generally *not at its absolute minimum values*. The latter is only approached as time increases. The absolute minimum would describe the long-term *stable state* toward which the entire society is moving. The stable state is, then, characterized by two conditions:

(a) Information level I maximally close to information level J (true at all times) and

(b) information levels I and J both at their absolute minimum values (true at large times), but not necessarily close to equal.

These conditions are discussed, in turn, next.

Condition (a) describes a level of sensate experiences that, at each time, are maximally close to satisfying the requirements set by the ideational principles of the ruling hierarchy. However, although the difference $I - J$ is minimized it is not yet necessarily small in absolute terms. In fact, there are multiple solutions to the problem $I - J =$ extremum, which are characterized by different attained values of the extremum. The situation is very much like that of ecological evolution, which obeys the same dynamical equations ([47], Chapter 8). Thus, for a given system, some solution differences $I - J$ are larger than others. Since each such solution represents a stationary solution, it also represents a locally stable (metastable) state in the evolution of the system. The system evolves from one such metastable state to another. It must be emphasized that, as in ecological evolution, this is not necessarily through ever-smaller minima. There could from time to time be a random transition to a state of larger minimum [65]. It is only the solution for the absolute minimum difference that is the absolute stablest state.

These trends suggest that knowledge of the curve $|K(t)| = |I(t) - J(t)|$ of information difference as a function of time might be useful as an indicator of future cataclysm for a society. Sudden rises in $K(t)$ would indicate changes toward *nonalliance* (and therefore toward instability). We later use the curve of $K(t)$ for this purpose in a simple example. In our application it turns out that $K(t) = -I(t)$ (Section 9.9.3) so that $|K(t)| = I(t)$. Predictive use of the curve $I(t)$ has previously been advanced for predicting environmental catastrophes [66, Chapter 7].

We can interpret these mathematical effects in sociocultural terms. Merely stationary solutions do not represent states of absolute stability since the information difference $I - J$ is not yet at its absolute minimum value. Also, as time progresses both information levels I and J tend to decrease, meaning, the complexities of both the ideational and sensate phenomena (structures and related behaviors) tend to decrease. This may occur gradually through the process of structural changes (since structure constrains and facilitates certain types of behavior), or morphogenically through emergence. The latter occurs, for instance, either when one culture suddenly dominates or when ideational and sensate cultures establish themselves into a new balance through a "joint alliance" that enables the formation of a new frame of reference that simplifies the way they are both seen. In fact we shall see such a shift as Figure 9.6 "emerges" from Figure 9.5 through transformation with the emergence of a new shared paradigm.

Also, randomly, transitions may be made to less stable systems, whereby the shared paradigm becomes less acceptable to the competing parties. A tenuous compromise is struck.

When sensate and ideational dispositions are each primitive, they take their simplest form, and therefore have minimum I and J values. Condition (b) (as above) describes the time at which both information levels have individually decreased to their minimal levels. They now describe a system that is both stationary and stable. The "simplest" form of living becomes dominant. This situation has a dramatic parallel in the growth of in situ cancers. Cancer is a life form with a minimal level of complexity, since it has lost the ability to function normally while maintaining the ability to reproduce. Hence its information level I is at an absolute minimum value for a living thing. As a verification of the latter, using it in an EPI approach for defining cancer growth gives the clinically correct law of growth [57, Chapter 3].

By comparison, a condition (b) society (as above), where the sensate information level I departs appreciably from the ideational level J, describes an unstable system where the potential for open warfare between different adherents becomes a recognized and feared potential. To use Sorokin's words, such a society is eclectic, self-contradictory, and poorly integrated logically. Their ideational and sensate elements remain adjacent and mechanically coexist, but without achieving genuine inner synthesis [58].

9.6.2. Consequences of Enantiomer Imbalance

One of the consequences of the imbalance between these enantiomer entities is that society moves in either an ideational or sensate direction. That is, one side or the other gains political power, and seeks to create stability and order by integrating the whole of society's external manifestations into its own aims, goals, and values. This constitutes Sorokin's conceptualization that the logical reasoning that underpins culture is integrative, representing an active dynamic force that creates dynamic stability as it achieves its goals. Ultimately, dynamic stability concerns structural phenomenal processes and the relationships of their associated behaviors.

In structurally *stable* societies the ever-existent potential for enantiomer conflict is suppressed by denying the debate between the cultural mentalities, thereby giving the impression of order and coherence. In this way the overt manifestation of violent conflict is inhibited.

On the other hand, a society may be dynamically stable while at the same time becoming increasingly structurally *unstable*. As a dominant cultural disposition evolves it continues to redefine, revalue, and reorder social structures and ideas by its worldview holders, but stable structures have to meet the needs of the society that they serve; otherwise they become unstable. The promotion of one enantiomer over the other will in the end be detrimental to the development of society and its needs. In such situations cultural disposition marginalization occurs [15]. As a result, one enantiomer faction or the other may find ways of expressing its distress phenomenally through violence.

9.6.3. A Quantification Using Information Parameter $\underline{K}$

In terms of EPI, the possible balance that can occur in a sociocultural situation in which an ideational/sensate cultural disposition arises is defined by $\underline{K} = \min|I - J|$, where $||$ denotes an absolute value and $\underline{K}$ is K at its positive extreme *minimum* value. Since the value of K indicates the *current condition of balance* of the culture, when K does not attain value $\underline{K}$ the conflict between enantiomers is maintained and no balance emerges.

When the minimum value $\underline{K}$ of K is attained, the indication is that either one cultural disposition has achieved complete domination over the other, or they have formed an "alliance," i.e., a balanced cultural disposition.

We have already considered the former case when one enantiomer disposition dominates, and have noted that the socioculture may not be durable because of its inherent inability to deal with either change or because of its impracticability. Let us now briefly discuss the latter alternative, and in the context of a third cultural disposition—the idealistic (Sorokin). This can be seen as balance of the two other dispositions working together. It is called idealistic because society requires a new political power base that is able to drive autopoietic processes that can manifest virtual images and logical processes phenomenally. The only way in which this new cultural disposition can arise, however, is as an emergent third cultural group. This is difficult to achieve because it involves new forms of proprietary enantiomer governance and involves mutual political processes across the two cultural dispositions. Thus it requires the dominant cultural dispositions to relinquish their existing power.

In corporate alliance theory, this is an essential component if joint alliances are to have a chance at surviving. As an illustration of this in a small-scale situation, the breakup of the US telecommunications group ATT (American Telegraph and Telephone) and the privatization of BT (British Telecom) stimulated an alliance process [67] that proved itself unstable. The joint alliance that formed was called Concert and resulted in failure after two years of operation at a cost of US$800 million annually before it shut down in 2001. One would typically associate this failure with the alliance's inability to develop an autonomous third culture with a proprietary disposition. The companies did not understand that the only way the alliance could survive was to develop a new autonomous alliance culture.

We return to the cultural disposition *alliance* state, quantifying it in terms of informations I and J. In the alliance formulation K gets extremized at its minimum value $\underline{K}$. This represents a minimum in the discrepancy between ideals and actions, or ideationality and sensatism. In other words, as with most real things, a compromise is struck. This could be seen to have occurred in the Western industrial revolution that began in the seventeenth century, when ideationality provided the ideate conceptualizations, and sensatism provided the capacity to manifest those ideas for the benefit of material returns. For two enantiomic agents to work well together, a *minimization principle* needs to be operative, and it is this that enables the partners in an alliance to structurally couple their ideates. A further requirement, of course, is for the worldviews of the enantiomic agents to be structurally coupled, since this provides for the principles and governance of the joint alliance. All these things are found by minimizing a difference of measures of the holon, which reflect the nature of the relationship between the yin–yang enantiomers. There is precedent in the physical sciences for what we are doing. As examples: in electromagnetism this compromise exists between electrical energy and magnetic energy; or in classical mechanics it is between mean potential and kinetic energies.

Since K can represent any balance that materializes, and I is of fixed form, information J may be used to provide a basis for learning the "representation of the current state of the culture". By the usual EPI model, the data are acquired by an observer who has no prior knowledge about the unknown structure. All he/she has to go on are the data and an idea consisting of an invariance principle that the structure obeys. Using these in the epistemological mechanism provided by EPI allows one to solve for the unknown structure. Thus, EPI is epistemological in nature, learning as it does about the ideational aspects of a sociocultural group.

9.7. The Sociocultural Propositions of EPI

EPI provides a *physical* description of a system. Therefore to use EPI, as in Section 9.8, requires a brief introduction to some necessary physical concepts.

It is assumed that the society is adequately described by the frequency of occurrence values $p_n(t)$ of its major population components $n = 1, \ldots, N$. Components n include the various peoples of the society, labeled in some arbitrary way, and also its capital resources. For purposes of a practical analysis, a minimal number N of such resources is chosen, specifically, those resources that have a significant effect upon the populations. Examples are arable land, retrievable oil, forests, and gold in reserve. With the components n so defined, an occurrence value $p_n(t)$ is defined to be the relative amount of substance of type n that is present in the society at the time t. This can be quantified as the number of (say) cubic meters of type n divided by the total number of cubic meters over all categories $n = 1, \ldots, N$. The p_n are then also probabilities, in the sense that if a cubic meter of the society (be it of a person, piece of land, sample of oil, etc.) is randomly sampled from it, it will of type n with probability p_n (the "law of large numbers": see [68]. The aim of the calculation is to find the time dependence of the probabilities $p_n(t)$.

Since we are only interested in establishing the occurrences $p_n(t)$ of the various population components, their internal spatial structures are ignored. In a sense, each component is regarded as a featureless "particle," as previously discussed.

9.8. An Illustrative Application of EPI to Sociocultural Dynamics

In this section we shall develop a general stochastic model of system change dynamics, using the principles of EPI. One illustration provides an analysis of population growth. A simple application of this is to consider the relationship between ideational and sensate populations. It presupposes that ideational or sensate cultural disposition in a society is determined by the relative *population size* of each enantiomer group, and there is a statistical relationship between any emergent cultural balance that may occur (and which is represented by $\underline{K}$) and the population sizes of each enantiomer represented by I and J. In a dynamically stable society it is the *balance*, where one develops, that becomes the dominant culture. Having said this, it may be naïve to consider that domination to be a statistical process.

Ideally, one might envisage that power is allocated to those who uphold the values and needs of the dominant culture. The naïve argument is that population dynamics can also establish the foundation for the dynamics of *political processes* from which power is assigned. This is especially when population is related to the capacity to mobilize power as, for example, occurs ideally in a democratic voting situation. But, however ideal such a proposition might be, there may be more complex aspects at work than this proposition would support.

First we can note that people who normally achieve consensual social power in a given culture do so when they have a personal cultural disposition (seen by those who support a given assignment) that is consistent with the dominant social culture. In other words "like appoints like," usually because this supports cultural inertia that contributes to the security for assignators. It is also likely to support stable political processes that contribute to social structural stability.

Second, it may be supposed that a balanced culture is determined statistically by enantiomer cultural components. However, it can also occur when members of a ruling class, who manages the political process, hold power independently of enantiomic population densities. This ruling class does not have to be maintained in their position by popular support that derives form processes of open semantic communication. Following Habermas [69] a society can be steered through other means (like the use of money or direct or indirect power) while suspending the meaningful or semantic communication processes. Military power acts as steering medium in that it encourages people to behave in certain ways with a penalty of physical or structural violence[25]. A form of indirect power can occur by creating powerful emotional incentives for people to agree.

Another medium of indirect power is the *mass media*. This appears to convince people of arguments (where they do not challenge their fundamental cultural values or beliefs) through emotive or other techniques. Whether the use of such steering media are able to enable the ruling class to maintain their power base for extended periods of time is not a question that will be considered here.

9.8.1. General Problem of Population Growth and Motion

Let a given society contain population components named $n = 1, \ldots, N$. At this point we refer the reader back to the defining concepts in Section 9.7. Our aim is to find the probabilities, or occurrence rates, $p_n(t)$ of these components. Note that

$$p_n(t) \equiv p(n|t), \tag{9.9}$$

where the vertical line means "if." That is, by definition $p_n(t)$ is the probability of randomly selecting from the society a population member type n, *if* the time is t. These $p_n(t)$ thereby define the dynamical evolution of the society, and our aim is to find them.

We will first work on a more general problem. This is to find the spatial motion and growth of each population component n. Hence, we first work with the general

---

[25] By structural violence we mean the constraints imposed by one's social structure, a concept originally proposed by Galtung [70]. This operates by limiting the development of the potential of an individual by not enabling access to the necessary resources through prejudicial or biased social structures.

probability $p_n(x, t)$ of the two-dimensional surface *position* x $= (x, y)$ of the nth population type at the time t. This by definition arises out of a generally complex probability amplitude function $\psi_n(x, t)$, as the amplitude times its complex conjugate,

$$p_n(x, t) = \psi_n(x, t)\psi_n^*(x, t). \tag{9.10}$$

By elementary probability theory [68], this probability relates to the ones we want, $p_n(t)$, as

$$p_n(t) = p(n|t) = \int dx\, p_n(x|t) \tag{9.11}$$

where

$$p_n(x|t) = p_n(x, t)/p_T(t), \quad p_T(t) = U(0, T). \tag{9.12}$$

Equation (9.11) states that the probability of finding the nth population component at a time t is its probability of being *anywhere* over space x $= (x, y)$ at that time. The first Eq. (9.12) is by definition of $p_n(x|t)$, and the second states that the a priori probability of a time value is uniform U over the total fixed time interval $(0, T)$.

We first establish the dynamics of the $p_n(x, t)$, and then use Eqs. (9.11) and (9.12) to get the dynamics of the desired $p_n(t)$.

Societies are very complex systems, containing a large number N of interacting "populations" (in the generalized sense above). Among these interactions, some are strong and some are weak. In order to keep the calculation of the dynamics tractable, *the dynamics are assumed to be defined to a good approximation by only those populations that strongly interact. This defines a smallest number N of effectively interacting populations.* Thus, the derived dynamics will only *describe* this smallest set of populations. Also, these dynamics are necessarily approximate, to the extent that the effects of other, more minor, contributors have been ignored. Finally, the size N of the number of effectively interacting populations is a function of the complexity[26] of the interactions as well as their individual "strengths."

9.8.2. Population Growth and Depletion Coefficients

Let us suppose that the dynamics are driven by N *change coefficients*, denoted as g_n and $d_n, n = 1, \ldots, N$. These describe, respectively, growth and depletion as a

---

[26] While we can talk of the number N indicating the complexity of a situation, we should note that according to Yolles [3] at least five types of complexity can be identified. These are (1) *computational complexity* is defined in terms of the (large) number of interactive parts, (2) *technical complexity* (also referred to as cybernetic complexity) occurs when a situation has a "tangle" of control processes that are difficult to discern because they are numerous and highly interactive. It also involves the notion of future and thus predictability, and technically complex situations have limited predictability, (3) *organizational complexity* is defined by the rules that guide the interactions between a set of identifiable parts, or specifying the attributes, (4) *personal complexity* is defined by the subjective view of a situation, and (5) *emotional complexity* occurs when a tangle of emotional vectors are projected into a situation by its participants (and can be seen as emotional involvement).

function of time t. These change coefficients are assumed to be *known* functions

$$g_n(p_1, \ldots, p_N; t) \quad \text{and} \quad d_n(p_1, \ldots, p_N; t) \tag{9.13a}$$

of the probabilities and of the time. Being a "growth" coefficient, g_n is positive, and likewise the "depletion" d_n must be negative,

$$g_n \geq 0 \quad \text{and} \quad d_n \leq 0. \tag{9.13b}$$

As examples of growth dependencies, the growth g_4 of population $n = 4$ could depend upon the level p_5 of population 5, as in the case where p_4 represents the relative number of fishermen in a developing society and p_5 represents the relative number of fish that is currently available to them. The more fish, the higher the *growth* of fishermen, so that g_4 would grow with p_5. The probabilities in Eq. (9.12) can (as indicated) depend upon t or, even, upon t at previous or future times, thereby exhibiting "memory" or "anticipation," a concept discussed by Dubois [71] and Yolles and Dubois [9].

9.8.3. EPI Solution

As shown in Chapter 5, the EPI–SWE formulation gives the equations of population growth as

$$\frac{dp_n}{dt} = (g_n + d_n)p_n, \quad p_n = p_n(t). \tag{9.14}$$

(See Eqs. 5.14 a, b or 5.32) This is a generalized Lotka–Volterra growth law (or a "replicator equation"), for properly constructed coefficients g_n, d_n in terms of the probabilities p_n. Planck's constant $\hbar$ has dropped out, as it should have since L–V growth is classical. The L–V law often well-describes biological systems [72]. Hence, the EPI approach predicts that *sociocultural* systems obey generalized L–V growth as well.

Mathematically, Eq. (9.14) is formally a simple, first-order differential equation. Such an equation can often be solved analytically, and is always soluble by numerical finite differences (although in many cases the solution is unstable). Regardless of the chosen approach to solution, the latter must always obey a condition of normalization,

$$\sum_{n=1}^{N} p_n(t) = 1 = \text{const.} \tag{9.15}$$

This can often be used as a check on a solution.

9.8.4. Simple Illustrative Example

The following is a simple solution to Eq. (9.14) that models a common class of sociocultural problems. Suppose that there are effectively only $N = 2$ populations competing. In this scenario it is simplest to express the change coefficients

g_n, d_n in terms of the usual "fitness" coefficients w_n of genetic population theory as

$$g_n = w_n, d_n = -\langle w \rangle = -(w_1 p_1 + w_2 p_2), \quad n = 1, 2, \quad \text{where } w_1 \geq w_2. \quad (9.16)$$

As a very simple case, we assume conditions to hold for which w_1, w_2 are approximately *constants* over time.

We have arbitrarily chosen population 1 to have the larger of the two change rates. Here the EPI solution to Eq. (9.14) is analytically known, as

$$p_1(t) = \frac{p_1(0)}{p_1(0) + p_2(0)\exp[-(w_1 - w_2)t]},$$

$$p_2(t) = \frac{p_2(0)}{p_2(0) + p_1(0)\exp[+(w_1 - w_2)t]}. \quad (9.17)$$

As a check, note that the left and right sides of both equations balance at $t = 0$, after normalization Eq. (9.15) is used. Note that equations (9.17) are a unique solution. Hence, the minimum that is achieved by the difference $I - J = K$ is here the absolute minimum as well, $K = \underline{K}$. However, as we shall see, this does not define an alliance minimum but, rather, a *domination-type* minimum.

The growth equations (9.17) show monotonic behavior with time, $p_1(t)$ ever growing and $p_2(t)$ ever falling. The monotonic growth is persistent. As long as w_1 exceeds w_2 by any amount, $p_1(t)$ eventually overtakes $p_2(t)$ and eventually achieves complete dominance ($p_1 = 1$, $p_2 = 0$) at $t = \infty$. Thus, when two species compete for the same resources one must eventually *dominate completely* over the other. This is called "Gause's law" in ecology. The law was recently used [73] to quantify the scenario of some 20,000 years ago when Cro-Magnon man wiped out Neanderthal man, apparently by this effect.

An example is shown in Figure 9.7. Here the population parameters are arbitrary growth rates $w_1 = 0.06$ (6% growth rate), $w_2 = 0.03$, and initial levels $p_1(0) = 0.25$ and $p_2(0) = 0.75$. The two solid curves of Figure 9.7 show that, although population 2 is initially twice that of population 1, it steadily falls toward zero while population 1 heads toward 1. Hence, population 1 eventually dominates completely. The sensate (and alliance) information curve $I(t)$ is also shown (dashed). The sociocultural significance of these trends is discussed in Section 9.9.

The theoretical time t_r at which p_1 attains any multiple $r \geq 1$ of p_2 is, from Eqs. (9.17),

$$t_r = \frac{1}{w_1 - w_2} \log\left[r\frac{p_2(0)}{p_1(0)}\right]. \quad (9.18)$$

For the population parameters of Figure 9.7, a case $r = 1$ of population equality occurs at a time of 36.6 years, as is confirmed by the curves.

These results for a lowest-order case $N = 2$ of populations in competition show that the effects of *constant* growth rates are *monotonic*. Similar trends are shown for a higher-order case $N = 3$ as shown in Figure 1 of [74]. These trends can only be reversed if the change coefficients *vary* appropriately with the time. Indeed,

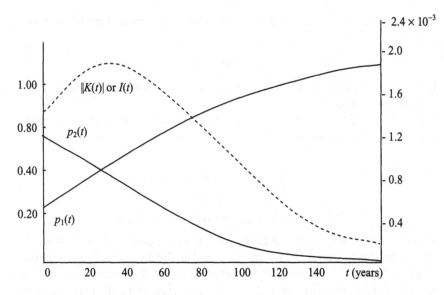

FIGURE 9.7. The two relative populations $p_1(t)$, $p_2(t)$ (scale values on far left) as a function of time t, and information $I(t)$ (scale values on far right) as a function of time in years.

if population 2 knew that its demographics were weakening as in Figure 9.7, it would probably take steps to accomplish such changes. This would entail not only increasing its change rate but also decreasing that of its competitor. This of course is a classic example where conflicts arise out of demographics, as population 1 would tend to resist such changes as detrimental to *its* future.

9.9. A Case Illustration: Postcolonial Iran

The preceding concepts, of the sensate and ideational dispositions, informations I and J as enantiomers, the significance of an alliance state, etc., can be applied to modern examples in the developing world. The reasoning is for the most part descriptive, qualitative, and open to debate. The aim is not to reach definitive conclusions but to show how real events can be fit within the framework. In doing this we provide an illustration of the possibilities of our approach. We have selected postcolonial Iran because its developmental dynamics is arguably seen as a starting point for current challenging international problems, among them terrorism.

9.9.1. Sensate vs Ideational Mindsets

The Western sensate dispositions adopted by the ruling classes of many post-colonial countries have failed in meeting the needs of much of their populations. As a result, these populations lose confidence in the ruling classes, and seek to empower the adherents of rival ideational dispositions. Often these are religious fundamentalists. Iran may be regarded as a good example here. The Iranian revolution placed in power those who adhered to an ideational disposition, creating stability and order. Adherents of sensate disposition were still there, but they now needed to realize their activities within an *ideational* framework. In the global economy [75], this movement spread to other postcolonial countries that were realizing increased internal support for ideational fundamentalism. Their posturing against Western sensatism manifested a relatively small faction of extreme fundamentalists who supported overt enantiomer conflict. To pursue this they adopted what they saw as their only global means, terrorism, which had been used successfully by Gavrilo Princip by 1914 ([76], p. 206) and others since then. In postcolonial countries, such instrumental use of violence often aggravates the crisis, in exacerbating a growing corresponding loss of confidence in the dominant cultural disposition. These push populations more toward empowering extreme adherents of opposing dispositions. Violence is not a product of cultural conflict but, rather, a promoter. *The more that violence is perceived as a crisis, rather than as one possible development that comes from cultural enantiomer conflict, the more likely it is that power can be seized by extreme adherents of one cultural disposition or another.*

In contrast to this ideational movement, Western society pursues its self-development in the global society that it dominates commercially, and its spiritual values and ideas are usually promoted, and valued, in terms of how they benefit the dominant sensate values.

A shift from the sensate to the ideational cultural mindset in postcolonial Iran occurred as the Shah (Iran's hereditary monarch), instituted as an eligible head of state from their perspective, did not provide the leadership appropriate to his population.

9.9.2. Cultural Instability

The insurgence of ideational cultural disposition resulted in a period of cultural instability. The impact of this instability has affected the autopoietic process, destabilizing the way in which phenomenal events are manifested. The outcome of this is that small random events can be exploded into significant ones. There is an argument that the emergence of Bin Laden, like the rise of Hitler before the Second World War, was one such event. The stable *maintenance* of such phenomena then becomes a function of the behavior of those in the social environment.

The change represents a sudden shift in direction, that sees society reordered and redirected in a time described in history as more revolution than evolution. It can be added that the adherents of the dominant disposition owe their position of power and status to it, which will ensure that there is a resistance to change, which tends to make that change more dramatic, potentially more violent, and with the need to be more fundamental when made. Equally, that shift results from a disposition's structured society increasingly unable to meet a major crisis, which will give urgency for the shift to be sudden rather than progressive. The catastrophic shift is accompanied during a period of chaos, of power struggles between collective mentalities that is maintained, until adherents of one or the other disposition gain dominance.

We can explain this formally. We use the present tense to simplify the language, and also because of parallels to today's conflicts in the region. The Iranian religious leaders, the ayatollahs, represent ideational disposition in Iran. Their ability to establish a stable alternative to the sensate disposition represented by the Shahites can only be attained if the population actually practices the ideational principles set by the religious rulers (through belief as opposed to coercion). Stability is not possible when the information difference $K = I - J$ is not yet at its absolute minimum value (i.e., $K \neq \underline{K} = \min|K|$). As time progresses, both information levels I and J tend to decrease, meaning, the complexities of both the ideational and sensate phenomenal structured behaviors tend to decrease. This may occur gradually through the process of structural changes, since structure constrains and facilitates certain types of behavior, or suddenly through morphogenic emergence that occurs (for instance) either when one culture suddenly *dominates* (an indication of this is given in Figure 9.7).

Or, instead, evolution toward a joint alliance can occur if ideational and sensate cultures evolve toward a new balance that simplifies the way both are seen. (We can see such a shift as Figure 9.6 emerges from Figure 9.5 through the emergence of a new shared paradigm.)

9.9.3. Quantitative Growth Effects

The growth Eqs. (9.14) were shown to hold for populations of people (and resources). Hence, they may well describe what happened to the populations of Ayatollahites (call it the ideational population 1) and Shahites (sensate population 2) during the era of the Shah. Can we also use the solutions (9.17) and (9.18) to (9.14)? These required constant growth coefficients w_1 and w_2. The demise of the Shah was so rapid that the growth coefficients might well have been approximately constant over that time interval. Also, the growth coefficient w_1 undoubtedly exceeded w_2, and possibly much exceeded it, so that even if initially the two populations were equal, it would not take long before the Ayatollahites much outnumbered the Shahites (Figure 9.7). For example, if $w_1 - w_2$ were 0.1 (10% growth advantage per year for Ayatollahites), by Eq. (9.18) it would only take $t = 10$ years for the Ayatollahites to achieve a population advantage of 2.72 (as a convenient number) times that of the Shahites.

A discussion of general trends is illuminating. The dominant effect in Eq. (9.18) is the logarithmic one. This means that the time required to achieve a general multiplier r goes only as the logarithm of r, a very slow dependence. Conversely, this shows a fast dependence—exponential—of the advantage multiplier r upon the time. For example, each subsequent multiplication by 2.72 of an already existing population advantage r only requires an added time of $1/(w_1 - w_2)$ years. This amounts to 10 years for the above case of $w_1 - w_2 = 10\%$ growth advantage/year. Or, 20 years give a population advantage multiplier of $2.72^2 = 7.4$,

30 years give $2.72^3 = 20$; etc. Of course these are moot points in the case of the Shah.

In other words, with constant growth rates present, one population rapidly achieves a state of complete dominance over the other. The effect is powerful in that it is independent of the initial population levels. If the increasingly recessive population resists the trend, this can result in war.

We discussed using the information alliance difference $K(t)$ as an indicator of impending danger for a system. In this problem there is a unique solution Eq. (9.17) so that $|K(t)| = \underline{K}(t)$, the minimized alliance difference. Also, if $\kappa = 1/2$, we have $I = (1/2)J$, so that $K = I - J = I - 2I = -I$, and consequently the absolute value $|K| = |I| = I$ since I is positive. We show the evolution of information $I(t)$ or $|K(t)|$ over time as the dashed curve in Figure 9.7. This shows a peak at about 36 years (when both populations are equal), after which the information monotonically falls toward zero. The curve indicates that 36 years is a critical time for the system. Before this time population 2 observes, probably in dismay, being overtaken by population 1. Turmoil might result as population 2 tries to reverse the trend. Beyond this time the system population 1 more and more dominates over 2. The contest is now over, and the system approaches absolute equilibrium, albeit without alliance or conciliation in this case.

9.9.4. Manifestations in Political Power and Dominance

Let us suppose that we are dealing with *power* as volume of human energy that is used viably through political processes. In other words, effort is put into a successful political process with the result that power is achieved. From this proposition, the population dynamics now represent a process of *power accumulation*. Interpreting this in terms of the EPI population results above, we can now say that where there are constant growth rates present in *the power* held by the two ideational factions, after a suitable amount of time (the order of t_r of Eq. (9.18)) *one faction rapidly achieves a state of complete dominance over the other*, and this is independent of their initial power levels. If resistance to this growth in power occurs, then conflict can result. This is more apt to occur before one faction catches up to the other, i.e., before a time of about 36 years in Figure 9.7. Beyond that time peace ensues since the contest is over and there is a decisive winner.

Thus, for general constant growth rates and initial population levels, the time t_1 as given at Eq. (9.18) defines a general state of parity of power. Beyond this time one faction becomes evermore powerful, indicating to all that the contest is over; consequently peace ensues. This scenario fits many former worldwide contests: The Ayatollahites vs Shahites, Axis powers vs Allies in WWII, USSR vs US, etc. Of course the peace we refer to may just mean physical nonviolence, as opposed to structural nonviolence from which social discontent may eventually follow.

A corollary of these numerical results is that if an alliance balance is to be struck, the growth rates *must change* with time. In particular, these changes must be ever in the direction of balancing the two populations. For example, if population 1 is currently tending to dominate its growth rate must be diminished relative to that of population 2. These rate changes can of course occur naturally or be imposed from the outside (the "third party" previously mentioned). If imposed, the art of achieving an alliance lies in continually adjusting them in the direction of Hegelian balance.

9.10. Overview

We have created a new paradigm of sociohistory, by linking EPI with SVS within the context of social change. It has been guided by the *ontological* specification of SVS and the *epistemological* imperatives of EPI. We refer to it as an emergent theory of sociohistory, whose aim is to provide a capability for explanation and prediction in social situations.

Part of this approach has involved the exploration of aspects of the sociocultural dynamics of Sorokin, and in doing so the need has emerged to elaborate on it in a way that follows the principles of both SVS and EPI theory. This theory also therefore enables us to use principles of joint alliance theory in SVS to explain the dynamics of yin–yang balance between cultural enantiomers and provides a capacity to explore this epistemologically through EPI. SVS has also illustrated a potential capacity to explore, through the notion of autopoiesis and with the facilities of EPI, the politics of sociocultural situations. This is of course something that Sorokin never considered.

The EPI-based analysis has accomplished a formulation of sociocultural dynamics. The three variables I, J, and $\underline{K}$, representing respectively sensate, ideational, and balanced cultural dispositions, fall naturally into the dynamic stochastic processes provided by EPI. New quantitative results have emerged, including the simultaneous description of both the evolutionary growth and dynamical motions of interacting populations. Application was made to a model, two-component problem.

The resulting predictions on population growth were found to represent simple, but self-evident, constructs. The fact that such self-evident propositions can be created from EPI demonstrates its innate capacity to produce sensible results from a set of basic propositions. It also shows the promise of the approach for future development as new forms of problem are tackled. This will be a learning process.

The importance of EPI within this context is not simply that it can produce self-evident propositions. It has also enabled us to more deeply question certain propositions in our understanding of Sorokin's sociocultural dynamics. It has shown in this regard that EPI provides *a framework* for measuring the nature of a culture under investigation. One elementary result, for instance, is in the examination of the emergence of a balanced culture when $K = \underline{K}$. Measurements could show whether this has emerged and complexity thereby reduced. This can be important not only within the context of large-scale sociocultures, but in small-scale ones too. Thus for instance we earlier referred to the failure of joint alliance ventures, and the use of EPI might well have been used in an attempt to quickly indicate whether a balanced or dominant culture had been formed, thereby determining the likelihood of imminent collapse. Other results from EPI allow one to quantify the level of complexity of the sociocultural system (see also Chapter 7).

We emphasize *that the ultimate purpose of the mathematical approach is limited to providing a general framework*, or paradigm, from which the logical projections of stochastic possibilities for a socioculture can be clearly identified. The trends that are so identified can also help an inquirer to pose sensible new questions about sociocultural dynamics. These can, in turn, provide a structural aid to the inquiry process. For example, in this way "What if...?" scenarios can be tested. The paradigm also acts in an unbiased manner, since it predicts likely outcomes without further partiality beyond the propositions created and their possible inconsummate interpretations.

References

Chapter 1

[1] G.A. Korn and T.M. Korn. *Mathematical Handbook for Scientists and Engineers, 2nd edn.* McGraw-Hill, New York, 1968.

[2] B.R. Frieden. *Probability, Statistical Optics and Data Testing, 3rd edn.* Springer-Verlag, Berlin, 2001.

[3] B.R. Frieden. *Science from Fisher Information, 2nd edn.* Cambridge University Press, Cambridge, UK, 2004. See also *Physics from Fisher Information, 1st edn.* Note that all references to [3] refer specifically to the 2nd edn. book, although some are also found in the 1st edn. Also, many are in the paper B.R. Frieden and B.H. Soffer. *Phys. Rev. E* **52**, 2274 (1995).

[4] D. Zwillinger. *Handbook of Differential Equations, 3rd edn.* Academic Press, Boston, MA, 1997, p. 120.

[5] H.L. Van Trees. *Detection, Estimation, and Modulation Theory, Part I.* Wiley, New York, 1968.

[6] A.J. Stam. *Some Mathematical Properties of Quantities of Information*, Ph.D. dissertation, Technical University of Delft, The Netherlands, 1959.

[7] B.R. Frieden. *Phys. Rev. A* **41**, 4265 (1990).

[8] M. Born and E. Wolf. *Principles of Optics.* Macmillan, New York, 1959.

[9] L.J. Savage. *Foundations of Statistics.* Dover, Englewood Cliffs, NJ, 1972, p. 236.

[10] L.H. Ryder. *Quantum Field Theory.* Cambridge University Press, Cambridge, UK, 1987, pp. 162–164, and 176–179.

[11] F. Wilczek. *Phys. Today* **53**,[1] 22 (2000).

[12] F.J. Yndurain. *Quantum Chromodynamics, 3rd edn.* Springer-Verlag, New York, 1999.

[13] J.A. Wheeler. In S. Kobayashi, H. Ezawa, Y. Murayama, and S. Nomura (eds.). *Proceedings of 3rd International Symposium on Foundations of Quantum Mechanics.* Tokyo, 1989, p. 354.

[14] J.A. Wheeler. In J.J. Halliwell, J. Perez-Mercader, and W.H. Zurek (eds.). *Physical Origins of Time Asymmetry.* Cambridge University Press, Cambridge, UK, 1994, pp. 1–29.

---

[1] This ansatz also fits in with the "knowledge game".

[15] I. Kant. *Critique of Pure Reason*, English translation of original German version of 1787. Willey New York, 1900.

[16] A.R. Plastino and A. Plastino. *Phys. Rev. E* **54**, 4423 (1996).

[17] E.R. Harrison. *Cosmology: The Science of the Universe, 2nd edn.* Cambridge University Press, Cambridge, UK, 2000.

[18] P.M. Morse and H. Feshbach. *Methods of Theoretical Physics*. McGraw-Hill, New York, 1953.

[19] B.R. Frieden, A. Plastino, and B.H. Soffer. *J. Theor. Biol.* **208**, 49 (2001).

[20] R.J. Hawkins and B.R. Frieden. *Phys. Lett. A* **322**, 126 (2004).

[21] R.A. Gatenby and B.R. Frieden. *Cancer Res.* **62**, 4675 (2002); *Mutat. Res.* **568**, 259 (2004); *Math. Biosci. Eng.* **2**, 43 (2005).

[22] M. Yolles and B.R. Frieden. *J. Organ. Transf. Soc. Change* **2**, 29–62; 63–77 (2005).

[23] B.R. Frieden and A. Plastino. *Phys. Lett. A* **278**, 299, (2001).

[24] B.R. Frieden, A. Plastino, A.R. Plastino, and B.H. Soffer. *Phys. Rev. E* **60**, 48 (2000); *Phys. Rev. E* **66**, 046128 (2002); S.P. Flego, B.R. Frieden, A. Plastino, A.R. Plastino, and B.H. Soffer. *Phys. Rev. E* **68**, 016105 (2003).

[25] B.R. Frieden and R.A. Gatenby. *Phys. Rev. E* **72**, 036101, 1–10 (2005).

[26] O. Morgenstern and J. von Neuman. *Theory of Games and Economic Behavior.* Princeton University Press, Princeton, NJ, 1947.

[27] K. Popper. *The Logic of Scientific Discovery.* Routledge, London, 2002.

[28] B.R. Frieden and A. Plastino. *Phys. Lett. A* **272**, 326 (2000).

[29] K. Shoulders and S. Shoulders. Charge clusters in action, In T. Valone (ed.), *Proceedings of the 1st International Conference on Future Energy*. Integrity Research Institute, Washington, D.C., 1999. ISBN 0-9641070-3-1; also see U.S. Patents by K.R. Shoulders: 5,018,180 (1991); 5,054,046 (1991); 5,054,047 (1991); 5,123,039 (1992); and 5,148,461 (1992).

[30] D. Bohm and B.J. Hiley. *The Undivided Universe.* Routledge, London, 1993.

[31] B.R. Frieden and B.H. Soffer. *Phys. Lett. A* **304**, 1 (2002).

[32] S.W. Hawking. The information paradox for black holes, In *17th International Conference on General Relativity and Gravitation*, Dublin, Ireland, July, 2004.

[33] M. Tegmark. *Class. Quantum Gravity* **14**, L69 (1997).

Chapter 2

[1]   J.E. Stiglitz. *Amer. Economist* **48**, 17–49 (2004).

[2]   J.E. Stiglitz. *Amer. Economist* **47**, 6–26 (2003).

[3]   J.E. Stiglitz. *Q. J. Econ.* **115**, 1441–1477 (2000).

[4]   S. Grossman. *J. Finance* **XXXI**, 573–585 (1976).

[5]   S. Grossman. *Rev. Economic Stud.* **64**, 431–449 (1977).

[6]   S. Grossman. *J. Economic Theory* **78**, 81–101 (1978).

[7]   S. Grossman and J.E. Stigltz. *Amer. Econ. Rev.* **LXVI**, 246–253 (1976).

[8]   S. Grossman and J.E. Stigltz. *Amer. Econ. Rev.* **LXX**, 393–408 (1980).

[9]   B.R. Frieden. *Science from Fisher Information: A Unification.* Cambridge University Press, Cambridge, UK, 2004.

[10] L. Bachelier. Theory of speculation. In P.H. Cootner (ed.), *Random Character of Stock Market Prices*. MIT, Cambridge, MA, 1964, pp. 17–78. English translation of Théorie de la Spéculation (Thesis), *Annales Scientifiques de l'École Normale Superieure* **III-17**, 21–86 (1900).

[11] M.F.M. Osborne. *Operations Res.* **7**, 145–173 (1959).

[12] P.H. Cootner (ed.). *Random Character of Stock Market Prices.* MIT, Cambridge, MA, 1964.

[13] F. Black and M.S. Scholes. *J. Polit. Economy* **81**, 637–654 (1973).

[14] R.C. Merton. *Bell J. Econ. Manage. Sci.* **4**, 141–183 (1974).

[15] R.N. Mantegna and H.E. Stanley. *An Introduction to Econophysics: Correlations and Complexity in Finance.* Cambridge University Press, Cambridge, UK, 2000.

[16] J.P. Bouchaud and M. Potters. *Theory of Financial Risks: From Statistical Mechanics to Risk Management.* Cambridge University Press, Cambridge, UK, 2000.

[17] J. Voit. *The Statistical Mechanics of Financial Markets.* Springer-Verlag, New York, 2001.

[18] J.Y. Campbell, A.W. Lo, and A.C. MacKinley. *The Econometrics of Financial Markets, 2nd edn.* Princeton University Press, Princeton, NJ, 1997.

[19] C.E. Shannon. *Bell Syst. Tech. J.* **27**, 379–423 (1948).

[20] E.T. Jaynes. *IEEE Trans. Syst. Sci. Cybern.* **SSC-4**, 227–241 (1968).

[21] R.J. Hawkins, M. Rubinstein, and G.J. Daniell. Reconstruction of the probability density implicit in option prices from incomplete and noisy data. In K.M. Hanson and R.N. Silver (eds.), *Maximum Entropy and Bayesian Methods*, Volume 79: *Fundamental Theories of Physics.* Kluwer Academic, Dordrecht, The Netherlands, 1996, pp. 1–80.

[22] R.J. Hawkins. Maximum entropy and derivative securities. In T.B. Fomby and R.C. Hill (eds.), *Applying Maximum Entropy to Econometric Problems*, Volume 12: *Advances in Econometrics.* JAI Press Inc., Greenwich, CT, 1997, pp. 277–301.

[23] E. Maasoumi. *Econometric Rev.* **12**, 137–181 (1993).

[24] J.K. Sengupta. *Econometrics of Information and Efficiency.* Kluwer Academic, Dordrecht, The Netherlands, 1993.

[25] A. Golan, D. Miller, and G.G. Judge. *Maximum Entropy Econometrics: Robust Estimation with Limited Data. Financial Economics and Quantitative Analysis.* Wiley, New York, 1996.

[26] T.B. Fomby and R.C. Hill, eds. *Applying Maximum Entropy to Econometric Problems, Volume 12: Advances in Econometrics.* JAI Press Inc., Greenwich, CT, 1997.

[27] J.A. Wheeler. It from bit. In S. Kobayashi, H. Ezawa, Y. Murayama, and S. Nomura (ed.), *Proceedings of the 3rd International Symposium on Foundations of Quantum Mechanics, Tokyo, 1989.* Physical Society of Japan, Tokyo, 1990, p. 354.

[28] J.A. Wheeler. Time today. In J.J. Halliwell, J. Perez-Mercader, and W.H. Zurek (eds.), *Physical Origins of Time Asymmetry*, Cambridge University Press, Cambridge, UK, 1994, pp. 1–29.

[29] B. Buck and V.A. Macaulay, eds. *Maximum Entropy in Action: A Collection of Expository Essays.* Oxford University Press, Oxford, UK, 1990.

[30] B.R. Frieden. *J. Mod. Opt.* **35**, 1297–1316 (1988).

[31] D. Edelman. The minimum local cross-entropy criterion for inferring riskneutral price distributions from traded option prices. Working Paper 47, University College Dublin Graduate School of Business Centre for Financial Markets, Ireland, 2003.

[32] B.R. Frieden, A. Plastino, A.R. Plastino, and B.H. Soffer. *Phys. Rev. E* **66**, 046128 (2002).

[33] B.R. Frieden, A. Plastino, A.R. Plastino, and B.H. Soffer. *Phys. Lett. A* **304**, 73–78 (2002).

[34] D.C. Brody and L.P. Hughston. *Proc. R. Soc. Lond. A* **457**, 1343–1364 (2001).

[35] D.C. Brody and L.P. Hughston. *Quant. Finance* **2**, 70–80 (2002).

[36] R.A. Fisher. Theory of statistical estimation. *Proc. Cambridge Philos. Soc.* **22**, 700–725 (1925).

[37] B.R. Frieden. *Physics from Fisher Information.* Cambridge University Press, Cambridge, UK, 1998.

[38] B.R. Frieden and B.H. Soffer. *Found. Phys. Lett.* **13**, 89–96 (2000).

[39] R.J. Hawkins and B.R. Frieden. *Phys. Lett. A* **322**, 126–130 (2004).

[40] J.H. McCullough. *J. Finance* **XXX**, 811–829 (1975).

[41] O.A. Vasicek and H.G. Fong. *J. Finance* **XXXVII**, 339–356 (1982).

[42] G.S. Shea. *J. Finance* **XL**, 319–325 (1985).

[43] M. Fisher, D. Nychken, and D. Zervos. Fitting the term structure of interest rates with smoothing splines. Technical Report FEDS 95-1, Federal Reserve Board, Washington, D.C., 1995.

[44] V. Frishling and J. Yamamura. *J. Fixed Income* **6**, 97–103 (1996).

[45] K.J. Adams and D.R. Van Deventer. *J. Fixed Income* **4**, 52–62 (1994).

[46] S. Kullback. *Information Theory and Statistics.* Wiley, New York, 1959.

[47] B. Tuckman. *Fixed Income Securities: Tools for Today's Markets, 2nd edn.* Wiley, Hoboken, 2002.

[48] C.R. Nelson and A.F. Siegel. *J. Bus.* **60**, 473–489 (1987).

[49] Bank for International Settlements, Monetary and Economic Department. Basel, Switzerland. *Zero-Coupon Yield Curves: Technical Documentation*, 1999.

[50] L. Krippner. The OLP model of the yield curve: A new consistent cross-sectional and inter-temporal approach. Victoria University of Wellington, New Zealand, Unpublished manuscript, 2002.

[51] F.X. Diebold and C. Li. Forecasting the term structure of government bond yields. PIER Working Paper 02-026, Penn Institute for Economic Research, Department of Economics, University of Pennsylvania, 2002.

[52] L. Krippner. Modelling the yield curve with orthonormalised Laguerre polynomials: An intertemporally consistent approach with an economic interpretation. Working Paper in economics 1/03, Department of Economics, University of Waikato, New Zealand, 2003.

[53] L. Krippner. Modelling the yield curve with orthonormalised Laguerre polynomials: A consistent cross-sectional and inter-temporal approach. Working Paper in economics 2/03, Department of Economics, University of Waikato, New Zealand, 2003.

[54] F.X. Diebold, G.D. Rudebusch, and S.B. Aruoba. The macroeconomy and the yield curve: A nonstructural analysis. Working Paper 18. Federal Reserve Bank of San Francisco, CA, 2003.

[55] K.D. Garbade. Modes of fluctuation in bond yields—an analysis of principal components. Technical Report 20, Bankers Trust Company, New York, 1986.

[56] R. Litterman and J.Scheinkman. *J. Fixed Income* **1**, 54–61 (1991).

[57] K.D. Garbade. *Fixed Income Analytics, 2nd edn.* MIT, Cambridge, MA, 1996.

[58] K.D. Garbade and T.J. Urich. Modes of fluctuation in sovereign bond yield curves: An international comparison. Technical Report 42, Bankers Trust Company, New York, 1988.

[59] W. Phoa. *Advanced Fixed Income Analytics.* Frank J. Fabozzi Associates, New Hope, PA, 1998.

[60] W. Phoa. Yield curve risk factors: Domestic and global contexts. In M. Lore and L. Borodovsky (eds.), *Professional's Handbook of Financial Risk Management*, Butterworth Heineman, Burlington, PA, 2000, pp. 155–184.

[61] J.-P. Bouchaud, N. Sagna, R. Cont, N. El-Karoui, and M. Potters. *App. Math. Finance* **6**, 209–232 (1999).

[62] P. Santa-Clara and D. Sornette. *Rev. Finan. Stud.* **14**, 149–185, 2001.

[63] R.J. Hawkins, B.R. Frieden, and J.L. D'Anna. *Phys. Lett. A* **344**, 317–323 (2005).

[64] P. Yeh, A. Yariv, and C. Hong. *J. Opt. Soc. Amer.* **67**, 423–438 (1977).

[65] S.P. Flego, B.R. Frieden, A. Plastino, A.R. Plastino, and B.H. Soffer. *Phys. Rev. E* **68**, 016105 (2003).

[66] M. Reginatto and F. Lengyel. The diffusion equation and the principle of minimum Fisher information. cond-mat/9910039, 1999.

[67] N.G. van Kampen. *J. Stat. Phys.* **17**, 71–88 (1977).

[68] H. Risken. *The Fokker–Planck Equation: Methods of Solution and Applications, Volume 18: Springer Series in Synergetics, 2nd edn.* Springer-Verlag, New York, 1996.

[69] R. Rebonato. *Interest-Rate Option Models.* Wiley, Hoboken, NJ, 1998.

[70] D. Brigo and F. Mercurio. *Interest Rate Models: Theory and Practice.* Springer-Verlag, Berlin, 2001.

[71] L. Sirovich. *Q. Appl. Math.* **XLV**, 561–571 (1987).

[72] L. Sirovich. *Q. Appl. Math.* **XLV**, 573–582 (1987).

[73] L. Sirovich. *Q. Appl. Math.* **XLV**, 583–590 (1987).

[74] K.S. Breuer and L. Sirovich. *J. Comput. Phys.* **96**, 277–296 (1991).

[75] B.R. Frieden, A. Plastino, A.R. Plastino, and B.H. Soffer. *Phys. Rev. E* **60**, 48–53 (1999).

[76] M. Aoki. *New Approaches to Macroeconomic Modeling: Evolutionary Stochastic Dynamics, Multiple Equilibria, and Externalities as Field Effects.* Cambridge University Press, New York, 1998.

[77] M. Aoki. *Modeling Aggregate Behavior and Fluctuations in Economics: Stochastic Views of Interacting Agents.* Cambridge University Press, New York, 2001.

[78] D. Bohm. *Phys. Rev.* **85**, 166–179 (1952).

[79] J. Tobin. *J. Money, Credit Banking* **1**, 15–29 (1969).

[80] A.B. Abel. Consumption and investment. In B.M. Friedman and F. Hahn (ed.), *Handbook of Monetary Economics*, Volume 2. Elsevier Science B.V., Amsterdam, 1989, pp. 725–778.

[81] M.L. Weitzman. *Income, Wealth, and the Maximum Principle.* Harvard University Press, Cambridge, MA, 2003.

Chapter 3

[1] H.J. Morowitz. *Bull. Math. Biophys.* **17**, 81–86 (1955).

[2] H.A. Johnson. *Science* **168**, 1545–1550 (1970).

[3] R.A. Fisher. *Philos. Trans. R. Soc. Lond.* **222**, 309–368 (1922).

[4] R.A. Fisher. In *Statistical Methods and Scientific Inference, 2nd edn.* Oliver and Boyd, London, 1959, pp. 1–112.

[5] C.E. Shannon. *Bell Syst. Tech. J.* **27**, 379–623 (1948).

[6] S. Kullback. *Information Theory and Statistics.* Wiley, New York, 1959.

[7] K.S. Trincher. *Biology and Information: Elements of Biological Thermodynamics.* Consultants Bureau, New York, 1965.

[8] J.R. Pierce. In *Information Theory and Physics in Introduction to Information Theory, Symbols, Signals, and Noise, 2nd edn.* Dover Publications, New York, 1980, pp. 184–207.

[9] E.N. Gilbert. *Science* **152**, 320–326 (1966).

[10] W. Ebeling and C. Frommel. *Biosys.* **46**, 47–55 (1998).

[11] L. Szilard. *Z. Phys.* **53**, 840 (1929).

[12] J.C. Maxwell. *Theory of Heat, 6th edn.* D. Appleton Co, New York, 1880.

[13] B.R. Frieden. *Probability, Statistical Optics and Data Testing, 3rd edn.* Springer-Verlag, Berlin, 2001.

[14] R.V.L. Hartley. *Bell Syst. Tech. J.* **7**, 535 (1928)[1].

[15] F.M. Reza. *An Introduction to Information Theory.* McGraw-Hill, New York, 1961.

[16] I. Prigogine. *Physica* **31**, 719–724 (1965).

[17] T.D. Schneider. *J. Theor. Biol.* **148**, 125–137 (1991).

[18] B.R. Frieden. *Science from Fisher Information, 2nd edn.* Cambridge University Press, Cambridge, UK, 2004. Alternatively, *Physics from Fisher Information.* Cambridge University Press, Cambridge, UK, 1998.

[19] E. Schrodinger. *What is Life?* Cambridge University Press, Cambridge, UK, 1944.

[20] A. Hariri, B. Weber, and J. Olmsted. *J. Theor. Biol.* **147**, 235–254 (1990).

[21] B.J. Strait, and T.G. Dewey. *Biophys.* **71**, 148–155 (1996).

[22] T.D. Schneider. *J. Theor. Biol.* **189**, 427–441 (1997).

[23] O. Weiss, M.A. Jimenez-Montano, and H. Herzel. *J. Theor. Biol.* **206**, 379–386 (2000).

[24] B. Zeeberg. *Genome Res.* **12**, 944–955 (2002).

[25] M. Dehnert, W.E. Helm, and M.T. Hutt. *Gene* **345**, 81–90 (2005).

[26] T.D. Schneider. *Nucleic Acids Res.* **28**, 2794–2785 (2003).

[27] D.R. Brooks, P.H. Leblond, and D.D. Cumming. *J. Theor. Biol.* **109**, 77–93 (1984).

[28] R. Wallace and R.G. Wallace. *J. Theor. Biol.* **192**, 545–559 (1998).

[29] C. Adami, C. Ofria, and T.C. Collier. *Proc. Natl. Acad. Sci. USA* **97**, 4463–4468 (2000).

[30] D. Segre, D. Ben-Eli, and D. Lancet. *Proc. Natl. Acad. Sci. USA* **97**, 4112–4117 (2000).

[31] B.D. Fath, H. Cabezas, and W. Pawlowski. *J. Theor. Biol.* **222**, 517–530 (2003).

[32] R.E. Ulanowicz. *Comput. Chem.* **25**, 393–399 (2001).

[33] R. Albert and A.L. Barabasi. *Rev. Mod. Phys.* **74**, 47–97 (2002).

[34] H. Jeong, B. Tombor, R. Albert, Z. Oltvai and A. Barabasi. *Nature* **407**, 651–654 (2000).

[35] A. Wagner and D. Fell. Technical Report 00-07-041. Santa Fe Institute, NM, 2000.

[36] E. Yeger-Lotem, S. Sattath, N. Kashtan, S. Itzkovitz, R. Milo, R. Pinter, V. Alon and H. Margalit *Proc. Natl. Acad. Sci. USA* **101**, 5934–5939 (2004).

[37] S.H. Yook, Z.N. Oltvai, and A.L. Barabasi. *Proteomics* **4**, 928–942 (2004).

[38] B. Grunenfelder and E.A. Winzeler. *Nat. Rev. Genet.* **3**, 653–661 (2002).

[39] J.D. Han, N. Berfin, T. Hao and D.S. Goldberg. *Nature* **430**, 88–93 (2004).

[40] N.M. Luscombe, M.M. Babu, H. Yu, M. Snyder and S. Teichmann. *Nature* **431**, 308–312 (2004).

[41] C. von Mering, R. Krause, B. Snel and M. Cornell. *Nature* **417**, 399–403 (2002).

[42] R. Albert, H. Jeong, and A.L. Barabasi. *Nature* **406**, 378–382 (2000).

[43] D.S. Callaway, M. Newman, S. Strogatz and D. Watts. *Phys. Rev. Lett.* **85**, 5468–5471 (2000).

[44] H. Jeong, S. Mason, A. Barabasi and Z. Oltvai. *Nature* **411**, 41–42 (2001).

[45] L. Zhao, K. Park, and Y.C. Lai. *Phys. Rev. E.* **70**, 1–4 (2004).

[46] R.A. Gatenby and B.R. Frieden. *Cancer Res.* **62**, 3675–3684 (2002).

---

[1] It is also known as the Hartley measure of information (or simply Hartley information).

[47]   R.A. Gatenby and B.R. Frieden. *Mut. Res.* **568**(2), 259–273 (2005).

[48]   T.D. Schneider. *J. Theor. Biol.* **148**, 83–123 (1991).

[49]   R. Lahoz-Beltra. *Biosystems* **44**, 209–229 (1997).

[50]   B. Alberts. *Cell* **92**, 291–294 (1998).

[51]   R.A. Gatenby and B.R. Frieden. *Math. Biosci. Eng.* **2**(1), 43–51 (2005).

[52]   B. Alberts, D. Bray, J. Lewis, M. Raff, K. Roberts, and J.D. Watson. *Molecular Biology of the Cell.* Garland Publishing Inc., New York, 1994.

[53]   J.P. Keener. *J. Theor. Biol.* **234**, 263–275 (2005).

[54]   J.D. Dockery and J.P. Keener. *Bull. Math. Biol.* **63**, 95–116 (2001).

[55]   D. Kaiser. *Ann. Rev. Genet.* **35**, 103–123 (2001).

[56]   L.H. Zhang and Y.H. Dong. *Mol. Microbiol.* **53**, 1563–1571 (2004).

[57]   M.E. Taga and B.L. Bassler. *Proc. Natl. Acad. Sci. USA* **100**, 14549–14554 (2003).

[58]   M.G. Surette, M.B. Miller, and B.L. Bassler. *Proc. Natl. Acad Sci. USA* **96**, 1639–1644 (1999).

[59]   F. Chamaraux, S. Fache, F. Bruckert, and B. Fourcade. *Phys. Rev. Lett.* **94**, 158102 2005.

[60]   J. Franca-Koh and P.N. Devreotes. *Physiology* **19**, 300–308 (2004).

[61]   J.A. Morris. *J. Biosci.* **26**, 15–23 (2001).

[62]   S.B. Garcia, M. Novelli, and N.A. Wright. *Int. J. Exp. Pathol.* **81**, 89–116 (2000).

[63]   L.A. Loeb. *Cancer Res.* **61**, 3230–3239 (2001).

[64]   W.S. Kendal. *Math. Biosci.* **100**, 143–159 (1990).

[65]   R.V. Sole and T.S. Deisboeck. *J. Theor. Biol.* **228**, 47–54 (2004).

[66]   M. Eigen and P. Schuster. *Naturwissenschaften* **64**, 541–565 (1977).

[67]   S.F. Gilbert. *Developmental Biology.* Sinauer Associates Inc. Sunderland, MA, 1994.

[68]   F. Gilbert and B. Marinduque. *Curr. Opin. Obstet. Gynecol.* **2**, 226–235 (1990).

[69]   M. Ilyas, J. Straub, I.P.M., Tomlinson, and W.F. Bodmer. Genetic *Eur. J. Cancer* **35**, 335–351 (1999).

[70]   J.W. Gray and C. Collins. *Carcinogenesis* **21**(21), 443–453 (2000).

[71]   F. Kerangueven, T. Noguchi, F. Coulier, F. Allione, V. Wargniez, J. Simony-Lafontaine, M. Longy, J. Jacquemier, H. Sobol, F. Eisinger, and D. Birnbaum. *Cancer Res.* **57**(24), 5469–5474 (1997).

[72]   F. Jiang, R. Desper, C.H. Papadimitriou, A.A. Schaffer, O.P. Kallioniemi, J. Richter, P. Schrami, G. Sauter, M.J. Mihatsch, and H. Moch. *Cancer Res.* **60**, 6503–6509 (2000).

[73]   M.A. Peinado, S. Malkhosyan, A. Velazquez, and M. Perucho. *Proc. Natl. Acad. Sci. USA* **89**(21), 10065–10069 (1992).

[74]   L.A. Loeb. *Cancer Res.* **54**, 5059–5063 (1991).

[75]   P.S. Rabinovitch, B.I. Reid, R.C. Haggitt, T.H. Norwood, and C.E. Rubin. *Lab. Invest.* **60**, 65–71 (1988).

[76]   J.M. Bishop. *Cell* **64**, 235–248 (1991).

[77]   H. Rubin. *Cancer Res.* **61**, 799–807 (2001).

[78]   C.C. Park, M.J. Bissell, and M.H. Barcellos-Hoff. *Mol. Med.* **6**, 324–329 (2000).

[79]   E.R. Fearon and B. Vogelstein *Cell* **61**, 759–767 (1990).

[80]   E.E. Shannon and W. Weaver. *The Mathematical Theory of Communication.* University of Illinois Press. Urbana, IL, 1949.

[81]   Li, S., et al. (47 other authors) *Nature* **303**, 540–543 (2004).

[82]   E.F. Mao, L. Lane, J. Lee, and J.U. Miller. *J. Bacteriol.* **179**, 417–422 (1997).

[83]   L. Brillouin. *Science and Information Theory.* Academic Press, New York, 1956.

[84]   B.R. Frieden, A. Plastino, and B.H. Soffer. *J. Theor. Biol.* **208**(1), 49–64 (2001).

[85]   B.R. Frieden and B.H. Soffer. *Phys. Rev E.* **52**, 2274–2286 (1995).

[86]   R.A. Gatenby and E.T. Gawlinski. *Cancer Res.* **56**, 5745–5753 (1996).

[87]   V. Ahuja, R.E. Coleman, J. Herndon, F. Edward, and E.P. Patz. *Cancer* **83**(5), 918–924 (1998).

[88]   J. Volker, H. Klump, and K. Breslauer. *Proc. Natl. Acad. Sci. USA* **98**(14), 7694–7699 (2001).

[89]   L. Tabar, G. Fagerberg, S.W. Duffy, N.E. Day, A. Gad, and O. Grontoft. *Radiol. Clin. North Am.* **30**(1), 187–210 (1992).

[90]   G. Fagerberg, L. Baldetorp, O. Grontoft, B. Lundstrom, J.C. Manson, and B. Nordenskjold. *Acta Radiol. Oncol.* **24**(6), 465–473 (1985).

[91]   B.A. Thomas, J.L. Price, P.S. Boulter, and N.M. Gibbs. *Recent Results Cancer Res.* **90**, 195–199 (1984).

[92]   H.J. De Koning, J. Fracheboud, R. Boer, A.L. Verbeek, H.J. Collette, J.H. Hendriks, B.M. van Ineveld, A.E. de Bruyn, and P.J. van der Maas. *Int. J. Cancer* **60**(6), 777–780 (1995).

[93]   P.G. Peer, R. Holland, J.H. Hendriks, M. Mravunac, and A.L. Verbeek. *J. Natl. Cancer Inst.* **86**(6), 436–441 (1994).

[94]   L.J. Burhenne, T.G. Hislop, and H.J. Burhenne. *Am. J. Roentgenol.* **158**(1), 45–49 (1992).

[95]   D. Hart, E. Shochat, and Z. Agur. *Br. J. Cancer* **78**(30), 382–387 (1998).

[96]   P. Ducommun, I. Bolzonella, M. Rhiel, P. Pugeaud, U. von Stockar, and I.W. Marison. *Biotechnol. Bioeng.* **72**, 515–522 (2001).

[97]   R. Tang, A.J. Cheng, J.Y. Wang, and T.C. Wang. *Cancer Res.* **58**(18), 4052–4054 (1998).

[98]   S.E. Artandi, S. Chang, S.L. Lee, S. Alson, G.J. Gottlieb, L. Chin, and R.A. DePinho. *Nature* **406**(6796), 641–645 (2000).

[99]   R.A. DePinho. *Nature* **408**, 248–254 (2000).

[100] R.A. Gatenby and B.R. Frieden. *Mutat Res.* **568**(2), 259–273 (2004).

[101] H. Buerger, F. Otterbach, R. Simon, C. Poremba, R. Diallo, T. Decker, L. Riethdorf, C. Brinkschmidt, B. Dockhorn-Dworniczak, and W. Boecker. *J. Pathol.* **187**(4), 396–402 (1999).

Chapter 4

[1]   M.A. Nielsen and I.L. Chuang. *Quantum Computation and Quantum Information.* Cambridge University Press, Cambridge, UK, 2000.

[2]   Special issue of *Europhysics-news.* **36** (2005).

[3]   M. Gell-Mann and C. Tsallis (eds.), *Nonextensive Entropy: Interdisciplinary Applications.* Oxford University Press, Oxford, 2004.

[4]   J.A. Wheeler. In W.H. Zurek (ed.), *Complexity, Entropy and the Physics of Information.* Adison-Wesley, New York, 1990, p. 3.

[5]   A. Plastino. *Physica A* **340**, 85 (2004).

[6]   B. Russell. *A History of Western Philosophy.* Simon & Schuster, New York, 1945.

[7]   M. Davies. *The Universal Computer.* W.W. Norton, London, 2000.

[8]   E.T. Jaynes. *Phys. Rev.* **106**, 620 (1957).

[9]   E.T. Jaynes. In W.K. Ford (ed.), *Statistical Physics.* Benjamin, New York, 1963.

[10] A. Katz. *Statistical Mechanics.* Freeman, San Francisco, CA, 1967.

[11] T.M. Cover and J.A. Thomas. *Elements of Information Theory.* Wiley, New York, 1991.

[12] R. Frieden, A. Plastino, A.R. Plastino, and B.H. Soffer. *Phys. Rev. E* **66** 046128 (2002).

[13] R. Frieden, A. Plastino, A.R. Plastino, and B.H. Soffer. *Phys. Lett. A* **304**, 73 (2002).

[14] S. Flego, R. Frieden, A. Plastino, A.R. Plastino, and B.H. Soffer. *Phys. Rev. E* **68**, 016105 (2003).

[15] E.T. Jaynes. In R.D. Rosenkrantz (ed.), *Papers on Probability, Statistics and Statistical Physics.* D. Reidel, Dordrecht, Holland, 1983, pp. 210–314.

[16] A. Plastino, A.R. Plastino, and M. Casas. In Stanislaw Sienutycz (ed.), *Variational Extremum Principles in Macroscopic Systems, Volume II* (Chapter 1). Elsevier, Amsterdam, 2005, pp. 379–394.

[17] B.R. Frieden. *Physics from Fisher Information.* Cambridge University Press, Cambridge, UK, 1998; *Science from Fisher information, 2nd edn.* Cambridge University Press, Cambridge, UK, 2004.

[18] R.B. Lindsay and H. Margenau. *Foundations of Physics.* Dover, New York, 1957.

[19] P.W. Bridgman. *The Nature of Physical Theory.* Dover, New York, 1936.

[20] P. Duhem. *The Aim and Structure of Physical Theory.* Princeton University Press, Princeton, NJ, 1954.

[21] R.B. Lindsay. *Concepts and Methods of Theoretical Physics.* Van Nostrand, New York, 1951.

[22] H. Weyl. *Philosophy of Mathematics and Natural Science.* Princeton University Press, Princeton, NJ, 1949.

[23] J.W. Gibbs. *Elementary Principles in Statistical Mechanics.* (in collected works), Yale University Press, New Haven, CT, 1948.

[24] B.R. Frieden and B.H. Soffer. *Phys. Rev. E* **52**, 2274 (1995).

[25] R. Frieden, A. Plastino, A.R. Plastino, and B.H. Soffer. *Phys. Rev. E* **60**, 48 (1999).

[26] Paul B. Slater. preprint (2005) [quant-ph/0504066], J. Math. Phys. **47** 022104 (2006).

[27] F. Pennini, A.R. Plastino, and A. Plastino. *Physica A* **258**, 446 (1998).

[28] F. Pennini, A. Plastino, A.R. Plastino, and M. Casas. *Phys. Lett. A* **302**, 156 (2002).

[29] Michael J.W. Hall. *Phys. Rev. A* **62**, 012107 (2000).

[30] Sumiyoshi Abe. *Phys. Lett. A* **254**, 149 (1999).

[31] R.N. Silver In W.T. Grandy, Jr. and P.W. Milonni (eds.), *E.T. Jaynes: Physics and Probability.* Cambridge University Press, Cambridge, UK, 1992.

[32] R.E. Nettleton. *J. Phys. A: Math. Gen.* **35**, 295 (2002).

[33] C. Villani. *J. Math. Pures Appl.* **77**, 821 (1998).

[34] A.R. Plastino and A. Plastino. *Phys. Rev. E* **54**, 4423 (1996).

[35] A. Plastino, A.R. Plastino, and H.G. Miller. *Phys. Lett. A* **235**, 129 (1997).

[36] H. Cramer. *Mathematical Methods of Statistics.* Princeton University Press, Princeton, NJ, 1946.

[37] F. Pennini and A. Plastino. *Phys. Rev. E* **69**, 057101 (2004).

[38] F. Pennini and A. Plastino. *Phys. Lett. A* **349**, 15 (2006).

[39] R.K. Pathria. *Statistical Mechanics.* Pergamon Press, Exeter, UK, 1993.

[40] F. Reif. *Fundamentals of Statistical and Thermal Physics.* McGraw-Hill, New York, 1965.

[41] A.J. Lichtenberg and M.A. Lieberman. *Regular and Chaotic Dynamics.* Springer-Verlag, Berlin, 1991.

[42] E.A. Desloge. *Thermal Physics.* Holt, Rinehart, and Winston, New York, 1968.

[43] C.E. Shannon, *Bell Syst. Tech. J.* **27**, 379 (1948).
[44] A.Y. Khinchin. *Mathematical Foundations of Information Theory*. Dover, New York, 1957.
[45] A.R. Plastino, M. Casas, and A. Plastino, Phys. Lett. A **246**, 498 (1998).
[46] A.N. Kolmogorov. *Dokl. Akad. Nauk SSSR* **119**, 861 (1958).
[47] J.G. Sinai. *Dokl. Akad. Nauk SSSR* **124**, 768 (1959).
[48] C. Beck and F. Schlögl. *Thermodynamic of Chaotic Systems*. Cambridge University Press, New York, 1993.
[49] C. Tsallis, Braz. *J. Phys.* **29**, 1 (1999); *J. Stat. Phys.* **52**, 479 (1988).
[50] A. Plastino and A.R. Plastino. *Braz. J. Phys.* **29**, 50 (1999).
[51] A.R. Plastino and A. Plastino. *Phys. Lett. A* **174**, 384 (1993).
[52] A.R. Plastino and A. Plastino. In E. Ludeña (ed.), *Condensed Matter Theories, Volume 11*. Nova Science Publishers, New York, 1996, p. 341.
[53] J.A.S. Lima, R. Silva, and A.R. Plastino. *Phys. Rev. Lett.* **86**, 2938 (2001).
[54] R.P. Di Sisto, S. Martinez, R.B. Orellana, A.R. Plastino, and A. Plastino. *Physica A* **265**, 590 (1999).
[55] S. Martinez, F. Pennini, and A. Plastino. *Physica A* **286**, 489 (2000).
[56] A.R. Plastino and A. Plastino. *Phys. Lett. A* **177**, 177 (1993).
[57] A.R. Plastino and A. Plastino. *Phys. Lett. A* **226**, 257 (1997).
[58] B.R. Frieden. *Phys. Rev. A* **41**, 4265 (1990).
[59] H. Grad. *Commun. Pure Appl. Math* **2**, 331 (1949); In S. Flügge (ed.), *Principles of the Kinetic Theory of Gases, Handbuch der Physik XII*, Springer, Berlin, 1958.
[60] Y.B. Rumer and M. S. Ryvkin. *Thermodynamics, Statistical Mechanics and Kinetics*. MIR Publishers, Moscow, 1980.
[61] D. Jou, J. Casas-Vázquez, and G. Lebon. *Extended Irreversible Thermodynamics, 3rd edn.* Springer, Berlin, 2001.
[62] H. Grabert. *Projection Operator Techniques in Nonequilibrium Statistical Mechanics*. Springer, Berlin, 1982.
[63] R. Zwanzig. *Lectures in Theoretical Physics (Boulder), Volume 3*. Interscience, New York, 1961.
[64] J.M. Cassels. *Basic Quantum Mechanics*. McGraw-Hill, New York, 1970.

Chapter 5

[1]   T. Deacon. *The Symbolic Species*. Penguin, London, UK, 1997, pp. 322–334.
[2]   E. Harrison. *Cosmology: The Science of the Universe, 2nd edn.* Cambridge University Press, Cambridge, UK, 2000.
[3]   J. Tauber. MAP blazes the way for Planck, ESA's probe into the birth of the Universe, *ESA Science and Technology*, April 13, 2005.
[4]   P.A.M. Dirac. *Proc. R. Soc. A* **165**, 199 (1938).
[5]   J.T. Bonner. *Size and Cycle*. Princeton University Press, Princeton, NJ, 1965.
[6]   J. Damuth. *Nature* **290**, 699–700 (1981).
[7]   T. Fenchel. *Oecologia* **14**, 317–326 (1974).
[8]   G.B. West, J.H. Brown, and B.J. Enquist. *Science* **276**, 122 (1997).
[9]   B.J. Enquist and K.J. Niklas. *Nature* **410**, 655 (2001).
[10] G.B. West and J.H. Brown. *Physics Today* **57**(9), 36 (2004) and reference therein.
[11] W.J. Jungers. *Size and Scaling in Primate Biology*. Plenum, New York, 1985.

[12] T.A. McMahon and J.T. Bonner. *On Size and Life*. Scientific American Library, New York, 1983.

[13] K. Schmidt-Nielsen. *Scaling: Why is Animal Size So Important?* Cambridge University Press, Cambridge, UK, 1984.

[14] W.A. Calder. *Size, Function and Life History*. Harvard University Press, Cambridge, MA, 1984.

[15] M.J. Rees. *The Allometry of Growth and Reproduction*. Cambridge University Press, Cambridge, UK, 1989.

[16] A. McKane, M. Droz, J. Vannimenus, and D. Wolf (eds.). *Scale Invariance: Interfaces and Non-Equilibrium Dynamics*. Plenum, New York, 1995.

[17] T. Nishikawa, A.E. Motter, Y-C Lai, and F.C. Hoppensteadt. *Phys. Rev. Lett.* **91**, 014101 (2003).

[18] L. Smolin. *Class. Quant. Grav.* **9**, 173 (1992).

[19] L. Smolin. *The Life of the Cosmos*. Oxford University Press, Oxford, UK, 1997.

[20] A. Linde. *Inflation and Quantum Cosmology*. Academic Press, Boston, 1990.

[21] A. Linde. *Sci. Amer.* **271**, 48 (1994).

[22] A.H. Guth. *The Inflationary Universe: The Quest for a New Theory of Cosmic Origins*. Perseus Books, New York, 1997.

[23] S.F. Hawking. *Black Holes and Baby Universes*. Bantam Doubleday Dell, New York, 1993.

[24] G. Damiani and P. Della Franca. *Riv. Biol.* **90**, 227 (1997).

[25] A. McKane, M. Droz, J. Vannimenus, and D. Wolf (eds.). *Scale Invariance: Interfaces and Non-Equilibrium Dynamics*. Plenum, New York, 1995.

[26] B.R. Frieden and R.A. Gatenby. *Phys. Rev. E* **72**, 036101, 1–10 (2005).

[27] M. Cignoni, P. Prada Moroni, and S. Degli. *Innocenti, Mem. S.A.It.,* Suppl. **3**, 143 (2003).

[28] M. Morris, K. Uchida, and T. Do, *Nature* **440**, (7082), 308 (2006).

[29] S.J. Brodsky and G.R. Farrar. *Phys. Rev.* **D11**, 1309 (1975).

[30] B.R. Frieden. *Science from Fisher Information: A Unification, 2nd edn.* Cambridge University Press, Cambridge, UK, 2004.

[31] R.H. Dicke. *Nat. Lett.* **192**, 440 (1961).

[32] B. Carter. Large number coincidences and the anthropic principle in cosmology. In M.S. Longour (ed.), *Proceedings of the IAU Symposium on Confrontation of Cosmological Theories with Observational Data*. Reidel, Dordrecht, The Netherlands, 1974, p. 291.

[33] B.J. Carr and M.J. Rees. *Nature* **278**, 605 (1979).

[34] J.D. Bjorken, *Phys. Rev. D* **67**, 043508 (2003).

[35] K.C. Kulander. *Phys. Rev. A* **36**, 2726 (1987).

[36] A. de Souza Dutra, M.B. Hott, and V. dos Santos. arXiv:quan-ph/0311044, **2**, 16, Sept. 2004.

[37] A. Mayer and J.-P. Vigneron. *J. Phys: Condens. Matter* **12**, 6693 (2000).

[38] J.B. Pendry. *Low Energy Electron Diffraction*. Academic Press, New York, 1974.

[39] V. Volterra. *Nature* **118**, 558–560 (1926).

[40] R. Stutzle, M.C. Gobel, Th. Horner, E. Kierig, I. Mourachko, M.K. Oberthaler. *Phys. Rev. Lett.* **95**, 110405 (2005).

[41] D.B. Newell, R.L. Steiner, and E.R. Williams. *Phys. Rev. Lett.* **81**, 2404–2407 (1998).

[42] M.I. Yolles and B.R. Frieden. Organisational Transformation and Social Change, **5**(2)103, 136 (2005).

[43]   D. Hume. *Dialogues Concerning Natural Religion* (1779) (republished Hafner Publ., New York, 1948) [20, 21].

[44]   G. Damiani and P. Della Franca. *Riv. Biol.* **90**, 227 (1997).

[45]   C. Darwin. *The Origin of Species*. Random House, New York, 1979.

Chapter 6

[1]   B.R. Frieden. *Science from Fisher Information: A Unification*. Cambridge University Press, Cambridge, UK, 2004.

[2]   S.-I. Amari and N. Nagaoka. *Methods of Information Geometry*. Oxford University Press, Oxford, 2000.

[3]   S.L. Braunsten and C.M. Caves. *Phys. Rev. Lett.* **72**, 3439 (1994).

[4]   A.R. Plastino and A. Plastino. *Phys. Lett. A* **181**, 446 (1993).

[5]   O. Morgenstern and J. von Neuman. *Theory of Games and Economic Behavior*. Princeton University Press, Princeton, NJ, 1947.

[6]   A.J. Stam. *Inform. Control* **2**, 101 (1959).

[7]   L. Brillouin. *Science and Information Theory*. Academic Press, New York, 1956.

[8]   P.J. Hüber. *Robust Statistics*. Wiley, New York, 1981.

[9]   L.D. Landau and E.M. Lifshitz. *Quantum Mechanics, 3rd edn*. Pergamon Press, Oxford, 1977.

[10]   T. Cover and J. Thomas. *Elements of Information Theory*. Wiley, New York, 1991.

[11]   E.T. Jaynes. *Probability Theory, the Logic of Science*. Cambridge University Press, Cambridge, UK, 2003.

[12]   D. Horn and A. Gottlieb. *Phys. Rev. Lett.* **88**, 018702 (2002).

[13]   R.C. Venkatesan. In K.L. Priddy (ed.), *Proceedings of SPIE Symposium on Defense and Security. Intelligent Computing: Theory and Applications II*, vol. 5421. SPIE Press, Bellingham, WA, 2004, p. 48.

[14]   P.E. Hydon and E.L. Mansfield. *Found. Comp. Math.* **2**, 187 (2004).

[15]   T.D. Lee. J. *Stat. Phys.* **46**, 843 (1987).

[16]   I.B. Rumer and M.S. Ryvkin. *Thermodynamics, Statistical Physics and Kinetics*. Mir Publishers, Moscow, 1980.

[17]   E.T. Jaynes. *Phys. Rev.* **106**(4), 620 (1957).

[18]   E.T. Jaynes. *Phys. Rev.* **108**(2), 171 (1957).

[19]   M. Casas, A. Plastino, and A. Puente. *Phys. Lett. A* **248**, 161 (1998).

[20]   S. Katzenbeisser and F. A. Petitcolas. *Information Hiding: Techniques for Steganography and Digital Watermarking*. Artech House, Norwood, MA, 2000.

[21]   J.A. Buchmann. *Introduction to Cryptography*. Springer, London, 2004.

[22]   H. Delfs and H. Knebel. *Introduction to Cryptography: Principles and applications*. Springer, Berlin, 2002.

[23]   B. Schneier. *Applied Cryptography, 2nd edn*. Wiley, New York, 1996.

[24]   C. Cachin. *Inform. Comput.* **192**(1), 41 (2004).

[25]   P. Moulin and A. Ivanovic. The Fisher information game for optimal design of synchronization patterns in blind watermarking. In *Proceedings of IEEE International Conference on Image Processing*, vol. 2. IEEE Press, Piscataway, NJ, 2001, p. 550.

[26]   L. Rebollo-Neira and A. Plastino. *Physica A* **359**, 213 (2006).

[27]   G. Sapiro. *Geometric Partial Differential Equations and Image Processing*. Cambridge University Press, Cambridge, UK, 2001.

[28] M.A. Nielsen and I. L. Chuang. *Quantum Computation and Quantum Information.* Cambridge University Press, Cambridge, UK, 2000.

[29] C. Bennett and G. Brassard. In *Proceedings of IEEE International Conference on Computers, Systems and Signal Processing. Bangalore, India.* IEEE Press, New York, 1984, p. 175.

[30] D. Bowmeester, D. Ekert, K. Artur, and A. Zellinger (eds.). *The Physics of Quantum Information: Quantum Cryptography, Quantum Teleportation, Quantum Computation.* Springer, Berlin, 2000.

[31] H.-K. Lo, T. Spiller, and S. Popescu (eds.). *Introduction to Quantum Computation and Information.* World Scientific, Singapore, 1998.

[32] G.H. Golub and C.F. van Loan. *Matrix Computations, 3rd edn.* Johns Hopkins University Press, Baltimore, MD, 1995.

[33] P.A.M. Dirac. *Principles of Quantum Mechanics (The International Series of Monographs on Physics).* Oxford University Press, Oxford, 1982.

[34] G. Arfken and H. Weber. *Mathematical Methods for Physicists.* Academic Press, San Diego, CA, 1995.

[35] A.F.J. Levi. *Applied Quantum Mechanics.* Cambridge University Press, Cambridge, UK, 2003.

[36] B.R. Frieden, A. Plastino, A.R. Plastino, and B.H. Soffer. *Phys. Rev. E* **60**(1), 48 (1999).

Chapter 7

[1] World Commission on Environment and Development. Our Common Future. Oxford University Press, Oxford, UK, 1987.

[2] B.L. II Turner, R.E. Kasperson, P.A. Matson, J.J. McCarthy, R.W. Corell, Lindsey Christensen, N. Eckley, J.X. Kasperson, A. Luers, M.L. Martello, C. Polsky, A. Pulsipher, A. Schiller. *Proc. Nat. Acad. Sci. USA* **100**(14), 8074–8079 (2003).

[3] M.L. Imhoff, L. Bounoua, T. Ricketts, C. Loucks, R. Harriss, and W.T. Lawrence. *Nature* **429**, 870–873 (2004).

[4] J.A. Foley, R. DeFries, G.P. Asner, C. Barford, G. Bonan, S.R. Carpenter, F.S. Chapin, M.T. Coe, G.C. Daily, H.K. Gibbs, J.H. Helkowski, T. Holloway, E.A. Howard, C.J. Kucharik, C. Monfreda, J.A. Patz, I.C. Prentice, N. Ramankutty, and P.K. Snyder. *Science* **309**, 570–574 (2005).

[5] H. Mooney, A. Cropper, and W. Reid. *Nature* **434**, 561–562 (2005).

[6] J.B. Schor. *Ecol. Econ.* **55**, 309–320 (2005).

[7] M. Scheffer, S. Carpenter, J.A. Foley, C. Folke, and B. Walker. *Nature* **413**(6856), 591–596 (2001).

[8] C. Folke, S. Carpenter, B. Walker, M. Scheffer, T. Elmqvist, L. Gunderson, and C.S. Holling. *Ann. Rev. Ecol. Evol. Syst.* **35**, 557–581 (2004).

[9] R.B. Alley, J. Marotzke, W.D. Nordhaus, J.T. Overpeck, D.M. Peteet, R.A. Pielke, Jr., R.T. Pierrehumbert, P.B. Rhines, T.F. Stocker, L.D Talley, and J.M. Wallace. *Science* **299**(5615), 2005–2010 (2003).

[10] D. Carment. Preventing state failure. In R.I. Rotberg (ed.), *When States Fail: Causes and Consequences*, Princeton University Press, Princeton, NJ, (2004), pp. 135–150.

[11]   J.A. Goldstone and J. Ulfelder. *Wash. Q.* **28**, 9–20 (2004).

[12]   P.A.T. Higgins, M.D. Mastrandea, and S.H. Schneider. *Philos. Trans. R. Soc. Lond. B. Biol. Sci.* **357**, 647–655 (2002).

[13]   A.L. Mayer, and M. Rietkerk. *BioScience* **54**(11), 1013–1020 (2004).

[14]   K.N. Suding, K.L. Gross, G.R. Houseman. *Trends Ecol. Evol.* **19**, 46–53 (2004).

[15]   B.D. Fath and H. Cabezas. *Ecol. Modelling* **174**, 25–35 (2004).

[16]   J. Patrício R. Ulanowicz, M.A. Pardal, and J.C. Marques. *Estuar. Coast. Shelf Sci.* **60**, 23–35 (2004).

[17]   B.D. Fath, H, Cabezas, and C.W. Pawlowski. *J. Theor. Biol.* **222**, 517–530 (2003).

[18]   M. Anand and L. Orlóci. *Ecol. Modelling* **132**, 51–62 (2000).

[19]   Y.M. Svirezhev. *Ecol. Modelling* **132**, 11–22 (2000).

[20]   R.E. Ulanowicz. Ecology, the Ascendent Perspective. Columbia University Press, New York, 1997.

[21]   H. Cabezas and B.D. Fath. *Fluid Phase Equilib.* **2**, 194–197 (2002).

[22]   R.A. Fisher *Philos. Trans. R. Soc. Lond.* **222**, 309–368 (1922).

[23]   B.R. Frieden, *Physics from Fisher Information: A Unification*. Cambridge University Press, Cambridge, UK, 1998, pp. 28–29.

[24]   B.R. Frieden. *Science from Fisher Information: A Unification, 2nd edn*. Cambridge University Press, Cambridge, UK, 2004, pp. 29–31.

[25]   A.L. Mayer, C.W. Pawlowski, and H. Cabezas. Ecol. Modeling **195**, 72–82 (2006).

[26]   R. Dilão and T. Domingos. *Ecol. Modelling*, **132**, 191–202 (2000).

[27]   N.J. Gotelli and G.R. Graves. Null Models in Ecology. Smith Sonian Institution Press, Washington, D.C., 1996.

[28]   F. Jordán, I. Scheuring, and I. Molnár. *Ecol. Modelling* **161**, 117–124 (2003).

[29]   V. Volterra. *Nature* **118**, 558–560 (1926).

[30]   S.E. Kingsland. Defining ecology as a science. Pp. 1–13 in *Foundations of Ecology: Classic papers with commentaries* (LA Real and JH Brown, eds.). University of Chicago Press, 1991.

[31]   T.M. Rocha Filho, I.M. Gléria, A. Figueiredo, and L. Brenig. *Ecol. Modelling* **183**, 95–106 (2005).

[32]   Y. Svirezhev. *Ecol. Modelling* **135**, 135–146 (2000).

[33]   J. Camacho, R. Guimerà, and L.A. Nunes Amaral. *Phys. Rev. Lett.* **88**(22), 228101.1–228102.4 (2002).

[34]   H. Cabezas, C.W. Pawlowski, A.L. Mayer, and N.T. Hoagland. *Clean Technol. Environ. Policy* **5**, 167–180 (2003).

[35]   P.M. Vitousek, P.R. Ehrlich, A.H. Ehrlich, P.A. Matson. *BioScience* **36**, 368–73 (1986).

[36]   H. Haberl. *Ambio* **26**, 143–6 (1997).

[37]   A. Balmford, R.E. Green, and M. Jenkins. *Trends Ecol. Evol.* **18**, 258–365 (2003).

[38]   Jenkins M. *Science* **302**, 1175–1177 (2003).

[39]   Millennium Ecosystem Assessment. *Ecosystems and Human Well-Being: Biodiversity Synthesis*. Island Press, Washington, D.C., 2005.

[40]   G.C. Daily (ed.). *Nature's Services: Societal Dependence on Natural Ecosystems*. Island Press, Washington, D.C., 1997.

[41]   D.U. Hooper, F.S. Chapin, III, J.J. Ewel, A. Hector, P. Inchausti, S. Lavorel, J.H. Lawton, D.M. Lodge, M. Loreau, S. Naeem, B. Schmid, H. Setälä, A.J. Symstad, J. Vandermeer, and D.A. Wardle. *Ecol. M.* **75**(1), 3–35 (2005).

[42]   J. Korhonen and J-P Snäkin. *Ecol. Econ.* **52**, 169–186 (2005).

[43]   H. Cabezas, C.W. Pawlowski, A.L. Mayer, and N.T. Hoagland. *Resour. Conserv. Recycling* **44**, 279–291 (2005).

[44] B.E. Beisner, D.T Haydon, and K. Cuddington. *Front. Ecol. Environ.* **7**, 376–382 (2003).

[45] B. Walker and J.A. Meyers. *Ecol. Soc.* **9**(2)3. [online] http://www.ecologyandsociety. org/vol9/iss2/art3. (2004).

[46] J.A. McGowan, D.R. Cayan, and L.M. Dorman. *Science* **281**, 210–217 (1998).

[47] T.D. Herbert, J.D. Schuffert, D. Andreasen, L. Heusser, M. Lyle, A. Mix, A.C. Ravelo, L.D. Stott, and J.C. Herguera. *Science* **293**, 71–76 (2001).

[48] D. Straile, E.H. van Nes, and H. Hosper. *Limnol. and Oceanogr.* **46**(7), 1780–1783 (2001).

[49] M. Scheffer and S.R. Carpenter. *Trends Ecol. Evol.* **18**(12), 648–656 (2003).

[50] C.H. Hsieh, S.M. Glaser, A.J. Lucas, and G. Sugihara. *Nature* **435**, 336–340 (2005).

[51] S.R. Hare and N.J. Mantua. *Prog. Oceanogr.* **47**, 103–145 (2000).

[52] C.C. Ebbesmeyer, D.D. Cyan, D.R. McLain, F.H. Nichols, D.H. Peterson, and K.T. Redmond. 1976 step in the Pacific climate: forty environmental changes between 1969–1975 and 1977–1984. In J.L. Betancourt and V.L. Tharp (eds.), *Proceedings of the 7th Annual Climate (PACLIM) Workshop*, April 1990, California Department of Water Resources. Interagency Ecological Studies Program Technical Report 26, 1990, 1991, pp. 115–126.

[53] D.R. Easterling and T.C. Peterson. *Int. J. Climato.* **15**, 369–377 (1995).

[54] J.R. Lanzante, *Int. J. Climatol.* **16**, 1197–1226 (1996).

[55] S.N. Rodionov, *Geophys. Res. Lett.* **31**(9), L09204 (2004).

[56] D.L. Rudnick and R.E. Davies. *Deep-Sea Res.* **I 50**, 691–699 (2003).

[57] N. Mantua, *Prog. Oceanogr.* **60**, 165–182 (2004).

[58] K.C. Taylor. *Amer. Sci.* **87**, 320–327 (1999).

[59] F. Gasse, R. Téhet, A. Durant, E. Gilbert, and J.C. Fontes. *Nature* **346**, 141–146 (1990).

[60] E.C. Grimm, G.L. Jacobson, Jr., W.A. Watts, B.C.S. Hansen, and K.A.A. Maasch. *Science* **261**, 198–200 (1993).

[61] P. deMenocal, J. Ortiz, T. Guilderson, J. Adkins, M. Sarnthein, L. Baker, and M. Yarusinsky. *Quat. Sci. Rev.* **19**, 347–361 (2000).

[62] C. Lorius, J. Jouzel, C. Ritz, L. Merlivat, N.I. Barkov, Y.S. Korotkevitch, and VMA Kotlyakov. *Nature* **316**, 591–596 (1985).

[63] J.M. Barnola, D. Raynaud, Y.S. Korotkevich, and C. Lorius. *Nature* **329**, 408–414 (1987).

[64] J. Jouzel, C. Lorius, J.R. Petit, C. Genthon, N.I. Barkov, V.M. Kotlyakov, and V.M. Petrov. *Nature* **329**, 402–408 (1987).

[65] J. Jouzel, N.I. Barkov, J.M. Barnola, M. Bender, J. Chappellaz, C. Genthon, V.M. Kotlyakov, V. Lipenkov, C. Lorius, J.R. Petit, D. Raynaud, G. Raisbeck, C. Ritz, T. Sowers, M. Stievenard, F. Yiou, and P. Yiou. *Nature* **364**, 407–412 (1993).

[66] J. Jouzel, C. Waelbroeck, B. Malaizé, M. Bender, J.R. Petit, N.I. Barkov, J.M. Barnola, T. King, V.M. Kotlyakov, V. Lipenkov, C. Lorius, D. Raynaud, C. Ritz, and T. Sowers. *Clim. Dyn.* **12**, 513–521 (1996).

[67] J. Chappellaz, J.M. Barnola, D. Raynaud, Y.S. Korotkevich, and C Lorius. *Nature* **345**, 127–131 (1990).

[68] P.U. Clark, N.G. Pisias, T.F. Stocker, and A.J. Weaver. *Nature* **415**(6874), 863–869 (2002).

[69] D.C. Esty, J. Goldstone, T.R. Gurr, P.T. Surko, and A.N. Unger. State failure task force report: Working Papers. Science Applications International Corporation, McLean, VA, 1995.

[70] D.C. Esty, J. Goldstone, T.R. Gurr, B. Harff, P.T. Surko, A.N. Unger, and R.S. Chen. State failure task force report: Phase II findings. Science Applications International Corporation, McLean, VA, 1998.

[71] J.A. Goldstone, T.R. Gurr, B. Harff, M.A. Levy, M.G. Marshall, R.H. Bates, D.L. Epstein, C.H. Kahl, P.T. Surko, J.C. Ulfelder, Jr. and A.N. Unger. *State failure task force report: Phase III findings* (in consultation with M. Christenson, G.D. Dabelko, D.C. Esty, and Parris McLean TM) Science Applications International Corporation (SAIC) McLean, VA, 2000.

[72] G. King, and L. Zeng. *World Pol.* **53**(4), 623–658 (2001).

[73] R.I. Rotberg, The failure and collapse of nation-states: Breakdown, prevention and repair. In R.I. Rotberg (ed.), *When States Fail: Causes and Consequences*. Princeton University Press, Princeton, NJ, 2004, pp. 1–50.

Chapter 8

[1] B.R. Frieden. *Science from Fisher Information: A Unification, 2nd edn.* Cambridge University Press, Cambridge, UK, 2004.

[2] J.R. Crow and M. Kimura. *Introduction to Population Genetics*. Harper and Row, New York, 1970.

[3] T.L. Vincent and J.S. Brown. *Evolutionary Game Theory, Natural Selection and Darwinian Dynamics*. Cambridge University Press, Cambridge, UK, 2005.

[4] B.R. Frieden, A. Plastino, and B.H. Soffer. *J. Theor. Biol.* **208**, 49–64 (2001).

[5] M.R. Raup. *Science* **231**, 1528 (1986).

[6] B.R. Frieden. *Probability, Statistical Optics and Data Testing, 3rd edn.* Springer-Verlag, Berlin, 2001.

[7] T. Koshy. *Fibonacci and Lucas Numbers with Applications*. Wiley, New York, 2001.

[8] J.H. Brown, J.F. Gillooly, A.P. Allen, V.M. Savage, and G.B. West. *Ecology* **85**, 1771–1789 (2004).

[9] G.B. West and J.H. Brown. *Phys. Today* **57**, 36–42 (2004) and references therein.

[10] B.J. Enquist and K.J. Niklas. *Nature* **410**, 655–660 (2001).

[11] G.B. West, J.H. Brown, and B.J. Enquist. *Science* **276**, 122–126 (1997).

[12] M.J. Rees. *The Allometry of Growth and Reproduction*. Cambridge University Press, Cambridge, UK, 1989.

[13] W.J. Jungers. *Size and Scaling in Primate Biology*. Plenum, New York, 1985.

[14] K. Schmidt-Nielsen: *Scaling: Why is Animal Size so Important?* Cambridge University Press, Cambridge, UK, 1984.

[15] W.A. Calder. *Size, Function and Life History*. Harvard University Press, Cambridge, MA, 1984.

[16] T.A. McMahon and J.T. Bonner. *On Size and Life*. Scientific American Library, New York, 1983.

[17] W.A. Calder. *J. Theor. Biol.* **100**, 275–282 (1983).

[18] J. Damuth. *Nature* **290**, 699–700 (1981).

[19] J. Fenchel. *Oecologia* **14**, 317–326 (1974).

[20] J.T. Bonner. *Size and Cycle*. Princeton University Press, Princeton, NJ, 1965.

[21] J.S. Huxley. *Problems of Relative Growth*. Methuen, London, 1932.

[22] B.R. Frieden and R.A. Gatenby. *Phys. Rev. E* **72**, 036101, 1–10 (2005).

[23] R. Badii and A. Politi. *Complexity: Hierarchical Structure and Scaling in Physics*. Cambridge University Press, Cambridge, UK, 1997.

[24] P.M. Binder. *Phys. Rev. E* **61**, R3303–R3305 (2000).

[25] A. McKane, M. Droz, J. Vannimenus, D. Wolf (eds.), *Scale Invariance, Interfaces and Non-Equilibrium Dynamics*. Plenum, New York, 1995.

[26] T. Nishikawa, A.E. Motter, Y.C. Lai, F.C. Hoppensteadt. *Phys. Rev. Let.* **91**, 014101 (2003).

[27] K. Nealson and W.A. Ghiorse. *Geobiology: Exploring the Interface Between the Biosphere and the Geosphere*. American Academy of Microbiology, Washington, DC, 2001.

[28] T.M. Blackburn and K.J. Gaston. *Trends in Evolution and Ecology* **9**, 471–474 (1994).

[29] T.M. Blackburn and K.J. Gaston. *Oikos* **70**, 127–130 (1994).

[30] J.H. Brown and P.F. Nicoletto. *Amer. Nat.* **138**, 1478–1512 (1991).

[31] V.J. Bakker and D.A. Kelt. *Ecology* **81**, 3530–3547 (2000).

[32] R.W. Sheldon and T.R. Parsons. *J. Fish. Res. Board Can.* **24**, 900–925 (1967).

[33] R.H. Peters. *The Ecological Implications of Body Size*. Cambridge University Press, Cambridge, UK, 1983.

[34] J.E. Cohen, T. Jonsson, and S.R. Carpenter. *Proc. Nat. Acad. Sci. USA* **100**, 1781–1786 (2003).

[35] J.H. Brown and J.P. Gillooly. *Proc. Nat. Acad. Sci. USA* **100**, 1467–1468 (2003).

[36] B.R. Frieden and B.H. Soffer. *Phys. Rev. E* **52**, 2274 (1995).

[37] B.R. Frieden. *Physics from Fisher Information*. Cambridge University Press, Cambridge, UK, 1998.

[38] B.R. Frieden and R.J. Hughes. *Phys. Rev. E* **49**, 2644 (1994).

[39] B. Nikolov and B.R. Frieden. *Phys. Rev. E* **49**, 4815 (1994).

[40] A.R. Plastino and A. Plastino. *Phys. Rev. E* **54**, 4423 (1996).

[41] B.R. Frieden and W.J. Cocke. *Phys. Rev. E* **54**, 257 (1996).

[42] B.R. Frieden, A. Plastino, A.R. Plastino, and B.H. Soffer. *Phys. Rev. E* **60**, 48 (1999).

[43] R.A. Gatenby and B.R. Frieden. *Cancer Res.* **62**, 3675 (2002).

[44] R.J. Hawkins and B.R. Frieden. *Phys. Lett. A* **322**, 126 (2004).

[45] S. Luo. *Found. Phys.* **32**, 757 (2002).

[46] S. Luo. *Phys. Rev. Lett.* **91**, 180403 (2003).

[47] H.L. Van Trees. *Detection, Estimation and Modulation Theory, Part I*. Wiley, New York, 1968.

[48] G.A. Korn and T.M. Korn, *Mathematical Handbook for Scientists and Engineers, 2nd edn.* McGraw-Hill, New York, 1968.

[49] R.V. Churchill. *Fourier Series and Boundary Value Problems*. McGrawHill, N.Y., 1941, pp. 114–118.

[50] R.A. Gatenby and B.R. Frieden. *Math. Biosci. Eng.* **2**, 43–51 (2005).

Chapter 9

[1] Galison, Peter Galison: 'Image and logic: a meterial culture of micro-physics', University of Chicago Press, 1997.

[2] P. Ball. Utopia theory. *Physics Web* (2003). http://physicsweb.org/articles/world/16/10/7/1, accessed August 2005.

[3] Yolles, M.I., *Management Systems: viable approach*. Financial Times Pitman, London, 1999.

[4] P. Ball. *Complexus* 1(4),190–206 (DOI: 10.1159/000082449) (2003). http://content.karger.com/ProdukteDB/produkte.asp, accessed December 2004.

[5]   A. Koestler. *The Ghost in the Machine*. Picador, London, 1967.

[6]   S.T. Kuhn. *The Structure of Scientific Revolutions*. University of Chicago Press, Chicago, IL, 1970.

[7]   M.I. Yolles. *Kybernetes* **33**(3), 726,764 (2004).

[8]   Yolles, M.I., Frieden, B.R., A Metahistorical Information Theory of Social Change: The Theory, *Organisational Transformation and Social Change*, **2**(2) 103–136, (2005).

[9]   M.I. Yolles, and D. Dubois. *Int. J. Comput. Anticipatory Syst.* **9**, 3–20 (2001).

[10]   Yolles, M.I., Guo, K., Paradigmatic Metamorphosis and Organizational Development, *Sys. Res.*, **20**: 177–199, (2003).

[11]   W. Whewell. *On the Philosophy of Discovery: Chapters Historical and Critical*, Parker & Son, London, 1860.

[12]   M. Crotty. *The Foundations of Social Research: Meaning and Perspective in the Research*. Sage, London, 1998.

[13]   Yolles, M.I., Organisations as Complex Systems: An Introduction to Knowledge Cybernetics, Information Age Publishing, Inc., Greenwich, CT, USA., 2006.

[14]   W. Ulrich. *Manage. Sci.* **23**, 1099–1108 (1977).

[15]   M.I. Yolles. *J. Oper. Res. Soc.* **51**, 1–12 (2001).

[16]   W. Ulrich. *J. Oper. Res. Soc.* **54**(4), 325–342 (2003).

[17]   J. Gleick. *Chaos*. Sphere Books Ltd., London, 1987.

[18]   Mandelbrot, B., *The Fractile Geometry of Nature*. Feeman, New York, 1982.

[19]   C.G. Jung. – Psychologishe typen, 1920, Rascher & Cie AG, Zurich, 1960. Tipos psicológicos, p. 389–467, Vozes, Brasil, 1991.

[20]   A. Aveleira. Consciousness and reality: A stable-dynamic model based on Jungian psychology. *Metareligion* (2004). http://www.meta-religion.com/Psychiatry/ Analytical_psychology/consciousness_and_reality.htm, accessed December 2005.

[21]   M.I. Yolles. *Cybern. Syst.* **31**(4), 371–396 (2000).

[22]   P. Iles and M. Yolles. *Int. J. Hum. Resour. Manage.* **13**(4), 624–641 (2002).

[23]   P. Iles and M. Yolles. *Hum. Resour. Dev. Int.* **6**(3), 301–324 (2003).

[24]   Sorokin, P., *Social and Cultural Dynamics*, in 4 volumes. Bedminster Press, New York, 1962.

[25]   R. Wollheim. *On the Emotions*, Yale University Press, New Haven, CT, 1999.

[26]   M.I. Yolles. *The Dynamics of Peace and War: A Mathematical Study*. Ph.D. dissertation, Department of Politics, Lancaster University, Lancaster, UK, 1980.

[27]   G. Kemp. *J. Conf. Proc.* **3**(1), 15–24 (1997).

[28]   Rosenblueth, A., Wierner, N., Bigelow, J., Behaviour, Purpose and Teology, *Philosophy of Science*, **10**(1) 18–24, (1943).

[29]   E. Schwarz. Anticipating systems: An application to the possible futures of contemporary society. Invited paper at CAYS' 2001, 5th International Conference on Computing Anticipatory Systems, Liege, Belgium, August 13–18, 2001.

[30]   S. Beer. *Cybernetics and Management*. English University Press, London, 1959.

[31]   S. Beer. *Decision and Control*. Wiley, Chichester, UK, 1966.

[32]   Espejo, R., 1997, Giving Requisite Variety to Strategic and Implementation Processes: Theory and Practice, *The London School of Economics Strategy & Complexity Seminar*, bprc.warwick.ac.uk/LSEraul.html, accessed 2000.

[33]   E. Schwarz. *Cybern. Hum. Knowing* **4**(1), 17–50 (1997).

[34]   Yolles, M.I., Iles, P., Guo, K.J., Culture and Transformational Change with China's Accession tro the WTO: The Challenge for Action Research, J of Technology Management in China. Accepted March 1st, 2006.

[35] A. Johansen and D. Sornette. The Nasdaq crash of April 2000: Yet another example of log-periodicity in a speculative bubble ending in a crash. (2000). Arxiv: Cond-Mat/ 0004263 V2, http://arxiv.org/.

[36] Z. Yu. 2004. Stock market crashes. http://guava.physics.uiuc.edu/~nigel/courses/ 463/Essays_2004/files/yu.pdf, April 2005.

[37] G.K. Still. *Crowd Dynamics*, Ph.D. dissertation, Mathematics Department, Warwick University, UK, 2000.

[38] Focus, p. 3, New insight into crowd behaviour—It's Fractal. (1994). http://www. geocities.com/Omegaman_UK/crowd1.html, accessed December 2004.

[39] S. Beer. *The Heart of the Enterprise*. Wiley, Chichester, UK, 1979.

[40] H. Van de Sande. *The Social Psychology of Crowd Behaviour*. (research book currently in process) 2004. http://www.ppsw.rug.nl/~vdsande/OC123.doc, accessed December 2004.

[41] M.I. Yolles and K. Guo. *Sys. Res.* **20**, 177–199 (2003).

[42] T.L. Brown. *Making Truth: Metaphor in Science*. University of Illinois Press, Urbana, IL, 2003.

[43] K.J., Guo, M.I., Yolles, P., Iles, Organisational Patterning, Knowledge cybernetics and viable systems theory in Chinese Banking System, 15th International Association for Management of Technology (IAMOT) conference, "East Meets West: Challenges and Opportunities in the Era of Globalization", Beijing, May 22–26, 2006.

[44] M. Schwaninger. *Sys. Res.* **18**, 137–158 (2001).

[45] M.I., Yolles. Organisational Intelligence, *Journal of Work Place Learning*, **17**(1&2), pp. 99, 114, (2005).

[46] B.R. Frieden. *Physics from Fisher Information*. Cambridge University Press, Cambridge, UK, 1998.

[47] B.R. Frieden. *Science from Fisher Information*. Cambridge University Press, Cambridge, UK, 2004.

[48] H.R. Maturana and F.J. Varela. *The Tree of Knowledge*. Shambhala, London, 1987.

[49] H.R. Maturana. *Int. J. Man-Mach. Stud.* **7**, 313–332 (1975).

[50] B.R. Frieden and B.H. Soffer. *Phys. Rev. E* **52**, 2274–2286 (1995).

[51] B.R. Frieden. *Phys. Rev. A* **41**, 4265–4276 (1990).

[52] A.R. Plastino and A. Plastino. *Phys. Rev. E* **54**, 4423 (1996).

[53] S.P. Flego, B.R. Frieden, A. Plastino, and B.H. Soffer. *Phys. Rev. E* **68**, 016105 (2003).

[54] B.R. Frieden, A.R. Plastino, A., Plastino, and B.H. Soffer. *Phys. Rev. E* **60**, 48–53 (1999).

[55] B.R. Frieden, A.R. Plastino, A. Plastino, and B.H. Soffer. *Phys. Rev. E* **66**, 046128, 1–8 (2002).

[56] R.J. Hawkins and B.R. Frieden. *Phys. Lett. A* **422**, 126 (2004).

[57] R.A. Gatenby and B.R. Frieden. *Cancer Res.* **62**, 3675–3684 (2002).

[58] P. Sorokin. *Social and Cultural Dynamics, Volumes 1–4*. Bedminster Press, New York, 1962. (Originally published in 1937–1941, American Book. Co., New York.)

[59] S.A. Kauffman. *The Origins of Order: Self-Organization and Selection in Evolution*. Oxford University Press, New York, 1993.

[60] P. Bak and K. Sneppen. *Phys. Rev. Lett.* **71**(24), 4083–4086 (1993).

[61] N. Eldredge and S.J. Gould. Punctuated equilibria: An alternative to phyletic gradualism. In T.J.M. Schopf (eds.), *Models in Paleobiology*. Freeman, Cooper, and Co., San Francisco, CA, 1972.

[62] E. Schwarz. A metamodel to interpret the emergence, evolution and functioning of viable natural systems. In R. Trappl (ed.), *Cybernetics and Systems '94*, World Scientific, Singapore, 1994, pp. 1579–158 (as Proceedings of the European Meeting on Cybernetics and Systems Research, Vienna).

[63] E. Schwarz. A trandisciplinary model for the emergence, self-organisation and evolution of viable systems. Paper presented at the International Information, Systems Architecture and Technology, Technical University of Wroclaw, Szklaska Poreba, Polland, 1994.

[64] E., Schwarz, Is Consciousness Reality of Illusion? A Non-Dualist Interpretation of Consciousness, *Computing Anticipatory System: CASYS'03 - Sixth International Conference*, Liege (Belgium), 11–16 August, 2003.

[65] T.L. Vincent. *Evolutionary Game Theory, Natural Selection and Darwinian Dynamics*. Cambridge University Press, Cambridge, UK, 2004.

[66] B.D. Fath, H. Cabezas, and C.W. Pawlowski. *J. Theor. Biol.* **222**, 517–530 (2003).

[67] P. Musso. 1998, Towards a worldwide communications oligopoly? *Le Monde Diplomatique*, MondeDiplo.com (http://mondediplo.com/1998/03/14telemus).

[68] B.R. Frieden. *Probability, Statistical Optics and Data Testing, 3rd edn.* Springer-Verlag, Berlin, 2001.

[69] J. Habermas. *The Theory of Communicative Action*. Volume 2. Polity Press, Cambridge, UK, 1987.

[70] J. Galtung. *Peace: Essays in Peace Research, Volume 1*. Christian Ejlers, Copenhagen, 1972.

[71] D.M. Dubois. *Int. J. Comput. Anticipatory Syst.* **10**, 3–18 (2001).

[72] J. Maynard Smith. *Models in Ecology*. Cambridge University Press, Cambridge, UK, 1974.

[73] J.C. Flores. A mathematical model for Neanderthal extinction, J. Theor. Biol. **191**, 295–298 (1998).

[74] B.R. Frieden, A. Plastino, and B.H. Soffer. *J. Theor. Biol.* **208**, 49–64 (2001).

[75] R. D'Aveni. *Hypercompetition: The Dynamics of Strategic Maneuvering*. Basic Books, New York, 1994.

[76] C. Dobson and R. Payne. *The Carlos Complex: A Pattern of Violence*. Book Club Associates, London, 1977.

Index